Inside the Communication Revolution

Evolving Patterns of Social and Technical Interaction

Edited by
ROBIN MANSELL

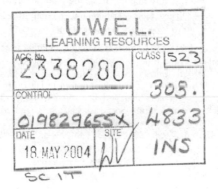
OXFORD
UNIVERSITY PRESS

OXFORD
UNIVERSITY PRESS

Great Clarendon Street, Oxford OX2 6DP

Oxford University Press is a department of the University of Oxford.
It furthers the University's objective of excellence in research, scholarship,
and education by publishing worldwide in

Oxford New York

Auckland Bangkok Buenos Aires Cape Town Chennai
Dar es Salaam Delhi Hong Kong Istanbul Karachi Kolkata
Kuala Lumpur Madrid Melbourne Mexico City Mumbai Nairobi
São Paulo Shanghai Singapore Taipei Tokyo Toronto

with an associated company in Berlin

Oxford is a registered trade mark of Oxford University Press
in the UK and in certain other countries

Published in the United States
by Oxford University Press Inc., New York

British Library Cataloguing in Publication Data

Data available

Library of Congress Cataloging in Publication Data

Inside the communication revolution: evolving patterns of social and
technical interaction/edited by Robin Mansell.
p. cm.
Includes bibliographical references and index.
1. Telecommunication—Social aspects. 2. Digital communications.
3. Information technology. I. Mansell, Robin.
HE7631 I468 2002 303.48'33—dc21 2001052079
ISBN 0–19–829655–X
ISBN 0–19–829656–8 (pbk.)

1 3 5 7 9 10 8 6 4 2

Typeset by Newgen Imaging Systems (P) Ltd., Chennai, India
Printed in Great Britain
on acid-free paper by
Biddles Ltd., *www.biddles.co.uk*

Contents

List of Figures   vii

List of Tables   viii

List of Boxes   ix

Abbreviations and Acronyms   x

List of Contributors   xiv

Preface   xvi

1. Introduction   1
   Robin Mansell

Part One. Mediating Social and Technical Relationships

2. Virtual Communities and the New Economy   21
   W. Edward Steinmueller

3. Cyberspace and Social Distinctions: Two Metaphors and a Theory   55
   David C. Neice

4. Knowledge Management Meets the Virtual Organization in the
   Newspaper Industry   85
   Jennifer J. Gristock

5. Mind the Gap: Digital Certificates, Trust, and the Electronic
   Marketplace   110
   Ingrid Schenk

6. The Colleague in the Machine: Electronic Commerce in the
   London Insurance Market   128
   Gordon Rae

7. Repersonalizing Data in the Banking Industry   144
   Andreas Credé

Part Two. Building Capabilities for Knowledge Exchange

8. Co-Design in Action: Knowledge Sharing, Mediation, and Learning   165
   Jane E. Millar

Contents

9. The Distribution of Spatial Data: Data Sharing and Mediated
   Cooperation   186
   Uta Wehn de Montalvo

10. Master of my Domain: The Politics of Internet Governance   206
    Daniel Paré

11. Missing Concepts in the 'Missing Links' for Brazilian
    Telecommunications   225
    Ana Arroio

12. Conclusion: Social Relations, Mediating Power, and Technologies   251
    Robin Mansell

    References   271
    Author Index   299
    Subject Index   305

List of Figures

4.1. Four hypothetical organizations in the organization-space   91
4.2. Division of the organization-space into virtuality 'zones'   100
4.3. Distribution of Internet newspaper organizations in the
     organization-space   102
8.1. Knowledge and capability in the zone of proximal development   170
8.2. Knowledge production during co-design   171
9.1. Relationships between data types, accuracy, and data complexity   191
9.2. Intermediating role of NSIF in the GIS community in South Africa   198
9.3. Model of willingness to engage in spatial data sharing in
     South Africa   201
10.1. UK Internet industry perceptions of Nominet operations   218
11.1. INPE annual investment, 1976–1997   232
11.2. INPE research and administrative employees, 1980–1997   233
11.3. Telebras system annual investment, 1974–1997   237
11.4. Embratel operating revenues and R&D investment, 1979–1997   239

List of Tables

3.1. Percentage of Internet-users for selected countries   67
4.1. Average scores of regional daily newspapers   97
4.2. Division of the organization-space into virtuality zones   99
4.3. Virtuality matrices of Internet newspapers: distance from
      London and average score of newspaper web sites   101
6.1. Size of London insurance market by income and number of firms   130
11.1. LEOS system overview   228

List of Boxes

3.1. Some common features of frontiers 64
5.1. Trust service market entry strategies in the UK 121
8.1. Factors responsible for outcomes of co-design in firms 184
9.1. Areas of GIS application in development projects 189
9.2. Types of spatial data 191
10.1. Summary of Naming Committee *co.uk* registration rules 213
11.1. Indicators of negotiating capability 230
12.1. Mediating social and technical relationships 263
12.2. Building capabilities for knowledge exchange 266

Abbreviations and Acronyms

AEB	Alliance for Electronic Business, UK
AEB	Agencia Espacial Brasileira (Brazilian Space Agency)
AICPA	American Institute of Chartered Public Accountants
Anatel	National Telecommunications Agency, Brazil
AOL	America Online
ARPANET	Advanced Research Projects Agency network
ATM	automated teller machine
ATS	advanced telecommunication satellite
BCC	British Chambers of Commerce
BSM	Business Systems Manager
CA	Certification Authority
CANARIE	Canadian Network for the Advancement of Research, Industry, and Education
CAST	Chinese Academy for Space Technology
CBC	Canadian Broadcasting Commission
CCI	Constellation Communications Inc.
CD-ROM	compact disc–read only memory
CEPR	Center for Economic Policy Research
CIAR	Canadian Institute for Advanced Research
CICA	Canadian Institute of Chartered Accountants
CMC	computer-mediated communication
CNPq	National Research Council, Brazil
COBAE	Brazilian Commission for Space Activities
CPqD	Centre for Telecommunication Research and Development, Brazil
CPS	certification practice statement(s)
CPSR	Computer Professionals for Social Responsibility
CSIR	industrial research organization in South Africa
DIGEST	Digital Geographic Information Exchange Standard
DKB	Dai-Ichi Kangyo Bank Ltd, Japan
DNS	Domain Name System
DTI	Department of Trade and Industry, UK
ECAF	European Certification Authority Forum
ECCO	Equatorial Constellation Communications Organization, Brazil
ECO	Equatorial Communications satellite project, Brazil
EDI	electronic data interchange
EEMA	European Electronic Messaging Association
EFF	Electronic Frontier Foundation
EPS	Electronic Placing System
ESRI	Environmental Systems Research Institute

ESSY	European Sectoral SYstems (of innovation)
FAQ	frequently asked questions
ftp	file transfer protocol
GIMS	Geographic Information Management System
GIP	Government Information Project, South Africa
GIS	geographic information system(s)
GOCNAE	National Commission for Space Activities, Brazil
GSDI	Global Spatial Data Infrastructure
HICSS	Hawaii International Conference on System Sciences
HNMC	Head of Non-Marine Claims
HSBC	HSBC Bank plc
HTTP	Hypertext Transfer Protocol
IBDT	Brazilian Institute for the Defence of Telecommunication
ICA	International Communication Association
ICANN	Internet Corporation for Assigned Names and Numbers
iCASE	integrated Computer-Aided Software Engineering
ICT	information and communication technology
IE	Information Engineering
IEC	Independent Electoral Commission, South Africa
IEEE	Institute of Electrical and Electronic Engineers
ILPF	Internet Law and Policy Forum
ILU	Institute of London Underwriters
INEGI	National Statistics, Geography and Informatics Institute, Mexico
INK	Information, Networks & Knowledge Research Centre
INPE	Brazilian National Institute for Space Research
INTELSAT	International Telecommunication Satellite Organization
IP	Internet Protocol
IPEC	Impacts and Perspectives of Electronic Commerce
IRC	Internet Relay Chat
ISOC	Internet Society
ISP	Internet service provider
ISS	Integrated Spatial Solutions
IT	information technology
ITU	International Telecommunication Union
JAD	Joint Application Development
JANet	Joint Academic Network, UK
JSD	Jackson Structured Design
LEOS	low earth orbiting satellite
LGT	General Telecommunications Act, Brazil
LIBC	Lloyd's Insurance Brokers Committee, UK
LIMNET	London Insurance Market Network
LINX	London Internet Exchange
MAT	Marine, Aviation, and Transport
MCI	Microwave Communications Incorporated

MECB	Brazilian Complete Space Mission
Minicom	Ministry of Communications, Brazil
MNGIS	Mexican National Geographic Information Systems
MIT	Massachusetts Institute of Technology
MPEG	Motion Picture Experts Group
MUD	multi-user dungeon
NASDAQ	National Association of Securities Dealers Automated Quotation
NATO	North Atlantic Treaty Organization
NCR	National Cash Register (company)
NIIA	National Information Infrastructure Agency
NLIS	National Land Information System, South Africa
NRS	Name Registration Scheme
NSIF	National Spatial Information Framework, South Africa
OECD	Organization for Economic Cooperation and Development
OFTEL	Office of Telecommunications, UK
OSS	Observatoire du Sahara et du Sahel
PABX	private automated branch exchange
PC	personal computer
PKI	public key infrastructure
PSAC	Policy Signing and Accounting Centre, UK
PSI	Performance Systems International Inc.
R&D	research and development
RAC	Royal Automobile Club
RAD	Rapid Application Development
RAM	random access memory
RDP	Reconstruction and Development Programme, South Africa
RFC	request for comment
RINET	ReInsurance Network, UK
RIPP	Rapid Iterative Production Prototyping
SACI	Advanced Interdisciplinary Communication Satellite
SACI/EXERN	Advanced Interdisciplinary Communication Satellite Project, Brazil
SAM	Southern African Metadata consortium
SANDF	South African National Defence Force
SBTS	Brazilian Satellite Telecommunication System
SME	small and medium-sized enterprise
SPRU	Science and Technology Policy Research
SSADM	Structured Systems Analysis and Design Methodology
STOA	Scientific and Technological Options Assessment (European Parliament)
STRADIS	STRuctured Analysis and Design of Information Systems
SWAT	Specialists with Advanced Tools
TCP/IP	Transport Control Protocol/Internet Protocol
TDMA	time division multiple access

TEN	Trans-European Network
TNO–STB	Netherlands Organization for Applied Scientific Research
TSER	Targeted Socio-Economic Research Programme
TSP	trust service provider
UBS	Union Bank of Switzerland
UKERNA	UK Education and Research Networking Association
UNCTAD	United Nations Conference on Trade and Development
UNESCO	United Nations Educational, Scientific, and Cultural Organization
UNITAR	United Nations Institute for Training and Research
URL	Universal Resource Locator
V2_OS	Operating System based on 32 bit assembler
VAN	value added network
VLS	satellite launch vehicle, Brazil
Web	World Wide Web
WELL	Whole Earth 'Lectronic Link
WIN	World Insurance Network
WMA	Windows Media Audio
WRI	World Resources Institute
WWW	World Wide Web
Y2K	Year 2000 (computer bug)
ZPD	zone of proximal development

List of Contributors

Dr Ana Arroio has a background in communications studies and completed her D.Phil. at SPRU (Science and Technology Policy Research), University of Sussex, in July 2000. She now works as an independent consultant in Rio de Janeiro, Brazil, and her research focuses on the technological and institutional capabilities required to harness the potential of new information and communication technologies and services.

Dr Andreas Credé is a research coordinator within the European Leadership and Organizational Practice at McKinsey & Company, London. He studied at SPRU, received his D.Phil. in 1997, and worked as a Research Fellow at SPRU. With a background in International Relations, he worked as a project finance consultant and had held senior positions with several banks before coming to SPRU. His research focuses on the nature of information and knowledge production in financial institutions.

Jennifer J. Gristock is a doctoral student at SPRU whose thesis is under examination. She trained as a Materials Engineer at the University of Wales Swansea. Her research focuses on the ways information and communication technologies are used in combination with meetings and traditional media to support virtual forms of working.

Dr Robin Mansell was Research Director and Professor of Information and Communication Technology Policy at SPRU, University of Sussex, where she also co-directed the Information, Networks & Knowledge (INK) research centre until December 2000. She now holds the Dixons Chair in New Media and the Internet in the Interdepartmental Programme in Media and Communication, London School of Economics and Political Science. Her research focuses on the role of information and communication technologies and services in the context of social and economic transformation, the implications for social organization and business strategy, and governance and regulation.

Dr Jane E. Millar is a Research Fellow in SPRU, where she works with the INK research centre on issues relating to skills and employment, gender, and the social implications of new information and communication technologies. She has worked as a consultant to the Higher Education Funding Council of England, the Department for Education and Employment, and the European Commission. She studied at SPRU and received her D.Phil. in 1996.

Dr David C. Neice was a senior research analyst with the Canadian government, where he worked on information highway and access to technology issues and he has undertaken consultancy research at several universities in Canada and in the United States. His D.Phil. thesis was successfully examined at SPRU in February 2001. His research interests include the dynamic relationships between the digital technologies and the signs and symbols of the social order.

Dr Daniel Paré has a background in political science and completed his D.Phil. at SPRU in December 2000. He is now employed at the London School of Economics and Political Science as a Research Fellow. His research interests focus on the political economy of governance in information and communication technology-based environments.

Gordon Rae is a part-time D.Phil. student at SPRU. He works as an independent consultant and has undertaken work for Barclays Bank, Lloyds of London, Global Asset Management, and the Ministry of Defence. With a background in philosophy and cognitive and computing science, his research focuses on the role of social presence in the application of information and communication technologies in the financial and insurance services sectors.

Ingrid Schenk is a D.Phil. student at SPRU. She has a background in political science and economics and has undertaken consultancy work for the NCR Financial Services Knowledge Lab, London. Her research interests focus on relationship building in electronic commerce environments and asymmetries in power relationships between customers and suppliers.

Dr W. Edward Steinmueller is Professor of Information and Communication Technology Policy at SPRU, University of Sussex, where he directs the INK research centre and is responsible for SPRU D.Phil. admissions. An economist, he was formerly Deputy Director of the Center for Economic Policy Research (CEPR), Stanford University (1986–94), and Professor of the Economics of Technical Change, MERIT, University of Maastricht (1994–7). His research draws upon industrial economics and focuses on science policy, advanced communications technologies and services, software, and other developments related to the industry.

Dr Uta Wehn de Montalvo has a background in computing science and her D.Phil. thesis was successfully examined in May 2001. She has worked as a research officer in SPRU and undertaken consultancy research for the International Development Research Centre, Ottawa, and the UN Commission for Science and Technology for Development. Her research focuses on the implications of information and communication technologies for social and economic development.

Preface

The authors and editor of this book worked and studied together at the University of Sussex's SPRU (Science and Technology Policy Research) Graduate Research Centre. In SPRU, the Information, Networks & Knowledge (INK) research centre focuses on the way social and physical communications networks are enabling the translation of information into potentially useful knowledge. SPRU's postgraduate programme in science and technology policy studies includes some eighty doctoral research students who come with diverse prior training in the social, management, natural, and engineering sciences. Students work closely with about forty full-time research staff in a distinctive institutional environment where the goal is to advance understanding of the relationships between developments in science, technology, and innovation and society through interdisciplinary research. Nine of the eleven contributors to this book either have been or currently are doctoral students who work closely with INK on the way information and communication technologies (ICTs) are implicated in social and economic transformation.

In the United Kingdom postgraduate research training in the social sciences has traditionally been characterized by intensive use of a tutorial system organized around a student–supervisor relationship. How that relationship operates to prepare a postgraduate student for a professional career is often the subject of mystification. In the process of writing this book, we sought to create an innovative postgraduate training experience for the students by introducing a group supervision process. The account of the experiment that led to this book begins in 1996 when pressures within SPRU to take on growing numbers of doctoral students increased substantially in line with the expansion of postgraduate education generally throughout the United Kingdom. We began to wonder whether the traditional form of individual doctoral supervision (consisting normally of face-to-face communication and comments on hard copies of written material) would become unsustainable as the numbers of students per supervisor increased. Was there an alternative, aside from reducing the growth of our doctoral research programme, that would not detract from the students' learning experiences at SPRU?

The process of learning that yielded this book displayed many of the features of transformation in the learning process that may be characteristic of changes in higher education in the social sciences in Britain. The learning processes needed successfully to complete a doctoral degree in the social sciences are becoming more formalized and involve more uniform processes of research planning, methodology selection, and implementation of research plans. The experiment underlying this book was, in part, about the extent to which such training can be formalized and, in part, about the increasingly intensive use of ICTs in the manifold activities involved in doctoral research and supervision. It is also very substantially about how social and physical communications networks are interacting and evolving within the student–supervisor relationship.

All the contributors to this book have a core interest in some aspect of the development and implementation of ICTs. Beyond this core concern, there was very little coherence initially between the various research projects of the student contributors. The student contributors had enormously diverse disciplinary backgrounds in cognitive studies, communications studies, economics, engineering, international relations and political science, sociology, philosophy, and psychology. They had undertaken their earlier studies in Brazil, Canada, Germany, and the United Kingdom. They included both young and mature students; they included full-time and part time students; and they included a nearly equal number of men and women. Their research plans encompassed finely grained micro-level work and institutional-level analysis. All had completed their first year of study, but they were at different stages in the research process. Their research methods spanned the qualitative to the quantitative; their research methodologies entailed single or multiple case studies and large-scale surveys. Their knowledge of epistemological and philosophical issues in the social sciences was uneven. When regular group supervision for these students was introduced, it was unclear whether this social innovation would contribute positively to the students' learning experience because of the breadth of their talents and interests.

In 1996 all these students were physically present at SPRU and it was feasible to initiate a process of discovery about what aspects of the learning process involved in doctoral research could be transferred to a new face-to-face group supervision process. Would such a process facilitate student interaction in ways that they would regard as productive in the light of their enormously varied interests and training needs? Would we be able to build connections theoretically and empirically that would succeed in extending and deepening our collective understanding of crucial issues for both the creators and the users of advanced ICTs? The experience of group supervision suggests that the answers to both these questions are affirmative. But the experience also offers some interesting insights into the limitations of this particular innovation. In particular, further efforts to create a more uniform learning process for students who are working in an intensively interdisciplinary environment would require an unacceptable level of constraint on the conceptual innovations junior researchers are likely to produce during their efforts to make an original contribution to knowledge.

During the first year of group supervision, we experimented with many kinds of seminar-style presentations on various issues. Communication about 'theory' and conceptual frameworks was a struggle initially for everyone who participated in the group supervision sessions. During monthly- -and sometimes more frequent—group supervision meetings, there were many awkward silences; there was hesitancy on the part of some students to participate; there was considerable eloquence on the part of others, particularly the more experienced students. All of the latter were men, creating a risk of gender inequality in the group in terms of the time their supervisor might spend interacting with the more vocal and experienced men than with the women students.

Social experimentation is always risky. My reflection as both supervisor and editor of this book is that the experiment proved to be very rewarding. This is due to the persistence and enthusiasm of the student contributors to this book. It is also due to the active participation in the group at various times of my SPRU INK colleagues,

Professor W. Edward Steinmueller and Dr Jane Millar, both chapter contributors; and Dr Richard Hawkins. Cynthia Little, INK's programme assistant, also contributed enormously, not only to the editing of this book, but also to managing and facilitating the face-to-face and virtual communication processes between all concerned.

In the second year of the group supervision experiment, the idea of producing a book emerged. The students wanted to show that we were learning to engage in fruitful interdisciplinary discussion and in research that had common themes and was yielding insights that went well beyond the original core interest in the same technologies and, more generally, in science and technology policy. A year later, we had uncovered an integrating research theme—inside the communication revolution: evolving patterns of social and technical interaction—which is the title for our book. We develop this theme by juxtaposing theories about changing processes of social and technical interaction and, specifically, processes involving mediation, with empirical evidence drawn from a variety of ICT application areas in the service sectors.

By the end of 1998, thanks to the confidence of our editor, David Musson, we had a contract for the production of this volume. In each of the chapters, there is an emphasis on social processes of mediation and, reciprocally, on their intermediation by innovative technologies. This emphasis emerged as a result of our experiences of learning to interact using ICTs; from the deepening of our understanding of the implications of these technologies in varied social and economic contexts; and from the insights generated by our theoretical discussions, which incorporated material from a substantial number of social science disciplines. The reader will find variation in the extent to which each of the contributors succeeds in applying the concept of mediation to reveal insights into the dynamics of social and technical interactions. Other perspectives and concepts may serve equally well. The key point, however, is that mediation was the theme by which individual research projects were brought into closer relationship with each other. The limitations in 'intellectual convergence' are one indicator of the inherent difficulties of stimulating originality while, at the same time, seeking a common conceptual framework. The general conclusions that we draw in Chapter 12 are an expression of both the success and limitations of this process.

The contributors to this volume were committed to enhancing the transparency of the processes of designing and using ICTs for all those who produce them and who may wish to use them in socially and economically beneficial ways. Despite our deep involvement in the study of ICTs, however, we had no idea how swiftly these technologies would come to play a substantial intermediating role in the student–supervisor relationship.

Group supervision involved an intense leaning process for all the participants. Apart from the subject matter of the chapters and the learning associated with the design, implementation, and management of research in diverse topic areas, there were a number of associated learning outcomes. For instance, students learned how to assess the strengths and weaknesses of different publishers and to prepare a synopsis for a book-length manuscript. They reached a consensus about who their intended readers were. They considered intellectual property rights issues, they drafted and redrafted their chapters, and they learned how to offer one another constructive scholarly criticism. They experienced the time-consuming process of retrieving missing bibliographic information that had been omitted from early draft chapters. They all became active and informed contributors to theoretical and empirical discussions that ranged considerably beyond the boundaries of their own research topics.

Group supervision provided a new forum for learning, but this forum eventually became a 'virtual' forum. By the time this manuscript was completed in 2001, the group (and indeed individual) supervision had changed radically for these students. ICTs were implicated in this process of change. For example, by 1998, the Internet and the World Wide Web were being used intensively by all of us. One student suggested that a web site should be created to facilitate our book-writing activities. We assumed at first that this site would be used intensively, because all the contributors were profoundly interested in using innovative technologies. Interestingly, however, this site was very rarely used to exchange material or ideas directly between the contributors. All the students favoured face-to-face communication or individual-to-supervisor/editor email communication throughout the book-writing process.

In addition, by 2001, nearly all the students had begun to work 'virtually' from their homes in the United Kingdom or in their respective countries. They used email more intensively to communicate and they preferred not to locate their work on the internal SPRU web site to enable comments to be provided by all contributors. Their comments with respect to each others' draft chapters would generally be sent to me by email, accompanied by questions such as 'is this appropriate comment/criticism of this author's draft chapter?' Providing assurance seemed to require private interaction between student and supervisor either by email or by telephone. I would comment on the student's observations and the student would then pass the message on to another contributor. In fact, all the contributors were sensitive to the potential for email communication to create social discord. Despite the potential of technology to support an open dialogue between all the contributors using the Web or other computer-supported collaborative writing tools, this was not pursued or championed, even by the students who were the most experienced ICT users.

The dynamics of the early period of face-to-face meetings produced many discussions that were about substantive matters. But they were also about how to conduct scholarly debate within a group of unevenly experienced peers, from different cultures and from diverse disciplinary backgrounds. These discussions were often both entertaining and productive. Nevertheless, they sometimes became inadvertently hurtful and risked undermining the confidence of a research student. On the occasions when this happened, face-to-face and one-to-one communication seemed essential. This book is unlikely to have been produced if the group supervision process had been conducted 'virtually' from the outset. Facilitating this type of discussion in an online environment so as to minimize this kind of risk and to redress any harm would be enormously demanding.

There were many surprises in the learning process that led to this book. For instance, there were times when I believed that codified knowledge could guide the students in the writing process and in preparing the manuscript for publication, but their experience suggested that I was wrong in many cases. I would suggest readings on various topics, for example, relating to the substance of our research or to the writing process itself, only to find that the students wanted to discuss in detail my experiences before they felt prepared to conduct their own part of the work. The 'tacitness' of the doctoral research process is something we rediscovered through the group supervision process. Conversely, there were occasions when we found that certain aspects of the students' tacit knowledge discovery process could be written as working notes—or 'codified'— and then conveyed to other SPRU students who were not part of this group. Some of these other students later claimed that this codified knowledge had been helpful to them in preparing their theses. In short, we all learned to re-examine our conceptions of 'best-practice' supervision in both face-to-face and electronic environments. We learned, too, that there is no 'right for all times and all places' means of using ICTs to support the learning processes that underpin the creative process.

We found—albeit on the basis of this one instance of group supervision—some answers to the questions that had been raised about the potential for innovation in the provision of postgraduate education. Group supervision does not substitute for individual supervision even at the margins. Group supervision creates double the amount of work for both supervisor and students. Individual supervision still needs to continue with the same frequency. Group supervision must be voluntary and students must be able to join in and withdraw depending on their individual experiences. Such groups must be allowed to mature and to wind up so that new groups can be created drawing upon the changing enthusiasms of students. To focus on a potentially unifying research theme in an interdisciplinary field with the scope of science and technology policy studies, a period of face-to-face interaction seems essential. It also seems essential for the student–supervisor relationship if this is to be productive for both parties.

It had become technically feasible by 2001, and, in many instances, affordable for students, to conduct the whole of the student–supervisor relationship virtually. Indeed, by the time this book was completed, almost all the students' supervision was being conducted on a 'distance learning' basis for one reason or another. This was often because the increasingly difficult financial circumstances of students meant that they could not afford to live close to the university or because they were forced to take up full-time employment to support their families.

Despite the potential for virtual learning offered by the new technologies, SPRU continues to require that new research students spend a minimum of one and often more terms in SPRU. Without this period of physically present time—and there are strong pressures to give way on this requirement as research degrees are offered in distance-learning format—there seems to be a high risk that the scope for interdisciplinary research will narrow. This is because of the substantial need to engage in communication processes that cannot (yet) occur successfully in virtual mode even using the richest means of electronically mediated communication. Face-to-face interaction is necessary to build trust and understanding in areas that must bridge the apparently deep chasms between the disciplines and also to communicate in ways that respect language, culture, and perceptions about how scholarly engagement should occur. Without the experience of face-to-face relationship building, there is a greater likelihood that supervisors and students will become less able to detect and manage the times when the stresses and strains of being a research student while managing all the other facets of life reach a point where the outcome desired by student—the award of the degree—is jeopardized. The process of writing this book, therefore, also revealed ambivalence towards the roles of ICTs in mediating relationships. This ambivalence is echoed in varying degrees in each of the chapters of this book. We hope our experiences and our research results will prompt further in-depth research and debate about the opportunities and limitations of the use of these technologies in support of postgraduate training and about the potential they offer for creating a basis for more systematic delivery of such training.

All the contributors to this book want to acknowledge and thank our SPRU colleagues for their insights and guidance on the conduct of the many research projects that gave rise to this collection. In particular, we thank Chris Freeman for his unceasing encouragement to SPRU research staff and students to persist in asking new questions about both the social and the economic implications of the new technologies. He has encouraged us to undertake difficult interdisciplinary research encompassing contributors whose training comes from both the social sciences and engineering and to devise and implement creative research designs that yield new empirical insights that deepen or challenge received theory in many instances. We also acknowledge the consistent support to the student contributors of the staff of the SPRU Graduate Studies Office and Director of Graduate Studies, Martin Bell, without whose leadership the creative learning environment that gave rise to this book in SPRU would simply not exist.

Finally, our research on patterns of technical and social interaction in the rapidly changing field of innovation in ICTs is intended to show why it is essential for all stakeholders in the causes and consequences of the 'communication revolution' to acquire an improved understanding of the dynamics of change. Without such understanding, technology designers and producers, citizens and consumers, businesses and policymakers have little chance of guiding the current manifestations of social transformation.

Robin Mansell

Cooksbridge, East Sussex
November 2001

1

Introduction

ROBIN MANSELL

1. INTRODUCTION

What does it really mean to live and work inside the ongoing communication revolution? Our focus in this book is on the nature and significance of newly emerging patterns of social and technical interaction as digital information and communication technologies (ICTs) and services become more pervasively present in our lives. We also examine how these new patterns overlap with and are often very closely connected to existing patterns of social and technical interaction. We challenge the idea that using digital technological tools is necessarily associated with improvements in society or in an individual's experience of his or her interactions within society. The new technologies offer vast potential for improvements, but there may, in some instances, be good reasons for not privileging the technological mediation of human experience over other forms of mediation. This book brings together a set of investigations into the places and spaces where the use of digital technologies is likely to be socially or economically advantageous. It also examines instances where privileging of the use of these technologies may, in fact, be resisted in favour of achieving social outcomes through communication in other ways.

We draw upon cognitive, economic, management, political, and sociological theories and the results of empirical studies to offer insights into emergent forms of interaction that are being mediated by the new technologies. We investigate the consequences of these interactions for a variety of different kinds of social, economic, and organizational structures and processes. We suggest that a stance of informed ambivalence is most appropriate with respect to the growing number of roles that ICTs are playing in our economy and society. The principal focus of the chapters in this book is on how, and in what ways, interactions between social actors are altered when digital technologies become available. The empirical studies that are brought together here are concerned with the learning processes associated with on- and offline interactions between individual actors. They are also concerned with collective attempts to govern and manage new forms of interaction through corporate strategy and government policy developments, which are examined at the institutional level of analysis. The research in these chapters, therefore, includes micro-level studies of how the new technologies are intertwined with changing social relationships and studies designed to illustrate how learning within firms and within policy and regulatory organizations produces expected and unexpected social and economic outcomes in parallel with the development and use of the new technologies.

We argue that a degree of ambivalence towards the new technologies is important because the social and economic development of societies that are infused with digital systems is far from certain. This uncertainty impels us to give a high priority to investigating the scope for social actors to alter the way humans will interact with each other in the future within the contours of the evolving information and communication revolution. We suggest that continual rapid changes in digital technologies are creating a growing need for assessments of the potential for incremental and very radical changes in people's lives.

The research in this book is located broadly within the framework of studies of techno-economic and socio-political change (Boden and Miles 2000; Freeman 1988, 1994*a*; MacKenzie 1996*a*) and specifically within those traditions that seek to explain the dynamics of developments at the sites of interaction between social and technical systems (Bijker 1993; Castells 2000; MacKenzie 1996*b*). Section 2 of this chapter indicates briefly how this book addresses the dominant discourses on the social issues arising from the growing use of ICTs. Section 3 explains why we have elected to study these dynamic processes through the lens of mediation. Section 4 provides the theoretical and empirical agenda for the book by highlighting the importance of the concept of paradigmatic change in understanding the social and economic environment in which technological innovation occurs. We draw particularly on arguments developed principally by Christopher Freeman and Carlota Perez (1988). Much of their work examines the economic determinants of the ICT revolution. Our work is intended to complement theirs by giving greater emphasis to the social processes that seem to be sustaining specific 'trajectories' of change within an overall framework of paradigmatic transformation that is increasingly being dominated by new ICTs. By developing insights into the particularities of the social processes at work within the communication revolution, we hope to demonstrate that there is remarkable scope for social actors to encourage the selection of alternatives to the predominant uses of the new ICTs.

Finally, Section 5 provides an overview of the structure of the book by highlighting the research themes and issues that have been addressed as a result of our research programme on the social and technical interactions embedded in the communication revolution. A principal feature of this research programme is its emphasis on the complementarity of the social processes that influence our lives in off- and online environments. This emphasis is intended to correct the prevailing tendency in many studies of the communication revolution predominantly or exclusively to consider the potential for online interactions to substitute for offline interactions.

2. ADDRESSING THE DOMINANT DISCOURSES ON INFORMATION AND COMMUNICATION TECHNOLOGIES

The selection of the topics that are addressed in this book was originally made in the light of two observations about the predominant characteristics of much of the research that has been and continues to be undertaken on the implications of digital ICTs for the economy and, more generally, for society. The first was that the majority of work is confined to single social science disciplines and there are few opportunities for the accumulation of insights based upon systematic empirical studies. The second observation was that a substantial amount of research offers a snapshot of various features of the ways that social systems are interacting with technical systems, but provides very little insight into the dynamics of the processes that are giving rise to widespread and continuous change.

With respect to the first issue, there are now a few interdisciplinary examinations of the social and technical features of the so-called communication or 'digital' revolution.[1] We are beginning to see the publication of studies that are grounded in systematic empirical research and that are informed by consideration of theoretical insights into developments within the communication revolution that are drawn from two or more disciplines (Gauntlett 2000; Jonscher 2000; Woolgar forthcoming). This is important in the light of the fact that the plethora of reports on the 'new economy' or on 'knowledge' economies or societies offer essentially descriptive accounts of the changes that are presumed to be underway.[2] They rarely entertain questions derived from theory; neither do they offer critical reflections or analyses that are informed by received theory. The studies in this book address questions about the nature of the transformations that appear to be associated with the new technologies in a way that is informed both by social science research methodologies, which enable a more systematic account of these changes, and by an effort to build upon or to critique existing theoretical perspectives on the digital revolution. For example, we consider questions derived from a reading of received theory in the growing field of knowledge management and several other disciplines. In each chapter we seek empirical verification of the claims that have been made about the nature of many different types of 'virtual' organization of social and economic life.

With respect to the second issue, we take the view that a dynamic perspective is essential for an informed analysis of the transformations in human behaviour that may be associated with the application of digital technologies (Castells 2000; Gibbons et al. 1994). Our focus on the underlying dynamics of change—that is, the social processes associated with learning and new knowledge creation and the settings in which these processes occur—is the foundation for assessing the potential for alterations in the information exchange process and in the processes that contribute to the generation of new meanings. Although the empirical research reported in this book is, of necessity, time-bound, each of the authors makes an effort to locate his or her work within a conceptual framework that acknowledges the emergent (and in some cases evolutionary) characteristics of the interactions between the social and technical systems that provide the focus for his or her work.

---

[1] See e.g. Boisot (1995, 1998), Castells (1996, 1997, 1998), Ciborra (1996, 2000), Dutton (1996, 1999), Mansell and Silverstone (1996), and Mansell and Steinmueller (2000).

[2] There are many examples of such work—e.g. Department of Commerce (1999, 2000), Dizard Jr. (1997), and Duff (2000).

3. BINDING NETWORKS THROUGH MEDIATED INTERACTION

The conceptual canvas of this book is located in the traditions of social science research that are concerned with mediation—or intermediation. Intermediation is to be understood for our purposes as a dynamic relational process that binds (or unbinds) networks of individual actors or institutions (Coleman 1994; Cooley and Nam 1998; Cosimano 1996; Meyrowitz 1994; Morris and Hopper 1980; J. Sarkar 1998; White 1995; Yanelle 1989). The term 'intermediary' is generally defined as an intermediate person or thing, or one who acts as a mediator, and is derived from the Latin word *intermediatus* (*inter* + *medius* (middle)) (Pearsall and Trimble 1996). The term can also refer to a medium or means and to acting between persons such as an intermediate agent or agency (Flexner and Hauck 1996).

This definition implies the presence of at least three parties with some entity or person providing a link between the other parties. Intermediation processes involve ongoing interactions within and between the social, political, economic, and technical realms. Social science research has addressed the way these interactions produce, reproduce, and transform the social and economic order when they occur in the physical presence of others (Goffman 1963; LeFebvre 1991; Levinson 1997; Meyrowitz 1985, 1994). An increasing number of studies focus on why social interactions might be expected to change significantly when they involve mediation by the new digital technologies and services (Bolter and Grusin 1999; Mitchell 1996; Shapiro and Varian 1998).

Bolter and Grusin (1999) argue that the 're-presentation' of one medium in another should be referred to as 'reintermediation'. They justify this argument in the following way. 'We have adopted the word to express the way in which one medium is seen by our culture as reforming or improving upon another' (Bolter and Grusin 1999). From the Latin *remederi* (to heal or restore to health), this term draws attention to the increasingly common expectation on the part of many commentators that the use of the 'new' or 'digital' media is unquestionably an improvement upon earlier generations of technologies in terms of their impacts. These authors recognize that this view needs to be examined critically—and empirically—to discern the specific features of innovative technologies and their implications for mediated interactions of all kinds.

In our work, we have sought to acknowledge that there may be fundamental (and positive) changes in individuals' preferred modes of social interaction when digital technologies are available. However, there may be no change, or only relatively little positive change, in individuals' preferred modes of social interaction. There may be change that is regarded as being regressive or harmful. These possibilities are central features of the insights offered by the authors of several of the chapters in this volume.

It is acknowledged by many contributors to studies of innovation in the science and technology field that behavioural and cognitive outcomes associated with the use of new technologies often diverge substantially from those that were 'planned' or 'intended' by participants in the technological design process or by those implementing digital applications and services (Bijker, Hughes, and Pinch 1987; Hughes 1987; Mansell and Silverstone 1996). This may occur because individuals resist the initially 'intended' uses of the new technologies and services. People may find unexpected ways of integrating the new technologies and services into their daily activities, or they may opt for non-use. Alternatively, even when the technologies are used in ways that appear to be consistent with initial expectations, the users themselves may have a variety of interpretations of their own behaviour and its consequences for themselves and others (Silverstone 1994, 1999; Silverstone and Haddon 1996).

When intermediation occurs through software, computing, or telecommunications systems, it involves much more than changes in the structural linkages between people and the technical interfaces supported by digital ICTs (A.J. Kim 2000; Kollock 1994, 1999; Mitchell 1996, 1999; Rheingold 2000; Zerdick *et al.* 2000). Intermediation in both physical and electronic space is giving rise to many new patterns and modes of communication and information exchange. Some observers claim that these new developments contain the seeds for revolutionary changes in all aspects of social and economic life, including the processes of knowledge creation and application (Brown and Duguid 1998; Gibbons *et al.* 1994; Shapiro and Leone 1999). Innovations in business processes and organizations, innovations in governance systems, and changes in perceptions of social status certainly do seem to be closely allied to the spread of ICT applications (Mansell and Wehn 1998). These technologies offer the potential for forming new types of network relationships that are not tied to physical places and that are not time-bound in the same ways as in the past (Cairncross 1998). However, the extent and implications of such changes are the subjects of much speculation (Leadbeater 1999; Negroponte 1995; Tapscott 1995; Tapscott, Ticoll, and Lowy 2000).

By focusing on what the coupling of social networks in the physically present world with the virtual world means for social actors in a variety of organizational and institutional settings, we provide insights into why people may resist certain types of technically mediated communication in some instances while, in others, they may seek to accommodate or shape the uses of these new technologies. Thus, for example, we examine the conditions that may sustain or limit the development of virtual communities (networks of individuals choosing to interact using the Internet and other forms of network technologies). We consider the distinctive perceptions of social ordering that seem to exist among very experienced Internet-users, but that do not necessarily inform their offline interactions. The dynamics of certain forms of electronic business (in the newspaper and banking industries and in the insurance market) are also examined in terms of their distinctive features and their similarities with social processes that are present in both online and conventional offline business settings.

We examine features of the learning processes that are involved in designing and using the new technologies. These learning processes can give rise to rejection of or accommodation to new technological systems. They appear to be closely related to the politics of interpersonal interactions and their associated power relationships, but this may have little to do with the characteristics of the new technologies themselves. Politics are also implicated in the interactions between social and technical systems that we examine in this book at the institutional level. At this level we consider changes in regimes for the governance of the new technologies and for sustaining new forms of virtual organization.

The studies in the chapters that follow are concerned with many of the determinants of the coupling processes between social and technical networks and their outcomes as interpreted by their developers and users. Most of the empirical data have been collected via in-depth interviews and through the design and application of original survey instruments administered either in face-to-face settings, by post, or via the World Wide Web. Our overriding goal is to go beyond many conventional accounts of the features of the communication revolution to assess why some ICT applications are being used and adapted enthusiastically while others are meeting with resistance or rejection. Section 4 explains how the research agenda that links the studies collected in this book was formulated.

4. FRAMING THE THEORETICAL AND EMPIRICAL RESEARCH AGENDA

We refer to the 'communication revolution' in this book.[3] We use this phrase as short-hand for the 'information and communication technology revolution'. Christopher Freeman and his colleagues have examined the profound significance of the changes that appear to be associated with an emerging techno-economic paradigm that rests upon innovations in ICTs (Freeman 1988). This tradition of analysis of the relationships between technological change, innovation processes, and the economy is based, in part, upon Joseph Schumpeter's examination of the causes and consequences of technological change for the economy (Schumpeter 1947, 1961). Schumpeter suggested that periodically a specific ensemble of enabling technologies emerges around which, through selective attention by economic actors such as entrepreneurs, a type of lock-in begins to develop. A new economic model or paradigm emerges that challenges the hegemony of the former predominant ways of organizing economic life. The replacement of one techno-economic paradigm by another involves pervasive and all-encompassing shifts in social and economic organization and these must be expected to affect every aspect of the economy (Schumpeter 1961).

The 'new economy' has been coined as a label that captures some of the features of the shifts associated with the communication revolution. This label directs attention to the increasing salience of services and immaterial transactions in the global economy. However, it provides little assistance in directing attention to the need to probe how the 'old' economy is likely to change as the 'new' economy expands and as the new technologies are used within 'old' economy sectors. Nor does it assist in highlighting the nature of the new forms of face-to-face social relationships that emerge alongside the more conspicuous developments in virtual relationships.

Christopher Freeman and Carlota Perez suggested that, as the application of these technologies becomes a key factor in economic and social development, the efficacy of existing ways of organizing both economic and social life is called into question (Freeman and Perez 1988; Freeman and Soete 1997; Perez 1983). And, as innovations

---

[3] The framing of the research agenda for the studies reported in this volume arose out of discussions among the contributing authors over an extended period of time. As editor, Robin Mansell has compiled this synthesis and bears principal responsibility for it. She acknowledges the contributions of all the chapter authors, and particularly David Neice for his contribution to this section of Chapter 1.

in technologies and social and economic organization occur, they argue that there are likely to be substantial costs in social, organizational, and political adjustment.[4] One adjustment scenario foresees that governance institutions that hitherto have provided the sites of social and economic power and societal control will be radically reshaped with very uncertain outcomes for groups within society including citizens, owners, workers, and managers. Another adjustment scenario envisages an accumulation of incremental changes in the technical, social, and economic spheres of activity as the new technologies are accommodated and, in some instances, resisted (Freeman 1994*b*, 1995; Freeman and Soete 1997; Mansell and Steinmueller 2000).

The new techno-economic paradigm that is at the heart of the communication revolution is characterized by a growing emphasis on the production and distribution of knowledge (Cohendet and Steinmueller 2000; Cowan and Foray 1997; David 1995; Eliasson 1990; Foray 1995; Neef 1998; Nonaka and Takeuchi 1995; OECD 1996; Shapiro and Varian 1998; Steinmueller 2000; World Bank 1998). This new paradigm is linked to the widespread use of microelectronics technologies. These technologies are achieving a position as *the* dominant factor influencing the latitude for social and economic development. As Freeman argues, ' "Intangible" investment in new knowledge and its dissemination are the critical elements, rather than "tangible" investment in bricks and machines' (Freeman and Soete 1997: 3). Romer (1995) has also suggested that hardware, software, and 'wetware' or human capital are becoming replacements for capital, raw materials, and production and non-production workers, in terms of their importance in the economy.

According to many recent accounts of the development of so-called knowledge economies, human capital, human organization and management, and human interaction, whatever the purposes of the agents, are increasingly informed by, or mediated by, some combination of hardware and 'wetware'. In some parts of the management studies, marketing, and technical trade literatures, managing knowledge is regarded as a process involving the 'harnessing' of advanced ICTs (Norris and West 2001; Ruggles 1997). Successful 'harnessing' is expected to produce many positive results including improved decision making and problem-solving capabilities on the part of human and technological agents (OECD 1997). Thus, the analysis of effective strategies for knowledge management and for the deployment of the new technologies, for example, to support electronic commerce, is often fused in the research that underpins many of the studies on the social and economic implications of the communication revolution (Anderson 1997; Garcia 1995; Information Infrastructure Task Force 1997; Kalakota and Whinston, 1997; Kalakota, Robinson, and Tapscott 2000; OECD 2000; Timmers 1999). There are frequently very strong assumptions in such studies about the synergies between social and organizational processes and the specific designs and architectures of the new technological systems. They are expected jointly to deliver improved decision-making and problem-solving capabilities (OECD 1997).

---

[4] European research in this vein has been complemented in Canada and the United States by the work of Richard Lipsey (1991, 1994) and others associated with new growth theory (Helpman 1998; Romer 1990, 1993, 1994; Stiglitz 1999).

Some proponents of the idea that the new paradigm will bring automatic benefits to all people argue that a global information society is on the verge of enveloping us all (Dizard Jr. 1997). This process, it is claimed, in which the world's social and economic order will become largely immaterial, will favour a vast number of virtual associations and networks from which no one will be excluded (Rheingold 2000; Romm, Pliskin, and Clarke 1997; Sarkar, Butler, and Steinfield 1998). This is a very positive view of an inclusive 'new economy' and global information society. Immaterial life is expected to favour ever-closer encounters between social and economic actors. These intensive networks of social and economic interaction, enabled by ICTs, are expected to become sources of long-term sustainable growth and development and a more inclusive social order.

In the social science community, the rapturous claims about the potential of the 'new economy' have been met with dissent (Garnham 1994, 1996; Mansell and Silverstone 1996; Robins and Webster 1999; Webster 1995). Some of these dissenters argue that there is an embedded bias towards exclusion and disadvantage for certain people as the paradigmatic shifts involving the application of digital technologies become more deeply integrated within the social and economic order (Angell 2000; Garnham 2000; Webster 2001). Others argue that the technological foundations of our social order do appear to be changing substantially and that there are both significant benefits and substantial risks (Mansell and Wehn 1998). The extent of the benefits and the risks and their distribution throughout the social and economic order is a matter for empirical investigation. Empirical research does in fact suggest that there are opportunities to shape and manage these changes in line with a broad range of preferences of technology designers and users (Mansell and Steinmueller 2000; Silverstone 1999), but that so far we have made very little progress in formulating what preferences are becoming predominant and embedded in the new technologies or in assessing whether these are socially desirable.[5] We suggest that these issues must be investigated and our research agenda has provided a small step towards that goal. The results of our research are intended to provide a basis for encouraging measures that may influence the deployment of the new technologies in ways that promote greater social inclusion and social and economic well-being. The benefits and the risks associated with the transformations that are giving rise to a new paradigm of techno-economic and socio-political organization, which seems to privilege virtual forms of communication, are also the focus of a relatively small number of detailed empirical studies about what the implications of the communication revolution are likely to be in practice (Dutton 1996, 1999; Woolgar forthcoming).

Some analysts have asserted that the implications of the communication revolution for the social, political, and economic organization that served in an earlier time to mediate between individual actors and other institutions including the market are that they will be swept away by the creative destruction of the 'new economy' (Armstrong and Hagel III 1996; Easterwood and Morgan 1991; Sarkar, Butler, and Steinfield 1998). The coordinating roles performed by intermediaries of many kinds will no longer be needed in an environment where direct electronic links can be established between

---

[5] To an extent this argument is similar to that offered by Lessig (1999).

social actors. However, these 'end-of-intermediation' proponents are now in retreat. This is largely because individual and organizational intermediaries can be observed to be redefining their roles and very few instances of sustained 'disintermediation' can be seen. The empirical challenge now is to determine which specific new forms of inter-mediation, or indeed reintermediation, are emerging and with what consequences for individuals and groups and their practices (Hawkins, Mansell, and Steinmueller 1999; Mansell, Schenk, and Steinmueller 2000; Verhoest and Hawkins 2000).

As indicated above, the conceptual framework that informs the research reported in this volume extends insights into the social features of recent paradigmatic transfor-mations by building on the work of Freeman and Perez.[6] Their work on technological innovation has sometimes been criticized for its 'determinist' viewpoint (MacKenzie 1996a; Williams and Edge 1992; Williams and Slack 1999).[7] We suggest, however, that their work on the nature and determinants of technological revolutions and their implications for the economic and social order makes no claim that there cannot be resistance to the predominant forms of paradigmatic change. In fact, Freeman (1992b) advocates actions that will promote efforts to refashion the emergent ICT paradigm so that it is more consistent with generating economic prosperity and with social benefits (including environmental sustainability) for all. Freeman and Perez's argument that is set out in their numerous works also makes no claim that the application of digital ICTs will lead, for example, to the complete dematerialization of social and economic con-duct. Nor do they argue that virtual social interactions should be privileged. In fact, Freeman and Soete (1994) advocate continuing emphasis on social interaction in the physically present moments of life in order to ensure that the goal of equity in social and economic development is achieved.

In this book, our position with respect to the role of innovative digital technologies and services is as follows. We argue that it is not technology *per se* that is responsible for a shift in the techno-economic paradigm. Rather, it is the interactions between emerg-ing technological forms and human beliefs, perceptions, and choices that, together, comprise a techno-economic paradigm. A techno-economic paradigm involves 'a new set of guiding principles' or common-sense practices (Freeman 1992a: 165). These evolve as a result of the dynamic interactions or mediations that occur within social and technical networks. These networks are forged mainly within the contours of the digital communication revolution but also within the contours of networks of rela-tionships that have not been touched by this revolution except perhaps by the conse-quences of their exclusion.

The emerging techno-economic paradigm rests on a new material factor of produc-tion—that is, relatively inexpensive microelectronics technologies and their associated digital embedding of codified knowledge. The declining price of both microchips and digital information means that the earlier paradigm, based principally on cheap oil and the movement of things, is giving way to a paradigm that privileges the rapid movement

---

[6] Other important contributors to these insights include Cawson, Haddon, and Miles (1995), Miles and Thomas (1995), Miles *et al.* (1999).

[7] And see Freeman (1987) himself on the theme of technological determinism.

of information. The expectation is that this will lead to the application of knowledge in ways that will generate sustained and global social and economic development, and especially poverty reduction (KPMG 2000; Primo Braga 2000). There is no doubt that the ascendancy of this new ICT paradigm is taking hold of the imaginations of entrepreneurs, managers, and engineers. It is beginning to grip the collective social psyche and to invade popular consciousness in the industrialized countries and, increasingly, in many parts of the developing economies. The new technologies and their associated services are being regarded by some people as a new means of achieving the long awaited 'catch-up' of many of the developing countries with the wealthy industrialized countries (Goldstein and O'Connor 2000; Mansell 2001). However, little is understood as yet about the detail of how such developments—and their potential for eliminating poverty—might be realized in practice. It is clear that the simple availability of the new digital technologies and services is no elixir for a 'catch-up' process. The prospects for a global economy, and for a more socially inclusive global information society, depend on a vast number of non-technical processes and developments.

In recent years, greater emphasis in the social science research community has been given to examining these non-technical features of the communication revolution. Researchers are investigating the social, cultural, political, and economic issues associated with this revolution in considerable detail (Castells 2000; Dutton 1999; Kling *et al.* 1999; Silverstone 1999; Woolgar forthcoming). The results of this work are beginning to reveal why it is important to challenge the initial exuberance and hype surrounding 'dot.coms' and assumed benefits of the 'e-society' that spring from largely undocumented and unsubstantiated claims about the role of ICTs in generating improvements in knowledge management processes (Davenport and Prusak 1998; Ruggles 1997). The accumulation of new empirical evidence suggests that a tempered view of the positive and negative features of change is appropriate (Woolgar forthcoming). This work also indicates that there is a major research agenda to be pursued to provide systematic analyses of how people's lives and livelihoods are being affected by the introduction of the new systems and services.

The future of the communication revolution is very uncertain. It might be argued, therefore, that social scientists can do little more than wait for its outcomes to examine the implications. We take a different view, however. The accumulation of theoretically grounded and empirically informed social science investigations of these developments can be an important component of the necessary foundation for deliberations on the social and economic value of developments in the social and technical realms and about whether they are consistent, or at odds, with the values of the social actors whose lives they are affecting. Social action can then be mobilized to guide and, in some instances, to reshape the character of the emergent technological order.

Each of the studies in this book investigates an aspect of the 'guiding principles' that seem to typify the new forms of interaction between social and technical systems that are mediated to some extent by the new ICTs. The authors consider whether or not these relationships, as perceived from the vantage points of different social actors, are encouraging social and economic relationships that are valued. There are indications of beneficial contributions from some of the new ways of organizing business and social relationships, but there are also problems with adjustment to the new ways of working and living. The potential for significant disruptions is also a possibility, as Christopher Freeman and Carlota Perez have suggested. Such disruptions are likely to be perceived by social actors as being inconsistent with emerging social values in various locations throughout the world. In our research, we try to be sensitive to an important distinction: that between the wide range of potential characteristics of the emerging techno-economic paradigm and the main features of the dominant technological and institutional regime. Many of the potential features of the ICT techno-economic paradigm are unlikely to become predominant in the absence of analysis of current trends and independent reflection on the implications of these trends. The possibilities for alternative regimes and choice within the broad contours of the emergent paradigm create opportunities for mobilizing social action to achieve a greater variety of means of mediating social and technical interactions.

The scope for choice and variety within the ICT revolution is influenced by the way that the technological innovation processes interact with humanly constituted selection factors. A given paradigm may come to be perceived as a dominant regime when it appears to have become very strongly entrenched (Dosi 1982; Dosi and Malerba 1996; Freeman 1988; Freeman and Perez 1988; Freeman and Soete 1997). This may happen when political and social actors come to have little or no capacity to imagine alternatives or to act upon them. Given the uncertainty even among the most ardent proponents of the benefits of the communication revolution about what the predominant features of the new paradigm will be, it is important to examine the interplay of power and of resistance to alternative outcomes that involve social actors with many different interests (Mansell and Steinmueller 2000). The results of this approach to the analysis of the potential and risks of the communication revolution may create opportunities for the selection of new pathways and alternative dominant technological regimes or perhaps permit greater latitude for smaller diversions.

The latitude for social actors to make choices about the design and use of ICTs is considered in a wide variety of contexts in this book. Castells (2000: 16) argues that:

networks, as social forms, are value-free or neutral. . . . They process the goals they are programmed to perform. All goals contradictory to the programmed goals will be fought off by the network components . . . once the network is programmed, it imposes its logic on all of its members . . . To assign different goals to the programme of the network . . . actors will have to challenge the network from the outside and in fact destroy it by building an alternative network around alternative values.

In this book, we focus on the distinct processes of mediation within the 'old' and the 'new' social and technical networks to discover the contours of the new patterns of social and technical interaction and to suggest a basis for policy action to correct features that appear to be inimical to the goals of achieving equitable and inclusive information societies. To an extent, therefore, our research begins from the premiss that there is potential for alternatives to the emergent dominant networks' 'logics' or regimes, but we suggest that these must be revealed through analysis of the dynamics at work within the communication revolution. Thus, we depart from Castells's admonition as we seek in many instances to inculcate alternative values by investigating the inside of the emerging network societies.

5. STRUCTURE OF THE INVESTIGATION

The book is organized in two main parts. The first focuses mainly on how digital technologies are increasingly implicated in mediating social and technical relationships. The second part is concerned mainly with the learning processes, both individual and collective, that seem to be essential to the building of capabilities for using new technologies to promote new forms of information exchange that contribute to wider processes of knowledge exchange.

5.1. *Part One. Mediating Social and Technical Relationships*

Part One offers six perspectives on mediation processes that infuse and inform interactions within increasingly complex social and technological networks. These chapters illustrate the diversity of these mediation processes. The authors draw upon theoretical perspectives on the nature of human choice, the articulation of ICT developer and user preferences, and the way that power relationships come to be embedded within emergent social and technical relationships. Empirical research on the social perceptions of intensive Internet-users and the developers of electronic commerce shows how both technical and social mediation processes are influencing perceptions of social status in the virtual world without necessarily destroying existing perceptions in the offline world. This research also illustrates how changes in the organization of work are exploiting certain features of technically mediated communication, but that, at the same time, these changes rely upon many kinds of non-virtual working practices. Building trusting relationships to support electronic business transactions also requires a detailed understanding of why users select certain information and communication systems to support or augment their decision-making or problem-solving capacities, and yet reject other technological systems. We explain such choices over technologies and organizational practices in the light of features of the mediation process itself.

Edward Steinmueller begins the exploration of these issues in Chapter 2 with a challenge to conventional views about the formation and sustainability of virtual communities in the 'new economy'. He focuses specifically on the 'technology' of social organization—that is, on whether virtual-community members who interact through the use of computer-mediated communication are likely to offer a major opportunity for organizational innovation or whether they simply represent a technological novelty that will not really succeed in generating major changes in the way society engenders the institutional and procedural authority that is necessary to sustain the economy. The aim of his chapter is to assess whether the new technologically mediated forms of communication can be expected to support innovative means of production and exchange in the 'new economy' that is regarded by many as being at the heart of the communication revolution.

In Chapter 3, David Neice focuses on the potential for intensive interactions between users of the Internet to give rise to new perceptions about social distinction and about appropriately mediated social interactions on- and offline. This chapter offers insights into how access to digital technologies is interwoven with perceptions of social status in society. His research was designed to elicit perceptions and interpretations of the social distinctions that really matter to experienced users of the Internet in their interactions online. This work reveals the very complex interactions between the technical and the social in the online and offline worlds. It suggests, for instance, that the attribution of social esteem occurs through systems of rules and values that amplify peer reciprocity and social exchange. The potential for Internet use to amplify certain values and behaviours associated with reciprocity does not appear, however, to be inconsistent with commercial forms of exchange that are colonizing the Internet, as has been suggested by some authors (Goggin 2000). Neice argues that policy-makers and entrepreneurs will benefit from further exploration of the complementarity of social systems in the on- and offline worlds. Given the high stakes for investors in new digital technologies—and the vicissitudes of the stock market—Neice's strategy for the analysis of social and technical mediation processes and the way they limit possibilities for human agency on- and offline offers a valuable foundation for further research.

In Chapter 4, we move to a consideration of how meanings are created by those who are requested to perform at least a portion of their work virtually, using different combinations of ICTs. Jennifer Gristock contests the oversimplified suggestion that there is a trend towards the emergence of completely virtual organizations. She suggests that too much effort has been devoted to formulating definitions of *the* 'virtual organization'. Instead of attempting to define the key features of an emerging organizational form, she analyses how the possibilities for organizing work activities across space, time, and organizational boundaries are actually being extended through the use of a variety of combinations of ICTs. In this chapter, the empirical evidence comes from a web-based survey of virtual teamworkers and detailed case studies of the UK newspaper industry. Gristock investigates the way that face-to-face meetings and the use of communication media are combined by virtual teams who write editorial copy and sell advertising space. She argues that insights into the varying degrees of 'virtual-ness' associated with different facets of work are vitally important for understanding how this kind of teamworking differs from the kinds of work in non-technology-enabled organizations. She demonstrates that the mere possession of a particular set of technologies does not guarantee their efficient use, nor does it ensure that they are put to effective use for any particular purpose. Knowledge creation in virtual teams is shown to vary with the extent to which team members are separated by geographical, temporal, and community barriers and the extent to which they are able to create shared contexts of meaning through their physically present and electronically mediated relationships.

Electronic commerce is one of the fastest-growing services to be supported by ICTs. It is gaining in status as it comes to encompass a wide range of electronically mediated commercial practices. In Chapter 5, Ingrid Schenk provides an analysis of how private- and public-sector actors are attempting to replicate trusting relationships that exist in offline commercial environments in support of their electronic commerce endeavours. The empirical study for this chapter focuses on whether the forms of mediation embodied in digital certificates that are used to authenticate the identities of those participating in electronic transactions are likely to create a sound basis for the expansion of business-to-consumer electronic relationships. Drawing on interviews with representatives of firms and policy-making organizations in the United Kingdom, the United States, and Canada, Schenk illustrates how alternative technical choices are contributing to the perceived legitimacy of traders who become involved in electronic transactions. She argues that the technical interfaces for electronic certification are interwoven with commercial practices and that this has major implications for the growth potential of the new electronic markets. This is because these interfaces influence, not only whether people will be attracted to cyber-trading, but whether they will persist in this form of commercial exchange.

Electronic commerce is also the focus of Chapter 6 and the sector examined is the insurance industry. Gordon Rae challenges a predominant set of ideas about how markets should function when they are linked to a rapidly evolving technical environment. In most cases, electronic commerce is expected to transform or refine business processes by helping purchasers to make better decisions. ICT systems are expected to allow them to amass, analyse, and control large quantities of specialized data. Rae suggests that it is important to consider the influence of the cognitive frames of electronic commerce developers and users to understand the situations in which the availability of the new technologies will be perceived as a positive development by their potential users. Rae investigates the phenomenon of 'presence', a concept that encompasses cultural and social perceptions, to draw attention to the processes whereby users seek and find social meanings, whether or not they were intended by electronic commerce system designers in the insurance services market. Rae shows just how different the construction of such meanings can be for participants in the same information system design process. He also offers a glimpse of the extent to which people in the London insurance market accept computerized systems as invisible intermediaries and as social agents capable of competent performance.

Social perceptions and the attribution of meaning to certain kinds of information that must be exchanged in the banking industry in the course of credit assessment is the subject of Andreas Credé's contribution in Chapter 7. He investigates the process of knowledge production and the role of ICTs in commercial banks. His examination of bank employees' selection of both older and newer ICTs confirms that the sector has experienced continuous technological change and major investment in computer hardware and software. Nevertheless, banks remain highly labour-intensive service organizations that are dependent on skilled staff. Banks are said to exist because they are efficient information processors and the use of ICTs in this sector is often expected to make information processing more efficient and less costly. Information can be digitized so that it can be readily stored, communicated, and processed through progressive automation. Credé's research suggests, however, that the principal role of the new technologies is to 'repersonalize' human communication rather than to process information. He examines how information is actually produced and exchanged within the banks in his sample for the purpose of credit evaluation. He finds that 'repersonalization' is an important mediation process, because bank employees depend on confidential, proprietary information that is made available to them by their clients. Bank employees, therefore, can be regarded as savings intermediators, performing their roles by adding validity, relevance, and significance to information so that it can be used for credit evaluation. Credé suggests that this social process has important implications for how information is exchanged and for the selection or rejection of different types of ICT systems.

5.2. *Part Two. Building Capabilities for Knowledge Exchange*

Part Two turns to investigations of the formation of capabilities for knowledge exchange when some part of the information exchange processes involve the use of ICTs. This section of the book focuses on how these capabilities are accumulated through a variety of social mediation or learning processes. It includes studies of learning processes at the micro-level within firms and at the institutional level. Empirical work undertaken in the United Kingdom, South Africa, and Brazil provides the basis for this section of the book. The emphasis is on the importance of organizational contexts in shaping learning outcomes associated with the design and use of information and communication systems. These organizational contexts influence social perceptions of barriers to the use of information systems, the social values that become embedded in technological systems such as the Internet, and whether social actors can acquire the necessary range of capabilities for the production of advanced information and communication systems.

In Chapter 8, Jane Millar compares the experiences of information systems developers and users within two firms in the United Kingdom to examine how different organizational contexts can influence perceptions of the effectiveness of information system design and organizational change processes that employ identical techniques and software tools. The analysis focuses on the relationships between efforts to stimulate knowledge sharing and learning among participants in the design process and perceptions about the usefulness of the learning outcomes. Collaborative involvement of technical and business specialists in the design of business processes and information systems—or co-design—requires the mobilization of employees' and systems designers' knowledge. Millar suggests that, when knowledge sharing and learning seem to occur, the specific characteristics of the organizational context influence the way that the new knowledge will be exploited and whether it will become embedded in new technical systems. This chapter highlights the importance of analysing the processes of knowledge production within firms that are situated in particular contexts.

Perceptions of barriers and resistances to using ICT systems in ways that developers envisaged are often investigated using research methods that elicit individuals' recognition of their effectiveness using participant observation or in-depth interviewing techniques. Barriers to the effective use of new systems are often attributed to technical problems, to skill deficits, or to a lack of user awareness of the potential benefits of adopting certain technologies and practices. In Chapter 9, Uta Wehn de Montalvo draws on the results of a comprehensive survey of users of geographic information systems (GIS) in South Africa to identify some of the reasons for resistance on the part of users to the use of these systems in ways that GIS developers argue would reduce their costs and encourage their wider use in developing countries. The complexity of GIS users' perceptions about the advantages and disadvantages of sharing spatial datasets, a practice that could extend the application of GIS, is investigated in this case. Wehn de Montalvo's analysis reveals a rich portrait of users' attitudes and intentions with respect to data sharing. She shows that the promoters of data sharing need to take account of the social and psychological determinants of data sharing if they are to succeed in reducing the users' resistances to developing a data-sharing culture, thereby facilitating the greater use of this ICT application, which could potentially support social and economic development goals.

Daniel Paré offers another consideration of the non-technical or social dynamics that influence the specific formation of ICT systems. In Chapter 10 the focus is on the political processes that influence the organization of the Internet's governance framework. Paré examines changes in the way the Internet Domain Name System is organized based on an analysis of the viewpoints of those who participated in the creation of a new governance regime in the United Kingdom. This chapter shows that the restructuring of the management and administration of the *.uk* name space was informed by technical features and by the way the Internet Service Providers (ISPs)—the new intermediaries—intermediated relationships between those wishing to register domain names and the domain name registry organization itself. Paré shows that the process of constructing new capabilities for governance of the Internet involves the management of values and choices. Political processes played a very important role in whether organizational innovations were perceived as a success or as a failure by different actors. The analysis demonstrates the value of coupling an investigation of the determinants of the technical architecture of the Internet and its governance arrangements with an analysis of the interests of its designers and users.

In Chapter 11, Ana Arroio examines the variety of capabilities that appear to be necessary for developing a new technological system, in this case, a domestic low earth orbiting satellite (LEOS) network. An analysis of the proposals of public and private actors in Brazil to develop their own LEOS system is used to demonstrate that social actors must accumulate certain specific capabilities if they are to succeed in developing and implementing new technological systems. Many of these capabilities are essential in order to strengthen the resolve of members of national governments and employees of firms to take actions that may enable them to produce an advanced technological system. Arroio traces the informal and formal learning processes over an extended period. She shows that, although LEOS were championed in the 1990s as a means of improving the communications infrastructure in the middle-income countries, and despite the accumulation by actors in Brazil of key technical and regulatory capabilities, substantial mismatches between stakeholder objectives in government, in the satellite manufacturing industry, and among the proposed communications service suppliers produced high levels of uncertainty and failure to construct an indigenous technology system. In the Brazilian case, the social actors were unable to produce a LEOS system that might have had the potential at least to be better geared to the social and economic needs of the country. Arroio suggests that too little attention was given to building up the actors' negotiating strengths and to creating regulatory practices that might have encouraged the use of LEOS services to extend communications services especially in rural areas in line with stated policy objectives.

In the concluding chapter (Chapter 12), Robin Mansell assesses the results of our research agenda, which has been aimed at exploring the social and technical mediation processes that are deeply embedded within the communication revolution. An important observation emerges from these results. Interactions between conventional and new forms of social and technical mediation are crucially important for the interplay between technical and social developments in the 'old' and the 'new' economy or society. The new patterns of mediation involving digital technologies are contributing to the ascendancy of a new dominant technological regime. However, this regime is the product of many interlaced combinations of virtual and physically present social processes. The regime is extraordinarily dynamic. The regime itself is the product of interdependent technical and non-technical features and processes, and its malleability is often not considered by those seeking either to explain or to predict the outcomes of the communication revolution.

Our investigations into mediated communication involving ICTs provide a point of departure. The insights revealed by our research confirm that there is no generic 'one-size-fits-all' model for the deployment of digital technologies. There is room for diversity within a digitally mediated global information society. However, this diversity must be constructed through continuous negotiation by well-resourced social actors. An emphasis on variety is important for ensuring that local articulations of values and preferences that are socially accepted do, in fact, become embedded in our social and technically mediated interactions. This is not to offer a postmodern account of a multiplicity of emergent paradigms encompassing the fragmentary experiences of individuals within their situated contexts.[8] Our results display emerging patterns of technical and social interactions between technology designers, citizens and consumers, and business people and policy-makers. They offer insights that provide a basis for enhanced learning and reflection by these stakeholders. They also provide a foundation for structuring social and technical interactions in ways that elicit the more desirable features of the ICT paradigm. They highlight a basis for alternative actions that may enable these features to become embedded in the technological regime that comes to be associated with the ICT revolution as it progresses. As Garnham (2000: 135) argues in the particular case of the consumption of media products,

It is this view that forms of consumption are in complex ways embodied in different forms and institutions, and that they in their turn reinforce certain personal and social character traits, that

---

[8] Such as the accounts offered by Baudrillard (1988). Garnham (2000: 114) argues for a distinction between mediated social communication involving, for instance, the media and dialogic interaction. He does so on the basis of the observation that, in the latter face-to-face contexts of dialogue, the interchange of meaning is constantly reciprocal and immediate. We apply the term 'mediated' to all forms of interaction, but, like Garnham, we emphasize that the two kinds of mediated experiences must be treated as instances of phenomena that have similarities as well as differences.

is the rational core of McLuhan's theories, which in their turn derive from studies of the ways in which the development of printing and reading and the shift from orality changed society and the individuals within it. If so much is granted then we also know that the production and distribution of cultural commodities, what is made available for consumption and to whom, is structured—and intentionally structured—in specific, determinate ways. If the connection to individual and group identity formation is granted, then how that power of structuring works and with what effects becomes a matter of legitimate interest.[9]

In order to understand these structuring processes in the broader context of emerging paradigms for ICTs, we argue for a multilevel analysis of structuring processes that embraces both the micro-level mediated experiences of individuals within their social contexts and the institutionally mediated experiences of people that are informed by specific aspects of the economic and political environment and their interactions with technologies.

---

[9] Garnham's reference is to Marshall McLuhan (1962: 31), who claimed, for example, that 'electronic interdependence recreates the world in the image of a global village'.

PART ONE

MEDIATING SOCIAL AND TECHNICAL RELATIONSHIPS

2

Virtual Communities and the New Economy

W. EDWARD STEINMUELLER

1. INTRODUCTION

This chapter is concerned with a new form of social grouping, the 'virtual community'. The most elementary defining characteristic of a 'virtual community' is that its members meet and interact through the use of computer-mediated communication—that is, the use of audio-visual display devices such as personal computers connected through a network such as the Internet. The virtual community is located at the boundary between organizations and other social groupings that exists despite the absence of procedural and institutional authority.

The principal question to be addressed is, does the virtual community have more to offer for issues of organizational innovation and change than its technological novelty? This question can be stated as a choice between two alternative propositions. This first is that virtual communities are simply organizations, with identifiable institutional and procedural authority, whose activities are conducted with the aid of a new technology, computer-mediated technology. The second is that virtual communities do, or have the potential to, provide novel approaches to procedural and institutional authority, and thereby are capable of generating new types of social structure. Deciding between these alternatives involves assessing claims that have been offered about the meaning and significance of virtual communities to their members as well as analysing how the 'enabling' or 'enhancing' possibilities of computer-mediated communication might support the latter of the two alternatives.

Answering the principal question involves four investigations that comprise the main headings of this chapter following a brief review of the economic and organizational terminology employed in the chapter. Each of these investigations is based upon a simple question. First, do virtual communities make it possible to achieve unique outcomes in voluntary association? Secondly, do these outcomes rely upon a fundamentally new ethos or principle of human association? Thirdly, is the basis for the

The research conducted for this chapter was supported under the Targeted Socio-Economic Research Programme (TSER) of the Directorate of Research of the European Commission in the project European Sectoral Systems of Innovation (ESSY) coordinated by Franco Malerba, Bocconi University. The author is grateful to Professor Malerba and the Commission Project Officer for ESSY, Ronan O'Brien, for their intelligent guidance of the project.

emergence and sustenance of virtual communities sustainable? Fourthly, do the factors influencing the competition between different types of virtual communities indicate a collapse in their variety? The short answers to these questions are, respectively, yes, no, yes, and no. The conclusion summarizes what these investigations contribute to answering the principal question.

2. ECONOMIC AND ORGANIZATIONAL CONTEXT AND TERMINOLOGY

A principal aim of economics is to explain how the social processes of organization, production, and exchange are best configured for utilizing limited resources to meet the material needs and desires of human beings. The two simplifying assumptions that allow economists to claim such a large canvas of social interpretation are, first, that human needs and desires are exogenous to economic analysis,[1] and, secondly, that organization, production, and exchange are governed dually by technological possibility and the search for individual advantage.[2] These assumptions impose effective constraints on the possible outcomes of the economic system and an analytical structure for examining its operation. The widespread use of information and communication technologies (ICTs) introduces a number of challenges to economic analysis, some of which may be readily accommodated, while others are more profound. This chapter focuses upon changes in the technology of social organization and upon the changes in ICTs that enable social innovation—that is, the reconfiguration of the social processes of organization, production, and exchange.[3]

These social innovations are one of the foundations of the 'new economy'. As the contributions in this book make clear, these innovations involve a complex mixture of opportunity and threat, which makes their critical examination an excellent candidate for social science research. This chapter focuses on broad changes in the technology of voluntary association that are enabled by computer-mediated communication. These changes are only one aspect of emerging organizational change (which also includes important developments in production and exchange within and between organizations). Further, organizational change is only one feature of the new economy. The emerging new economy also involves changes in the composition of human needs and desires and in the technology of production and exchange. These additional changes

---

[1] For some economists, the efforts to construct a theoretical and empirical basis for 'endogenous taste formation' represent a closure of this open loop in the system of economic analysis. Other economists and most social scientists remain sceptical.

[2] Although individual advantage may be broadly interpreted to include altruistic motives and behaviour, the extent of this altruism is clearly rather limited in the light of observable differences in income distribution in all market-based societies.

[3] The use of the word 'enable' is the result of the current fashion to reject 'technological determinism' in favour of 'social constructivism'. There are two short answers to this controversy. First, technological changes that produce major opportunities provide a powerful incentive for someone to become the social actor in their implementation. Secondly, producers of technology are social actors in their own right. They have the capability (and often the intent) to produce social change by creating opportunities that will be taken up. While some may argue that enacting social changes by mobilizing opportunism and self-interest is problematic, the superiority of deliberative alternatives is uncertain.

may not only be equally important; they may also underlie the possibilities for changes in the 'technology' of human organizational, production, and exchange endeavours.[4]

Changes in the technology of production often involve the peculiarities of information as an intermediate and final 'good'. The non-rival properties of some types of information—that is, its ability to be reused by an indefinitely large number of users— when coupled with ubiquitous and highly interconnected systems for information distribution, suggest new possibilities for achieving increasing returns in production (Romer 1986). The extent to which these possibilities are empirically important depends upon the location and magnitude of these increasing returns and on whether the value in use of new information is expanding at a more rapid rate than in the past. Improvements in ICTs are likely to enhance the distribution of useful information to the extent that potential users are able to distinguish the useful from the useless (David and Foray 1995).

Correspondingly, the use of ICTs appears to present substantial opportunities for improving efficiency in the exchange process. Improvements in the quality and availability of information about the qualities and prices of goods and services will reduce the costs of exchange. These cost reductions are a distinct and additional source of 'increasing returns' to those available from the reuse of information.[5] This conclusion, however, is built upon the premiss of improvements in the quality of information. It is possible that problems in locating relevant information or ascertaining its reliability may offset any improvements in its quality or efforts to make it more available.[6] Thus, it cannot be expected that greater dissemination of information, *per se*, will improve economic outcomes.[7] Instead, it is likely to be an impetus to new forms of intermediation devoted to authenticating, critiquing, and repackaging information, efforts that themselves will vary in quality and accessibility.

Fully exploiting the opportunities that ICTs offer for improving production and exchange involves transformations in the technology of human organizations. ICTs enable new configurations of control, cooperation, and coordination. These new configurations initially favoured extensions in the scope and depth of control that

---

[4] Neice's chapter illustrates clearly how human needs and desires are the outcome of social processes that interact with, but also transcend, economic analysis. The chapters of this book (Gristock, Rae, Credé, Millar, Arroio) that examine changes in the technology of production make important criticisms of the views promoted by advocates of the 'virtual organization' such as Davidow and Malone (1992) or Tapscott, Ticoll, and Lowy (2000). In assessing electronic commerce, Mansell and Steinmueller (2000: ch. 7) provide a similar critique of technological change in the means of exchange in the new economy. Several chapters in this book (Schenk, Wehn de Montalvo, and Paré) further illustrate the complexities of the evolution of exchange relationships and their governance.

[5] It is the case, however, that the extent of these gains may be enhanced by the unusual features that information has as an economic good. The very low costs of replicating information by making it available in the form of World Wide Web pages, for example, augment the value of producing it in the first place. This is so not only because of the reduction in the costs of transmitting the information, but also because of the economies of scale in providing searching and sorting methods for locating relevant information.

[6] The cheapness of 'publication' on the Internet virtually assures that efforts to improve the availability of information will lead to problems in locating information that is relevant and of high quality.

[7] There are numerous possibilities for coordination failure in the distribution of information, an issue that is briefly examined in the latter half of this chapter.

might be exercised by those occupying the top levels in hierarchical organizations. Between 1950 and 1975, ICTs, with a few notable exceptions such as the ubiquitous telephone, were scarce and expensive. Their growing capabilities for acquiring, storing, processing, and reporting data were directed towards augmenting the trajectories of mass production and distribution inherited from prior industrial revolutions (Beniger 1986). In the years since 1975, technological change has produced a flood of new technologies and applications that has deposited ICTs ubiquitously throughout the organizations of industrialized societies.

The computing and telecommunications platforms that were once the tools of the managerial elite have become accessible to vast numbers of individuals in the industrialized world in the form of the personal computer connected to a data telecommunications network, the 'information appliance'. New ways to communicate with other individuals were among the many new 'features' of these information appliances. Within organizations, these new communication capabilities opened new opportunities for control, cooperation, and coordination. They created the opportunity for new technologies of organization such as tracking systems for logistics and inventory control, call centres for order taking and customer support, and production-flow controls supporting an unprecedented degree of flexibility and coordination.

This chapter examines the applications of new communication capabilities outside predefined organizational contexts in establishing and maintaining voluntary associations. The focus on voluntary associations allows examination of the use of ICTs outside the complexities and particularities of specific instrumental purposes. While many of the features of voluntary associations may be different in the ICT-mediated environment, the human needs, aspirations, and purposes that they serve are familiar. The building blocks for voluntary association will involve the basic issues that have always shaped human organizations. These include the need to mediate between individual and collective interests, to resolve conflicts between individuals to achieve a semblance of cooperation, and to attain shared goals and aspirations through some measure of coordination of individual initiative.

The defining principles or *raison d'être* of organizations are that they are able to establish some measure of 'procedural authority' by which to acquire and process information and that they are able to coordinate the actions of individuals through a complex mixture of incentives, social norms, and power.[8] Regardless of whether this procedural authority takes the form of a hierarchy, a participatory democracy, or of some other authority principle, its existence will make it possible to make decisions about procedures (for example, practices, routines, or customs) that 'ought' to be followed within the organization. That is, procedural authority makes it possible either to set or to validate norms and standards for individual behaviour in important domains of the organization's activities. In practice, of course, there is great variety in the extent to which procedural authority effectively governs the variety of individual behaviour

---

[8] See Ancori, Bureth, and Cohendet (2000), Cowan, David, and Foray (2000), and Lave and Wenger (1991) for different approaches to the issues raised by governing the 'codification' of information or the conduct of cooperative learning.

within organizations. Nor can it be said that increases in the power of procedural authority to *impose* such norms and standards in an organization necessarily lead to improved performance. What can be said, however, is that the distinguishing feature of organizations is their engagement with the creation and exercise of procedural authority.

Organizations also exist to provide a sustained means of coordinating the actions of individuals through the exercise of institutional authority. Institutional authority is exercised when organizations control the processes of selecting their members, negotiating the responsibilities or roles of individuals, and censuring or excluding dissenters' expressions by governing the conditions of membership or expression. *Ad hoc* or spontaneous social groupings are vulnerable to the discontinuities and disruptions from contests for authority and from the volatility of their membership. As such groupings do not have a clear recourse to institutional and procedural authority, they experience problems in recruiting and governing membership, in negotiating individual roles and responsibilities, and in reaching closure about means and ends for collective endeavour.

Procedural and institutional authority defines the boundary between organizations and other social groupings. A specific implementation of these types of authority constitutes a 'technology' of social organization. For example, in hierarchical organizations such as companies, governments, schools, or churches, procedural and institutional authority are vested in an 'ultimate authority', an individual or governing body, which may choose to devolve some measure of its authority to lower levels of the hierarchy. Past technological innovations in hierarchical social organization have included new methods for selection and review of the organization's ultimate authority as well as new means for devolving authority within organizations. Although most human organizations are hierarchical in nature,[9] the ultimate authorities' powers may be extensively devolved and their tenure may be delimited in time and subject to prompt review and termination. The economic and organizational issues identified in this section provide a foundation for analysing the emergence of virtual communities. The specific nature of these communities is the subject of the next section.

3. VIRTUAL COMMUNITIES AND CYBERSPACE: THE NEW AND THE OLD

'Computer-mediated communication', the somewhat arid term for the use of computers and data communications networks to exchange written messages, to converse, to establish tele-visual linkages, or to carry on all of these processes at once, is the long-heralded advent of the convergence of computing and telecommunications. The new 'converged' systems seem capable of translating many, if not all, of the known ways for humans to communicate with one another into streams of digital data that are

[9] Exceptions include associations that collectively make decisions, such as town-hall meetings, Anabaptist congregations, and various collectives or cooperatives. In all of these cases, the conditions of membership are strongly regulated by collective processes and, in many, procedural authority is often devolved to an individual or governing group by collective decision.

conveyed almost instantaneously between two locations to be reproduced by suitable 'interfaces' or stored for later access. Visual and auditory interfaces are the best developed, while tactile interfaces are still crude and olfactory and taste interfaces are largely speculative. These interfaces, and the networks connecting them, open up a new 'cyberspace' frontier for human interaction. 'Cyberspace', aptly named by William Gibson (1984), has its own geometry, better described by topology than by geometry, where distance matters less than connection and where time can be elastic, lengthened through the use of intermediate storage and compressed through the capacity to 'scan', 'jump', or 'hyperlink' through transmissions. In this twilight zone, the imagination is set free to produce new meeting places in which individuals may reproduce their interactions in ordinary space and time or, potentially, create new ways to interact with one another or with their constructions.

Lest the reader suspect that the conclusion is foregone because new ways of interacting in cyberspace can be imagined, it is important to keep two ideas in mind. First, new means of interaction do not necessarily suggest new approaches to the issues of procedural or institutional authority. While the telephone had a major influence on how authority might be exercised, it enhanced rather than transformed hierarchical relationships. Similarly, although individuals were not able previously to exchange multiple windows of audio-visual content over which the receiver could exert interactive control, it is not clear that these capabilities will transform the technology of organizations. In short, the novelty of the communication modality says little about the process of constructing or managing human associations. Secondly, the existence of new methods of communication surely does not guarantee the ability or desire to utilize them. The vast majority of information exchanged today through the use of computer-mediated communication rather drearily, but also imperfectly, mirrors well-established forms of communication such as the letter or phone call. Even if new modalities of communication made it possible to enable new approaches to the construction of organizations, they must be utilized for these purposes and this requires an extended process of learning and adaptation.

The existence of a cyberspace as one of the features of modern societies provokes many people to draw distinctions or boundaries. For some, this boundary is signalled by the identification of a 'real' and a 'virtual' world, one in which human beings are flesh and blood and the other in which thoughts and actions are disembodied by representation. For others, the boundaries are less certain, since, so far, people are the principal inhabitants of cyberspace and communication processes are not less real because machines mediate them. Drawing distinctions between these views need not be the exclusive province of philosophy. One may formulate and answer questions within a social science framework about the relation between behaviour and context or experience. For example, what role do physical meetings play in stimulating 'online' or computer-mediated communication? Some have suggested that virtual social interactions are strongly dependent upon physical meetings, which would suggest that there are inherent limits in computer-mediated communication for establishing independent 'communities'.[10] It is, however, important to be cautious about the situated nature of

---

[10] See e.g. Kling *et al.* (1999) or Rice (1993).

such investigations. Computer-mediated communication is relatively new for most people and it is not obvious that the children playing interactive video games with each other rather than watching television will behave, when they become adults, like contemporary adults, who spent *their* childhood watching television.

The domain of cyberspace that is explored in this chapter is 'virtual community', a term for social interactions occurring in cyberspace where multiple individuals communicate with one another. In cyberspace, the capacity for achieving connections between users transcends the concept of locale. Virtual communities may have 'locations' such as web sites or chat rooms but are best conceived as a social construct in which participating users establish links with one another.[11] For 'virtual communities' that do have locations, they are distinguished from other locales in cyberspace by the communicative behaviour of their users. When users do not communicate with one another at a cyberspace location, the cyberspace 'site' can be classified either (1) as interactive broadcast, where the owner engages in communication with users or, (2) as simple broadcast, where the owner undertakes to distribute information to potential visitors. In both of these cases, the traditional term 'broadcast' is employed to recognize that the nature of the communication process is from a single originator to multiple receivers. Both simple and interactive broadcasts involve exchanges between users that are mediated by the *owner* of a cyberspace location, a person or group that is capable of exercising procedural and institutional authority.[12] The boundaries between the virtual community and broadcast models for cyberspace locales are subject to social and technological innovation. For the purposes of this chapter, however, the direct provision of communication links between users is taken as a defining condition for the creation of a virtual community.[13]

'Virtual communities' are an instance of voluntary association, the nascent form of almost all types of social groupings, including organizations. Economists maintain that individuals engage in exchange, including the sale of their labour, voluntarily. It is important, therefore, to distinguish the employment relationship, which may not be entirely 'voluntary', from other types of voluntary associations. Throughout this chapter, the term voluntary association is used to describe social groupings arising from interactions that are not established or maintained as a requirement of employment. Associations, such as those that occur in work groups or teams, may employ some of the techniques and language of 'virtual communities' but are not analysed in this

---

[11] A 'chat room' is a capacity for switching text messages among connected users so that individuals may interact with one another.

[12] For example, amazon.com provides a means for its users to contribute reviews of the books it offers for sale. In effect, one user can communicate with others by using the interactive broadcast features of the amazon.com site and provide a means for reply. This is an example of the recognition of the value of user-specific information, a characteristic that Steinmueller (1992) has previously identified as a useful means for building subscriber loyalty to information services.

[13] The exclusion of sites where user interaction is 'mediated' by the site-owner is an arbitrary choice taken for the purpose of simplifying the preconditions for exchange between users, the chapter's central theme. Future research on the construction of social networks from user interactions with broadcast sites would be a valuable addition to the approach undertaken here. An interesting example, which is also excluded from the present analysis, is software that allows users simultaneously visiting the same World Wide Web site to send messages to one another (e.g. 'gooey', see **www.getgooey.com** (accessed 19 July 2001)).

chapter because the 'directed' element in an individual's participation would make the analysis unwieldy.[14]

In principle, all social groupings have the latent potential to evolve into organizations with a well-defined procedural authority and some functional purpose. However, actual social groupings also encompass a much broader class of social interactions, including those that are ephemeral, *ad hoc*, transitory, exploratory, or experimental. Voluntary associations may be formed simply for the purpose of facilitating social interaction among their members and involve few rules or institutional structures. It is therefore appropriate to maintain a degree of agnosticism about the 'purpose' of a voluntary association, even discounting various statements of purpose that may be made on behalf of its members, without exploring the actual content of interactions among members and what meaning they derive from these interactions.

In short, the definition of virtual community employed in this chapter is minimal, but precise. A virtual community exists when it is possible for a group of individuals to meet and interact with each other in cyberspace and these individuals voluntarily choose to participate in these meetings and interactions.[15] This minimalist definition sets aside issues of the purpose and various standards of performance that might be applied to the functioning of the virtual community. Incorporating the voluntary nature of association in the definition serves to distinguish virtual communities from other social interactions among computer-users such as those that occur in work groups or teams. A final advantage of this minimalist definition is that it provides a unit of analysis for social science analysis and a solid departure point for introducing issues of performance and purpose that might serve further to distinguish the processes and outcomes in specific virtual communities.

The next step is to establish a few basic features on which the 'performance' of virtual communities might be assessed. Many ideas about performance require a pre-definition of some instrumental purpose, intent, or objective. Some of these will be considered in this chapter. With a very minimum set of assumptions about purpose, however, it is possible to define three features of performance that might be applied to voluntary associations generally, and to virtual communities in particular. This starting assumption is simply that voluntary associations are comprised of members who perceive a benefit in interacting with one another. In other words, without assigning any sort of purpose or instrumentality to the grouping, it is possible to assess performance in terms of the value, broadly defined, that individuals perceive as deriving from their association with one another. This value is revealed through their communication behaviour (whether this can be observed or not).

This simple and minimal definition of purpose leads to three simple features or dimensions by which virtual communities or other voluntary associations might be

[14] Virtual communities among employees within an organization are included so long as membership or participation is not a condition of employment.

[15] As examined in the next section, several writers in the field would prefer a definition of virtual communities that has a more expansive meaning than voluntary associations of individuals in cyberspace. For reasons discussed further in that section, however, such a definition is neither needed nor particularly helpful in examining the basic issues of virtual communities introduced in this section.

examined. The first is 'recruitment'—that is, the issue of how individuals enter into association. Recruitment is meant to encompass both passive and active processes by which individuals come into association with one another, through their own efforts to connect with a virtual community or through efforts either of individuals within, or of the virtual community collectively, to seek linkages with new individuals.

The second is 'governance', which encompasses the issues of how members may protect themselves collectively from individuals who may wish to 'join' or to participate, but who are collectively perceived as reducing the value of the grouping by their behaviour. The challenges surrounding governance may be difficult to resolve because of the lack of consensus regarding what behaviours reduce the value of the group to other members.

The third feature is 'sustainability'. A voluntary social grouping should persist for as long as those participating in it find it of value. Failures of sustainability can occur for two different reasons. First, the requirements of convening or maintaining a virtual community may impose costs on particular individuals in excess of their willingness to contribute to the sustenance of the community. Secondly, even if some individuals were willing to bear these costs, it is possible that these individuals will be unable to assume them. Economists call the latter a 'coordination' failure. Coordination failures are particularly likely in the absence of institutional and procedural authority and are one of the reasons for establishing organizations. Because of the possibility of inter-user communication in virtual communities, it is difficult to make generalized statements about coordination failure. A particular network of interpersonal connections facilitated by a coordinator may reconstitute itself with the coordinator's demise, much as the closure of a popular pub may lead to its patrons cooperating in finding another locale. It is also possible, however, that the role of the coordinator is essential for the functioning of the virtual community, much as the popularity of a pub may be the result of its publican.

Each of these issues is 'scalable' in the sense that it will also be relevant for assessing the performance of groups that decide on purposes or goals or move towards better defined organizational structure. The features of recruitment, governance, and sustainability are, however, sufficiently basic in concept that they can be applied to a very wide range of voluntary social associations ranging from individuals who regularly visit a pub to large international voluntary associations.

To achieve a better understanding of these features, it is useful to consider three specific examples: the USENET, the 'open-source' software, and the peer-to-peer file-exchange communities. Although each of these communities involves a very large number of participants, it is their value in illustrating the features of recruitment, governance, and sustainability that is the primary reason for choosing them.

3.1. *Instances of Virtual Communities: USENET*

USENET is a system comprised of a standard for inter-computer exchange of a collection of user messages and a user interface for examining, originating, or replying to messages. The stream of USENET messages is organized in interest groups, each of which has the characteristics of a virtual community. Participation is voluntary and those who do contribute messages are required to provide their email address for their messages to be posted. Recruitment to USENET involves a user discovering the availability of the facility, currently provided as part of popular Internet browser software, but originally provided as an application program on time-shared computer systems.[16] USENET was established with, and continues to maintain, very few rules or restrictions on the contribution of material by users. It provides a way for users to participate in discussions on controversial issues such as political and cultural extremism (alt.anarchy.rules), downloading pornographic images (alt.binaries.pictures), and contributing off-beat humour (alt.urban.folklore), as well as to discussing conventional topics of interest (sci.archaeology).[17]

Because creating a USENET server requires acquiring and frequently updating files stored on many different 'server' computers, it could, in theory, be regulated by the owners of these computers. In practice, 'network administrators', the individuals who maintain and update USENET information resources on a particular computer, have not established a uniform set of practices. Moreover, if a user does not like the practices of a 'network administrator' on a particular server, it is possible to connect to another and, in part for this reason, there appears to be little practical censorship of the material available. The informality of the governance of the USENET as a whole does not necessarily extend to individual groups, some of which are 'moderated'; individual messages are examined before they are made publicly accessible and may not be included.[18] The practices within groups, such as moderation, for using USENET have evolved over many years and offer specific means of controlling or disciplining those users who violate the norms within the individual virtual communities on USENET.[19] In effect, some USENET virtual communities have established part of the mechanism of institutional authority (regulating the expression of participants) while others have not. The exercise of procedural authority (attempting to resolve disagreements among participants) does not appear to be present in USENET interest groups, although some groups create one or more frequently asked questions (FAQ) messages to reduce the repetitiveness of messages asking for the same information.

USENET continues to attract a large number of participants who are responsible for 'posting' and viewing messages and other material, a clear demonstration of its sustainability from the users' perspective. From the viewpoint of 'system administrators', USENET's sustainability may be somewhat more problematic. Pressures to regulate Internet content and legal liability issues may eventually make the current method of USENET distribution non-viable and require the development of new methods of distribution. Such a change could provoke a coordination failure, since the USENET currently relies upon the contribution of disk storage and communication capabilities

---

[16] For an analysis of the USENET as a virtual community, see Mansell and Steinmueller (2000: ch. 2), and Overby (1996).

[17] With the advent of widespread access to the Internet and its commercialization, many of the messages are posted with the objective of encouraging users to connect to sites offering advertising (and thereby generating revenue for their owners).

[18] Although several studies have been made of the content of particular USENET virtual communities, there is no practical way to monitor access, so it is not possible to trace the relationship between different governance regimes and actual usage, including passive viewing of content. It would be possible, however, to examine the relation between subject matter, governance, and participation (i.e. 'posting' of messages).

[19] Overby (1996) considers this issue in detail.

of large computer facilities. An alternative would require substantial resources and, probably, a revenue stream to pay for them.

3.2. *Instances of Virtual Communities: Open Source Software*

Voluntary associations are responsible for a very wide range of purposive activity, including the production of goods and services. The most conspicuous contemporary example of such activity is the open-source software movement, which has been responsible for producing a number of software products, including Linux, a computer operating system, and Apache, the software application most commonly employed for operating World Wide Web servers.[20] The aim of the 'open-source' virtual communities has been the production of a 'collective good' that is freely available to anyone with access to a high-quality Internet connection.[21]

The recruitment of individual participants in open-source software development involves the construction of a 'contributor' culture in which individuals provide documented software code for a specific project. Governance involves the determination by self-appointed project coordinators of the value of the contribution and its incorporation into the next 'release' of the software. Although serious disputes about contributions may lead to a branching of development paths and competing development efforts, the extent of 'branching' appears to be limited and does not seem to be a serious hindrance to the advance of open-source projects. The reasons for this have only begun to be examined by social scientists but are likely to be a consequence of the pace of development efforts coupled with the relatively clear design objectives of open-source software products. Both of these features suggest that most contributions will be incremental improvements that can be assessed by relatively uncontroversial criteria such as whether it works, is useful for the purpose of the effort, and is an efficient software implementation. It is, nonetheless, surprising that the second of these criteria does not appear to create major controversies.

A principal basis of the sustainability of the open-source movement is that many of its participants gain appreciable knowledge of the software tools that they intend to employ professionally. They also form professionally useful relationships. These benefits, combined with the extent of access of participants to computing resources, make it reasonably certain that recruitment will continue and sustainability will not become an issue.

3.3. *Instances of Virtual Communities: Peer-to-Peer File Exchange*

The third example of a virtual community illustrating the principles of recruitment, governance, and sustainability is the peer-to-peer file-exchange movement. Initiated by the software program Napster (**www.napster.com**), this movement involves the sharing of data files among users. The Napster program establishes a means for users to participate in an exchange of music files compressed according to the Motion Picture Experts

---

[20] Servers are networked hardware and software systems that distribute computer files (such as World Wide Web pages) or provide other services to connected users. For some observations about the growth of the open-source software movement, see Raymond (1999).

[21] In addition, companies such as Red Hat Inc. have emerged to provide 'user-friendly' versions of the Linux product—e.g. SuSE GmbH.

Group (MPEG) 3 format (referred to as MP3 files) devised by the Fraunhofer Institute or the Microsoft Windows Media Audio (WMA) format. A Napster-user connects to a file server maintained by Napster that executes a search for the desired MP3 files residing on connected users' computers. In effect, Napster is operating a 'super list' of the contents of its users' hard disk storage capacity and managing the transfer of these files. Napster meets the definition of a virtual community with the capacity to communicate with connected users through instant messaging (the message appears on the users' display) and 'chat rooms', where users may meet one another to negotiate exchanges.

Napster is a highly controversial virtual community because it is a relatively simple matter for users to convert the copyrighted contents of their music collection to MP3 files and exchange them with other users without payment of royalties to the publishers or artists.[22] Thus, many of Napster's users are engaged in the exchange of musical recordings that have been copied ('ripped' in the parlance of the community) from commercial recordings protected by copyright. Efforts to suppress this activity involve 'banning' users that are found to have violated copyright restrictions. However, the 'banned' user can readily overcome this exclusion.

Issues of recruitment and governance of the Napster community are very complex. While the user's true identity or email address is not publicly disclosed, information about his or her contributions and technical capacities (for example, their connection speed and online status) is available to other users along with an 'alias' allowing other users to communicate with them. Users may block or disrupt the downloading of files by monitoring the process and may set their computers to limit the extent of outgoing file transfer. Direct communication among users makes it possible to establish a variety of other linkages, including the possibilities of negotiating additional transfers (using other file-exchange methods) of material. Several tens of millions of downloads of Napster software have occurred, making Napster among the largest virtual communities on the Internet.

Within this community there are thousands of subcommunities established by interpersonal communication and the exchange of email addresses. Some of these are beginning to use new types of peer-to-peer exchange software such as 'Poinster',[23] which provides a means for users to define a community of friends with whom they wish to exchange files. Poinster operates as an extension of the America Online (AOL) Instant Messaging Service, which has millions of users. Even though substantial legal action for copyright infringement has been launched against virtually all of the companies involved in the peer-to-peer file-transfer movement, it is highly unlikely that these efforts will overcome users' interests in establishing exchanges for systematically violating copyright laws and for pursuing legitimate information exchange activities. Thus, while Napster or its rivals may succumb to legal challenges, the model of peer-to-peer file exchange is likely to be sustainable despite its frequent use for the transfer of copyright protected material.

[22] The *Los Angeles Times* (www.latimes.com (accessed 30 Dec. 2000)) has documented the Napster controversy, which involves continuing legal action, in considerable detail.
[23] See www.poinster.com (accessed 30 Dec. 2000).

3.4. *Virtual Communities as a Basis for the New Economy*

These examples make it straightforward to conclude that the Internet as a new communication medium offers a new means for interpersonal communication and social interaction and that it will offer new ways to establish voluntary associations and recruit new members. Each of the three examples illustrates the advantages and disadvantages of the virtual community organization. USENET clearly provides a means for users with specialized interests and considerable computer knowledge to exchange useful information with one another. Negotiating the 'chaotic' governance environment of USENET is, however, cumbersome and users often must have patience with irrelevant or objectionable 'postings'. The problems experienced by USENET are likely to be faced by any virtual community that does not enact the 'moderated' format utilized within some USENET groups.

The much stronger governance structure implemented within the 'open-source' software community comes a great deal closer to enactment of a clear 'procedural authority' with regard to virtual community activities. It is clear that, while highly specialized, the open-source virtual community is highly effective in recruitment, governance, and sustainability. Part of its effectiveness may be attributed to the cohesion of its users, who are among the most sophisticated computer programmers and are operating in this virtual community with a well-defined purpose, the creation of software. The apparent cohesiveness of this community suggests that a mechanism for peer interaction in 'qualifying' users is helpful for addressing recruitment, governance, and sustainability issues.

The Napster example indicates the potential power of new virtual community tools. Providing a means for the exchange of music is clearly a service that many users value. Although a substantial amount of user enthusiasm may derive from the capacity to exchange copyright-protected music, this is a better illustration of the inherent problems of maintaining copyright protection in cyberspace than it is of a fundamental weakness in the viability of virtual communities based upon information sharing. The size and enthusiasm of the Napster community suggests the need for recorded music publishers to move towards new methods of music distribution that will be even more convenient for users than creating *ad hoc* exchange markets.

It should now be apparent that virtual communities in cyberspace establish a basis for new economic activities. Moreover, each of these communities is able to achieve outcomes that would not be possible using traditional methods of voluntary association. Voluntary associations do exist for uniting individuals with very specialized interests for the purpose of information exchange, software construction, or exchange of musical recordings. They are, however, strongly limited by the costs and other problems of bringing the individuals together in the same physical space.

The examples of virtual communities considered so far, however, reflect a rather narrow portion of the spectrum of human interaction related to information exchange. While it is certainly likely that individuals engaged in any of these communities may establish social ties and friendships, these are likely to be dyadic in nature. Although some may argue that each of these virtual communities has a particular ethos or community spirit, there is also strong indication that individuals meeting in this 'space' find other means to interact more directly. In other words, the virtual communities considered so far are essentially public forums. They are relatively well adapted, either to their primary purpose or as a 'meeting place' for the formation of dyadic relationships. The next section considers other, less instrumental, aspects of virtual communities that also present issues of recruitment, governance, and sustainability.

4. IDEOLOGY AT THE CYBERSPACE FRONTIER: DISTINGUISHING MOTIVE AND BEHAVIOUR

The invention of the term 'virtual community' is often attributed to Howard Rheingold, who has offered one of the most engaging evocations of its potential from his own experience:

My seven-year old daughter knows that her father congregates with a family of invisible friends who seem to gather in his computer. Sometimes he talks to them, even if no one else can see them. And she knows that these electronic friends sometimes show up in the flesh, materializing from the next block or the other side of the planet.   (Rheingold 1993: 1)

Rheingold is both social chronicler and analyst. On the same page he reports: 'I'm not alone in this emotional attachment to an apparently bloodless technological ritual. Millions of people on every continent also participate in the computer-mediated social groups known as virtual communities, and this population is growing fast.' One of the virtual communities from which Rheingold draws his experience is Whole Earth 'Lectronic Link (WELL), located in the physical world in San Francisco, California, and in cyberspace at the address **www.well.com**. Although it began as a dial-in bulletin-board service, The WELL has, in recent years, made a complete transition to the Internet in performing its 'gathering magic'.[24]

Regardless of the technological method by which people are gathered, their interpersonal exchanges evolve into relationships to which the term 'community' is often applied. The crux of this section is a critical examination of the preconditions for, and the implications of, a more intimate form of community than the previous examples. The definition of 'virtual community' used in this chapter is minimalist. It assumes that the fact of people choosing to return to the same site to engage in interpersonal communication is sufficient to define 'community'. Is something more needed? In Rheingold's account, the emphasis intended in the use of the word 'community' rather than words such as clubs, groups, or associations is that the community is a means of achieving a level of intimacy that can support his use of the term 'family of invisible friends'. This intimacy is, however, only one of several different types of intensive interaction that may give rise to a collective perception of the existence of a community relationship among people, as Rheingold acknowledges in his original work and in its revised edition (Rheingold 2000).

---

[24] Mitchell (1996) analyses the architecture of these meeting sites, examining the historical continuities and discontinuities between non-virtual and virtual spaces.

Information services with a large subscriber base, such as AOL, that originated before the widespread use of the Internet offered inter-user communication as a key element of their service provisions.[25] The creation of 'user-forum' activities that allow individuals to find others with similar interests is a characteristic that AOL and The WELL share with many other efforts to create virtual communities. This 'architecture' is similar to the discussion-group structure pioneered by USENET. Similar capacities are incorporated in a vast number of political and community activism web sites as well as in more specialized sites within the research and education communities. From a review of the broad array of sites in which inter-user communication is an important feature, it may be concluded that such capabilities are valuable in recruiting users and encouraging them to revisit the site.

The Internet is evolving towards being an 'advertiser-supported' medium in which attracting an audience provides the basis for generating revenue from 'banner ads' (small displays advertising goods and services that can be activated by the user to serve as entry points to commercial sites). Under these conditions, if creating inter-user communication will increase recruitment or revisitation, there is an economic incentive to do it.[26] Even commercial ventures such as amazon.com have provided a limited means for establishing inter-user communication.[27] It does not appear to be appropriate to exclude the variety of motives and mechanisms involved in establishing inter-user communication from the virtual-community definition. Some advocates of 'intimate' communities may claim that these groupings lack the identity, trust, or commitment of their preferred form of involvement. At this stage of Internet development, however, it appears that many users are seeking social architectures in larger meeting places. This may or may not be a transitional phase. Among these relationships, the aim of building 'intimate' communities is, however, of considerable significance. The WELL virtual community chronicled by Rheingold's own experience clearly defined itself around this goal, embracing it as a differentiating feature. Even though its owner ship structure has changed, as Rheingold (2000) notes, the individuals responsible for The WELL indicate in their 'About The WELL' description that:

The WELL is an online gathering place like no other—remarkably uninhibited, intelligent, and iconoclastic. For an action packed fifteen years, it's been a literate watering hole for thinkers from all walks of life, be they artists, journalists, programmers, educators or activists. These WELL members return to The WELL, often daily, to engage in discussion, swap information, express their convictions and greet their friends in online forums known as WELL Conferences.[28]

---

[25] Problems with the 'moderation' of user communication during the pre-internet era by services such as Prodigy contributed to their failure, despite the enormous investment made in their establishment. Steinmueller (1992) offers an analysis of the role of inter-user communication and specific criticisms of the Prodigy service, as does Rheingold (1993).

[26] This is a central tenet of Hagel III and Armstrong (1997).

[27] During 2000, amazon.com established a means for users to establish an identity on their site, augmenting the previous ability of users to submit book reviews to a 'moderator'. Users can now provide personal information including their email address to allow other users to contact them. While this capacity is less complete than the 'forum' or interest-group model where exchanges between users are viewable by others, it is arguably an effort to establish a virtual community among amazon.com users.

[28] From **www.well.com/aboutwell.html** (accessed 1 Oct. 2000).

The signals provided by this message are relatively clear. It probably is not possible to find like-minded individuals on The WELL if an individual is inhibited, stupid, or rigidly traditionalist. In other words, the appeal is meant to be broad. The message, however, also signals that the acceptance of any particular individual by this community may represent a challenge. Although from all walks of life, WELL members are depicted as 'thinkers', a category that might cause more self-selection.

Rheingold also indicates that another characteristic of certain 'virtual communities' is that the relationships established within them need not be *confined* to the virtual domain. Rheingold observes that individuals *materialize* from the virtual into the real, even though they may normally be separated in actual space. Since, at present, the available 'partners' for online interaction are animated by human beings, the potential for non-virtual exchange is latent in all virtual relationships.[29] Individuals may have the intention to resist or seek out non-virtual exchanges, and the analysis of these intentions may prove to be useful for analysing the nature of different virtual communities. In the case of The WELL, a particular appeal is made to potential members that The WELL, itself, has the feel of a physical location regardless of the member's location:

The WELL is a cluster of electronic villages on the Internet, inhabited by people from all over the world. A discussion on the great eateries of Paris might include playful banter from people typing to one another from San Jose, Tokyo, Boston . . . as well as the Left Bank. Yet the ambience is all WELL. More than just another 'site', The WELL has a sense of place that is nearly palpable.[30]

Systematic research on the behavioural links between participation in virtual communities and physical meetings is needed. At present, anecdotal evidence suggests that participation in some virtual communities is motivated by this possibility, including 'sites' organized around the search for romantic or sexual partners.[31]

4.1. *Is Intimacy a Requirement for Defining Virtual Communities?*

Rheingold highlights the willingness of individuals to engage with one another outside or alongside the kind of exchange relationships that characterize much of their lives. This is a particularly important feature of some of the virtual communities that Rheingold considers and it illuminates his concern with the intimacy of personal exchange. He notes:

Reciprocity is a key element of any market-based culture, but the arrangement I'm describing feels to me more like a kind of gift economy in which people do things for one another out of a spirit of building something between them, rather than a spreadsheet-calculated quid pro quo. When that spirit exists, everybody gets a little extra something, a little sparkle, from their more practical transactions; different kinds of things become possible when this mind-set pervades.

[29] Of course, it may be blocked by anonymity, or by the financial or physical capacity to 'materialize' at a distant location.

[30] From www.well.com/aboutwell.html (accessed 1 Oct. 2000).

[31] The use of 'sites' in this context must be taken to include the facilities for real-time conversation provided by Internet Relay Chat (IRC). Casual observation suggests that both sexuality and proximity are important in the establishment of IRC communities.

Conversely, people who have valuable things to add to the mix tend to keep their heads down and their ideas to themselves when a mercenary or hostile zeitgeist dominates an online community. (Rheingold 1993: 59)

This rhetoric of anti-commercialism assumes a bifurcation between exchanges that are a contest in which one party gains at the other's expense and exchanges that are cooperative and therefore devoid of the potential for exploitation. It obscures the nature of transactions by refusing to acknowledge that individuals may offer their time, money, or emotional commitment to an outcome they regard as positive (whether it accrues to them personally or serves goals or interests to which they are sympathetic). The evocation of a 'gift economy' may be useful as a way to focus attention on the conditions supporting such transactions. It is difficult, however, to support the claim that it represents a fundamental departure from existing types of human relationship. If it were such a departure, we would expect participants to be indifferent about the outcomes and to regard the process of participation as sufficient.[32]

A broader base for constructing a theory of virtual communities is the acceptance that both human motivations and the social institutions for satisfying these motivations are complex. Neither the view that enrolment in a virtual community is a form of market exchange, nor the view that such enrolment must offer a refuge from the market-based culture, needs to be a starting point for defining virtual communities.

The minimalist definition of virtual communities as voluntary social associations based upon computer-mediated interpersonal communication encompasses the possibility that some of these communications will lead to close personal ties and that some of these relationships may be further developed through physical meeting. It also includes the possibility that these voluntary associations may undertake a variety of collective actions. The last of these elements is particularly important in helping to differentiate the virtual community from electronic commerce in which the most common aim is dyadic economic exchange.

The electronic commerce environment may, however, overlap with virtual communities and may even involve collective action. For example, producers or consumers may join in collective bargaining arrangements, thereby furthering their interests in market transactions. Similarly, individuals may joint in a collective effort to produce goods or services for their own benefit or for some wider social purpose on which they agree.[33] It is reasonable to expect that such collective endeavours will provide the stimulus for deep and lasting personal relationships, even if this is not an explicit aim of recruitment or maintenance of members in the association.

Just as interpersonal relationships may derive from common participation in collective action, they may also emerge from more casual interactions. Chat rooms and

---

[32] One situation in which people contribute to an organization without concern about outcome is membership in apocalyptic religious belief systems. Since a condition of membership is subscription to the belief that the outcome is preordained and outside members' control, there is no basis for questioning their contribution of time, money, or emotional commitment. The consequences of removing the 'reference points' by which individuals gauge their contributions can be quite dramatic.

[33] Those who would disparage these 'commercial' communities may not have had the experience of participating in a producer or consumer cooperative that can have considerable social content.

discussion forums are forms of virtual community that function in a way that is analogous to non-virtual gathering places such as coffee houses or pubs. In such establishments (virtual and non-virtual), some regular habitués are content to watch and perhaps to listen to the interactions taking place around them. The mildly pejorative term 'lurkers' has been coined to refer to these individuals. Others seek attachments with specific individuals (who may be unknown to them previously) and return with the intention of reviving and extending these attachments. Still others may be attracted to transient interactions, developing and leaving relationships as suits their fancy. In principle, a single meeting place may satisfy each of these individual needs, particularly if its architecture supports individual control of others' access. In these examples, the intention to engage in collective action is well outside the context established by the meeting place. While deep or lasting interpersonal relations or collective action may occur as a by-product of these social interactions, they need not be the intent of the gathering. Moreover, because the tools that exist to implement such facilities can be employed by a single person, their establishment and management do not require collective action either.

4.2. *Intimate Relationships are Dyadic, not Collective*

In short, many examples can be produced of social gatherings on the Internet that are not inherently devoted to the development of interpersonal relationships that would lead participants to describe those who interact with them as a 'family of friends'. In gaming environments, individuals may interact as friend or foe, but are united by a common enthusiasm for the experience. In interest-group forums, the initial focus is that of a shared interest rather than a personal interest in the participant. Even in meeting places devoted to particular health or personal issues, the scope for interaction may, in the first instance, be bounded by the subject of concern. All of these examples simply indicate the obvious fact of social life that one may meet and interact with others in a variety of personas and with diverse motives. Whether anything more intimate or personal evolves from these meetings depends upon the participants and is fundamentally a dyadic rather than a collective product of the association.

This reductionism to dyadic relationships may be disappointing, particularly to those who would like to believe, as casual readers of Rheingold may, that virtual communities necessarily embody a communitarian potential. Instead, it is reasonable to presume that virtual communities will exhibit the range of individual behaviours characteristic of their participants—that is, human beings. Exploitative, selfish, and abusive behaviours may be expected to coexist with those that are cooperative, altruistic, and supportive. These possibilities indicate that the way in which participation in the virtual community is structured may be very important to the value of the outcomes it produces for participants.

The establishment of norms, rules, and standards is no less significant in virtual environments than in other features of everyday life. These social institutions comprise the constraints and incentives necessary for successful exchange relationships to emerge. They may be implemented through authoritarian, democratic, or anarchistic methods. That is, they may reflect or dictate majority rule or collective assent and they must somehow be enforceable against the will or preference of dissenting and deviant individuals. Enforcement may involve disenfranchisement of enrolment, ostracism, or retribution, depending upon how enrolment in the community is organized.

It is necessary to create social institutions for a virtual community to persist and, in the first instance, these institutions can be linked to the conditions and possibilities of joining in the virtual community. These institutions cannot, alone, create either a meaning or a purpose for a virtual community. Nor will these institutions necessarily ensure that the community will persist in enrolling members or making progress towards realizing its meanings or purposes, however defined.

The purpose of this section has been to establish a basis for concluding that virtual communities as instances of human voluntary association broadly share, rather than radically depart from, the characteristics of other voluntary associations. In particular, individual participation in such communities may be expected to represent a type of exchange in which commitments of time, money, and emotional attachment are linked to immediate or speculative expectations about beneficial outcome accruing either to the individual or to purposes valued by that individual. Establishing an ethos or philosophy for a virtual community based upon sharing, cooperation, and other communitarian values may be an effective means for distinguishing a community. It should not be taken, however, as a defining characteristic either of virtual communities or of the possibilities of forming deep and lasting interpersonal relationships based upon virtual-community participation. Separating the ideological issues of virtual-community ethos and philosophy from the issues of voluntary association and exchange highlights the need for institutional norms, rules, and standards for individual behaviour in order to achieve productive outcomes and avoid opportunistic or malicious social behaviours.[34] These institutional features of the community interact with the processes of member recruitment and retention that are considered in the next section.

5. FINDING A HOME ON THE CYBERSPACE FRONTIER: THE MARKET FOR VIRTUAL COMMUNITIES

Cyberspace as implemented on the Internet/World Wide Web is a vast construct in which an incredibly large number of physical locations can be established, each having the potential to create a unique architecture for interpersonal interaction. The formation of virtual communities can easily founder because of the inability of potential members to find the cyberspace location of the community. At present, none of the existing 'search engines' provides a means for distinguishing sites at which inter-user communication is possible. For example, using the Northern Lights search engine and searching for 'community' returned 20,194,576 different web sites in the autumn of

---

[34] The critical focus necessary to make these distinctions has been achieved by referring to a portion of Rheingold's exposition. It is important, however, to stress that Rheingold clearly recognizes that virtual communities involve a much broader range of associations than those typified by his experiences with The WELL and that he provides many other examples of the formation and operation of virtual communities in his path-breaking work.

2000. Even if the user is capable of scanning the list (and ignoring the time needed to access pages) at the rate of one per second, a complete examination would require over thirty years of full-time work. Even the more restrictive search for the simultaneous occurrence of the words 'join', 'virtual', and 'community' (which assumes the term 'virtual community' is used by the web site) leads to 229,535 different listings and sixty-four hours of search. Moreover, this ignores the very large number of alternative groupings that might better suit a particular user. For example, on 1 October 2000, U-Net, a leading Internet Service Provider in the United Kingdom, listed some 61,100 topical discussions in USENET-type bulletin boards.[35]

The prospects for developing reliable means of identifying potential communities of interest is fading rather than growing over time. This is not only because of the pro-liferation of web and other cyberspace locations, but also because of efforts to com-mercialize the 'community' concept with a view to attracting audiences. There are important parallels between the promotion of virtual communities and the creation of other types of 'audiences' for media by responding to audience interest. For example, promoting viewer emotional attachment and expectation is an important feature of developing certain types of broadcast television programmes and the existence of 'fan clubs' indicates the potential for 'passive' viewers to become active in sharing their experience with others. In each case, the possibility of advertising revenue exists either if substantial numbers of users make use of the site or if the characteristics of the audience can be clearly identified for more targeted advertising.[36] Because inter-user communication has been identified as one of the features that builds audience, a grow-ing array of sites offers this capability, as noted earlier in the case of amazon.com.

5.1. *Congestion Costs, Coordination Failure, and Graffiti Attacks*

While these developments favour the relative growth of virtual-community sites over the simple or interactive broadcast models of the sites described above, they also serve to spread potential participation in virtual communities more thinly among the user population. From an economic viewpoint, because the additional costs associated with creating inter-user communication are relatively low, there is continuing growth in sup-ply. The continuing growth of inter-user communication by site developers produces a negative externality or spillover effect for users, which is an example of congestion costs.

The response to this type of congestion cost is the development of better capabilities for user selection through filtering, sorting, and searching. While these activities can be accomplished for web-based content based upon indexing schemes, it is much more

---

[35] Many of these bulletin boards do not have active users and are a cyberspace ghost town of previous developments.

[36] Advertising prices are based upon the value the advertiser attributes to exposing an advertisement to a relevant audience. The demand of advertisers seeking to reach broad market groups drives up the price of advertising and often makes media with broad market reach (such as television or newspapers) uneconom-ical for advertisers seeking specialized markets. A much smaller 'circulation' of the advertising message can command a substantial price if the medium carrying it has a high share of the audience that is relevant to a particular advertiser.

difficult to provide guides that are responsive to the wide variety of user preferences for interactive communication. A principal problem is that cyberspace has no geography. Proximity is defined by linkage, and the structure of linkages is neither discernible to the casual user nor integrated with the usage data that would facilitate 'mapping'. When this absence of structure is combined with the large number of broadcast-oriented sites (hundreds of thousands of which are personal and of the variety 'Hello world, my name is . . .') that cannot be distinguished from interactive sites, the problems of searching and filtering are magnified.

Efforts to address these problems have been based upon strategies of 'localization' and 'branded' networks of sites. Localization strategies have been pursued by a number of actors. For example, local Internet Service Providers (ISPs) have provided a means for users to identify one another. However, in many countries, these ISPs are under considerable competitive pressure from national or global companies, often offering lower connection charges for identical connection quality. For similar reasons, all localization strategies are under pressure. Because the link between physical and virtual location is largely fictitious, a 'localized' site can be created at a national, or even international level, with features that are indistinguishable from sites that are actually physically co-located with their local content.[37]

The second strategy being pursued by 'brand-name' companies such as AOL, Yahoo, Amazon, and Lycos is to offer 'channel guides', many of which are affiliated in some way to the company promoting the linkage. These channels can and often do include reference to virtual-community sites organized by, or in affiliation with, the 'brand' companies. The principal problem with these trends is that the strategies of building large audiences are at least partially inconsistent with the architecture for establishing interpersonal communication. In effect, users may be so overwhelmed by the wealth of opportunities to meet others that they do not know where to begin. Moreover, in the larger public spaces that are being created there is likely to be a greater incidence of opportunistic or exploitative behaviour that serves to discourage interpersonal communication.

In effect, the congestion problem is heightened by a coordination failure that has two axes. First, there are no technological means for identifying 'virtual communities' as a distinct category because of the inability to create consensus about the defining features of the virtual community beyond inter-user communication capabilities (since *some* users will have meaningful experiences in almost any environment in which they can communicate with others). Secondly, in the absence of a standard definition based upon a broad consensus of producers and users, the actors that have stepped forward to sponsor a definition have failed to reduce the congestion costs substantially. This, of course, is an opinion that may not be shared by the promoters or the thousands of users of the currently available services offered by the brand-name Internet sites. As previously noted, some of the branded Internet companies, especially Yahoo and Lycos, have clear incentives to direct users to sites that can enhance their revenues and are prepared to organize virtual communities across a wide spectrum of interests.

---

[37] Of course, any organization adopting this strategy will need to solve the problem of gathering sufficient current local content. A local 'correspondent' may, however, suffice for this purpose.

Accepting that virtual-community site creators, the community convenors, persist only by responding to user needs and expectations, it remains questionable whether the incentives adequately incorporate user needs and expectations. In the meantime, no other actor has stepped forward to contest the legitimacy or value of the commercial approach. This may be, in part, because a number of virtual communities, including, for example, The WELL or bianca.com, have their own incentives to build audience based upon either direct subscription (The WELL) or advertising (bianca.com) revenue.[38] From the user's perspective there is a coordination failure in the provision of user-centred information about virtual communities that heightens the problems of 'searching' for ones that are appropriate.

So far, the focus in this section on the promotion of cyberspace locations has been treated in a general way. In effect this has meant assuming a representative and, therefore, featureless user confronting the vast featureless plane of cyberspace. In practice, however, actual users have a wealth of social connections and substantive interests that provide reference points in their search for like-minded individuals. At present, it is primarily through these social reference points that virtual communities are likely to form. For example, and not surprisingly, many virtual communities are devoted to issues surrounding major human life events including birth, marriage, debilitating or terminal illness, job seeking, retirement, and so forth. In many cases, these virtual communities are organized with the specific intention of supporting interpersonal exchange about the life event. For example, in the case of illness, there are many broadcast sites devoted to medical information and the questions of patients. Finding relevant virtual communities that are devoted to specific illnesses is, however, not difficult (entering the name of the disease as the subject to a search engine) and there are often many linkages to other sites that may further reflect the interests of the user.

A striking feature of web development is that sites with direct user interaction are seldom organized by larger non-profit or commercial organizations. Thus, for example, the sites of non-profit organizations as diverse as the Boy Scouts of America, Greenpeace, and Computer Professionals for Social Responsibility may be visited without finding the implementation of a virtual community. Less surprisingly, major corporations such as General Motors, Kellogg, or British Airways do not provide a means for interpersonal communication at their sites, although all these organizations do provide interactive rather than simple broadcast capacities (sometimes using only a generic email return path). The reason for this is starkly illustrated by the following posting from a 'localized' site devoted to Oxfordshire in the United Kingdom:

Name: Colostomy News
Location: UK, UK UK
Date: Tuesday, May 9, 2000 at 14:05:15
Comments:
Smell the aura of excrement as it escapes from the colostomy bag and wafts over to your desktop, leaving the smelly remains that is Colostomy News spread over your screen like a proverbial skidmark. Smell, Read, Enjoy . . . [followed by a Web address].

---

[38] The site bianca.com is a well-established 'lifestyle' site organized as a house party with 'rooms' devoted to chat and information resources. It has substantial adult, i.e. sexual, content.

Organizations operating virtual communities may establish appropriate use policies and take actions to delete such messages or attempt to enact sanctions against users who post them. It is not possible, however, to prevent such messages from appearing without 'moderating'—that is, intercepting—interpersonal communications. It is therefore not surprising that larger established organizations are reluctant to set up inter-user communication that is subject to such graffiti attacks. The simple existence of such messages not only might cause distress to 'legitimate' users, but might also be seen as reflecting poorly upon the organization hosting the exchange. Larger organizations are more likely to create 'closed' communities in which a member must provide authenticated information about his or her identity, allowing would-be transgressors to be barred from anti-social actions.

Given the possibility of graffiti attacks, how are the larger branded organizations able to organize virtual communities? First, these organizations have greater technical resources for pursuing the progenitors of graffiti that may provide some deterrence. Secondly, and most importantly, the larger branded organizations disclose that inter-user communication has this potential. They have been successful (so far and for the most part) in convincing users that they cannot both host a site offering relative freedom of expression and be held responsible for the content appearing at that site.[39] Thirdly, these organizations have achieved sufficient credibility that they may intervene in inter-user communications in which there is a time delay between message transmission and the 'posting' when it is made available to all readers, without this being seen as contravening the process of inter-user communication.[40] All of these abilities carry substantial costs, which these branded virtual-community convenors are willing to absorb because it is their principal business. This is not the case for non-profit and commercial organizations and, therefore, these organizations are unlikely to convene virtual communities, even if this would benefit their principal mission.

Congestion costs, coordination failure, and graffiti attacks limit the 'market' for virtual communities. There is little doubt that users seek inter-user communication capabilities or that having these capabilities provides them with individual benefits and may lead to persistent social groupings. In other words, there is a clear demand. The problems that have been identified in this section are in the supply of such services. Congestion costs increase the difficulty of organizing and promoting a virtual community. To attract users to a virtual community, the convenor must find features that distinguish it in a crowded cyberspace that has no physical features allowing users to seek out appropriate neighbourhoods to pursue their interests. These congestion costs are worsened by the absence of a means to filter out sites exclusively based upon the simple or interactive 'broadcast' model and by the incentives to build new sites that will attract an audience and thereby generate advertising or other revenues.

---

[39] The latter claim has been challenged in the case of messages related to Nazism in Germany, where national laws prohibit the dissemination of this type of content. While these actions create a problem for organizations hosting such sites that are seeking to conduct business in Germany, effective means for suppression simply do not exist (Smith System Engineering 1997).

[40] These organizations also host 'chat' sites in which there is no effective means of censoring content. Such sites are now well recognized as problematic for individuals seeking freedom from intrusive or abusive behaviours.

A remedy for congestion costs would be the creation of information resources that perform the 'sorting and searching' function on behalf of the user. This has proved problematic, however, for several reasons: (1) users have diverse interests and it is difficult to define a strategy that will have broad appeal; (2) actors with sufficient size and credibility to undertake this task have incentives to create related networks in order to capture audience; and (3) other actors often act competitively because of their desire to retain users. No effective solution to these three problems has yet emerged. Other means of establishing linkages between the social world of the user and cyberspace are plagued by the problem of graffiti attacks. Graffiti attacks make it unattractive for organizations with a desire to defend their established reputations to provide inter-user communication and also raise the costs of doing so. Organizations with established reputations may build 'closed virtual communities' in which membership validation is a condition of entry. A limited number of Internet-related companies that have promoted their 'brand' identity have, so far, been able to distance themselves from the graffiti-attack problem by a combination of actions. The most important of these actions is persuading users that they cannot simultaneously experience free expression and be free from anti-social intrusive behaviour.

5.2. *A Typology of Virtual-Community Convenors*

These developments suggest that four types of actors will be the dominant convenors of virtual communities. First, the brand-name Internet companies are likely to organize a proliferating array of places to meet people with the facilities to support various types of inter-user communications. These meeting places will remain subject to graffiti attacks, although the technological capabilities of these companies will provide both a deterrence and a sanction potential that may reduce the severity of this problem. The issue of institutional authority is partially resolved in these communities by the promulgation of explicit norms and standards and the mediation of user communication. The sites of 'brand-name' companies that involve user subscriptions will maintain a stronger form of institutional authority in their ability to set terms and conditions of membership that will allow them to exclude non-complying members. Users are thus able to select a level of procedural and institutional governance suiting their tastes, although it will remain difficult for users to assess the value of these virtual communities to their interests. These problems are partially mitigated by the prevalence of 'trial subscriptions' that allow a period of free membership. Their ability to fund not only the governance activities, but also a variety of user resources, suggests that these virtual communities will continue to be attractive to a large number of users.

Secondly, smaller closed virtual communities whose principal purpose is voluntary association and that cover their costs through advertising or user subscription fees are likely to be a viable means of addressing the graffiti attack and governance problems. These virtual communities can heighten procedural and institutional authority to whatever level will be accepted by their members including the authentication of user identities. Their closed nature and absence of a brand-name identity will, however, make them vulnerable to the growing congestion of similar virtual communities. Users will find it difficult to identify which of these communities will meet their needs because of their inability to 'see' community life or casually participate in it to see how it would meet their interests or expectations. The problems that this second group of virtual communities has in recruiting members in the light of congestion costs suggests that many of these virtual communities will create an explicit linkage with non-virtual locations or associations. These ties may be an effective substitute for the advertising of brand-name Internet companies. User authentication—that is, restrictive registration procedures—may be employed to build trust among users that they are interacting with individuals sharing a common non-virtual association.

Thirdly, a number of small 'open' virtual communities is likely to persist. These communities' primary defences against graffiti attacks and other trust problems arising from their open membership will be their relatively small size, which reduces the scale of such problems, and the recognition of their members that being open involves these problems. Their small size and open membership will prevent them from generating significant revenue and they will continue to face congestion cost and coordination problems. They are, therefore, likely to be based upon a voluntary commitment by an individual or a group that is small enough to retain coherence. In effect, the convenors of such a virtual community may be motivated by the satisfaction that they achieve by being identified as the 'founders' or leaders of a particular community.

Fourthly, 'purpose-built' virtual communities dedicated to creating or maintaining club or public goods (shared within the group or more generally) are likely to persist. Their purposes partially solve the problem of congestion costs and coordination failure by providing an objective method of differentiating them from other cyberspace locations. While they are subject to graffiti attacks and related problems stemming from their relative openness, their focus on a particular purpose or task orientation make them less attractive targets. In addition, these communities can, if needed, quickly establish institutional authority and are likely to employ procedural authority not only to govern interpersonal exchange, but also to further the purposes of the community.

Each of these four types or models of virtual communities is likely to persist. The number and size of these communities are likely to fluctuate over time as user experience in inhabiting cyberspace increases and as these communities evolve and so need to compete to recruit and retain members. Competition between these types of communities is the result of the limited time that an individual spends in cyberspace and of the limits of his or her engagement with a particular virtual community. Some communities will grow, others will disappear as the result of this process. In addition, technological innovations will have a further influence on the growth and decline of communities. New types of user interfaces (for example, affordable video conferencing) or new tools for user interaction (for example, virtual worlds based on visualization such as those offered by various computer gaming environments) are likely to stimulate the formation of new virtual communities and to provoke adaptation or fission among existing community types. The next section examines some of the key economic and social issues that are likely to affect the size and growth of these different types of virtual community.

6. THE DYNAMICS OF VIRTUAL-COMMUNITY TYPES

The four types of virtual communities identified in the previous section involve three different, but potentially overlapping, types of 'service' to community members. The 'brand-name' virtual communities are principally identified with providing *informa-tion services and resources* to members that justify their subscription fee or the regular-ity of their use that produces advertising revenue. The closed and 'open' virtual communities based primarily upon user interaction are offering *affinity services* to their members. The purpose-built virtual communities engaged in producing club (confined to members) or collective (available to the public) goods are offering *collec-tive production services* to their members. While each of the types of communities appears primarily to specialize in the associated service, there is ample room for over-lap. For example, brand-name virtual communities clearly also offer affinity services to their members in addition to the provision of specific information services and resources such as downloadable software, game-playing capabilities, or recipe collec-tions. Similarly, the smaller closed and open virtual communities, or affinity-based communities, may offer information services and resources to their members. Although they will be likely to have fewer funds for the acquisition or provision of these information services and resources, what they do provide may be better tailored to the needs and interests of their members. Affinity-based communities may also engage in the production of club or public goods as an incidental purpose or sub-activity of their membership.

There are, however, reasons for the primary specialization of the communities. Although the brand-name communities may encourage affinity or collective produc-tion services, these activities are not subject to the same economies of scale as the pro-vision of information services and resources. The result is that they are unlikely to devote as much attention to the governance of affinity groups or to managing collective production services as communities that are principally devoted to these purposes. Similar logic supports the specialization of the other types of community in the services identified with them.

It is important to recognize, however, that these various types of communities can transform themselves over time. Affinity-based communities may grow and diversify in the direction of becoming brand-name communities. In addition, it is useful to ask what the limits to growth of existing brand-name communities might be. The first half of this section is devoted to these issues. The remainder examines the issues surround-ing the opportunities and limits to growth of the purpose-built virtual communities offering collective production services and devoted to producing club and public goods.

6.1. *The Dynamics of Brand-Name and Affinity-Based Virtual Communities*

While it might be possible to differentiate between brand-name and affinity-based virtual communities by interrogating the organizers of these communities about their intentions, it is also possible to focus on the interaction between behaviour with regard to user recruitment and the economics of virtual-community size. The principal revenue sources available to brand-name virtual communities that are not engaged in electronic commerce such as amazon.com are subscriber revenues and advertising.

Subscriber revenues are the result of the number of paid subscribers (if any) and, in the most common pricing model, the monthly duration of connection after a 'free' minimum allowance. The relatively few affinity-based virtual communities that have subscriber charges are based upon unlimited connection time, a fixed monthly or annual subscription charge. The size of these affinity-based virtual communities with subscriber charges primarily will be driven by the social engagement of their users in community life. For the brand-name virtual communities that collect subscriber revenue such as AOL, a similar process will account for a portion of their revenue. In the case of services like AOL, however, both types of subscriber revenue are also likely to be responsive to increased investments in information services and resources. These investments will tend to be made in those information services and resources that are most popular, the common denominator of user interests as long as the additions to subscribers and subscriber connection time respond positively. It is also important, however, to offer new features and services that may initially have a relatively low level of user engagement but that are believed to have the potential for significant growth. These service and resource creation investment decisions are very similar to the content creation decisions made by newspaper publishers or broadcasters and will not be analysed further here although they suggest important new areas for research.

The issues arising from advertising revenue for both affinity and brand-name virtual communities are somewhat more complex. Advertising rates are based either upon page impressions (the number of times an advertisement is part of a page 'served' by the Internet) or upon click-through counts. A typical income that a medium-sized virtual community organized as a web site in the United States might expect in 2000 is $1 per 1,000 page views. Thus, a virtual-community site recording 125,000 page impressions per month would generate revenue of $125.[41] Unless the site contains extensive content of substantial interest to users, 500–1,000 active users would be required to generate this many page views. If such a site is an active part of an 'affiliates' network (in which additional payment is made if users make a purchase as a result of clicking on an advertisement), it might be able to realize double the revenue, $250.[42] Thus, an affinity-based virtual-community convenor has some very clear incentives. If he or she elects to create a relatively small affinity site, the revenue from advertising will certainly not pay a living wage. A small site may be started with the intention of attracting a larger audience over time. In doing so, however, it will be difficult to achieve the necessary level of recruitment without incurring various marketing, advertising, and promotion costs. Moreover, this strategy is likely to reduce the intimacy of the community, require investments in content and design (at least of the site author's time), and increase the problems of anti-social behaviour including graffiti attacks.

---

[41] Based upon an actual site represented as typical in Hanson (2000).
[42] As reported by the same webmaster (who is responsible for several sites) in Hanson (2000).

In short, the affinity-site convenor has two choices. First, the affinity-site convenor can adopt the strategy of aiming to attract a large audience and hope that this audience will materialize. Secondly, he or she can focus on a small audience and either forgo advertising (which may discourage users in any case) or expect that, at best, the advertising revenue might pay the costs of hosting the web site. The prospects for the first strategy are rather discouraging. The top fifty web sites are believed to receive 95 per cent of the web advertising revenue.[43] While total web advertising revenue crossed the $1 billion-per-quarter level in 2000, the number of sites competing for the $200 million residual from the top fifty sites is very large. From the viewpoint of advertising-supported media, print journalism, where the bottom 80 per cent circulation newspapers and magazines share 20 per cent of advertising revenue, is a much more attractive market for an aspiring publisher who needs to start small. The very low entry costs, however, suggest that there are likely to be many aspiring publishers among affinity-group convenors. There is likely to be a relatively high rate of exit among these affinity-group convenors as it becomes clear that their primary goal will not be realized. It remains to be seen whether entry rates will remain high after experience has accumulated and the shortcomings in expectations are more widely recognized.

At the other end of the spectrum, among the top fifty web sites (some of which meet the virtual-community definition), the issues are what limits growth and what governs strategy? As noted earlier, investment in content can increase subscribers and subscriber connect time, both of which increase revenue. There is, however, a trade-off as the number of subscribers grows. The subscriber community will develop subcommunities with more intense and specialized interests that need to be served in some fashion. As these subcommunities grow, there is a real possibility that they will 'fission', departing to join or establish other virtual communities that better serve their interests. Thus, over time, investments in content creation will diversify.

The more important issue, however, is that the large subscriber base in a brand-name virtual community can begin to reproduce the congestion problem characterizing the Internet in which it is difficult for the user to navigate or search among the variety of information services and resources on offer. This results in periodic efforts to re-engineer the user interface to provide a better 'map' or navigation guide and in the possibility of innovations that would be more broadly useful within the larger community of the Internet. The other, related, problem experienced by the largest brand-name virtual communities is that it becomes increasingly difficult to differentiate the common denominator information services and resources, the publishing of interactive and simple broadcasts, from broadcast sites already available on the Internet. Thus, an individual interested in news may navigate locally to the news section of the brand-name virtual community or may navigate across the Internet to a major site such as the *New York Times* or *The Economist* and similarly meet his or her needs and interests at other general-interest sites. In other words, the very large brand-name virtual community faces the problem of having to reproduce the capabilities of the Internet and having to compete with other very large content producers to attract subscriber interest and commitment.

---

[43] This and the following discussion of advertising revenues derives from *Ad Resource* (2000).

The generation of advertising and subscriber revenue, however, is not the only mechanism for sustaining virtual-community organizers. Many affinity-group convenors forgo advertising or expect only nominal revenue from carrying advertisements. These individuals must have other motivations than financial return for devoting their time to web-site creation and maintenance activities. In many cases, they will also need to subsidize their activities from their own financial resources. What might motivate these individuals? There are at least three possibilities.

First, affinity-based virtual-community convenors may have a strong interest in the subject matter of their sites and convene the virtual community in order to share those interests with others. There are many possible sources of such strong interest. They include specialized knowledge such as a fascination with the literary works of Jane Austen or Georges Simenon, the need to share life experiences of a particular type such as cancer survival or hip replacement, or a hobbyist topic such as Roman coins or Model T automobiles.[44]

Secondly, affinity-based virtual-community convenors may be motivated by their interest in finding like-minded individuals to interact with on a broad array of subjects. Affinity groups based upon nationality, age, sexual preference, religious belief, and dozens of other characteristics exist. Although many of these virtual communities are organized within the brand-name web sites, there are also many that are organized by an individual or small group of convenors.

Third, affinity groups may be organized in order to create shared experiences or interaction 'spaces'. At present, this category is dominated by multi-user dungeon (MUD)-style role-playing games in which users are confronted with, and may choose to enter into, an environment structured by the convenor in the form of the personal computer connected to a data telecommunications network. Many, but by no means all, of the sites hosting these virtual communities are subscription based. The larger context of these sites can be thought of using the framework, proposed by Mitchell (1996), of virtual architecture. For example, a current experience in minimalist architecture is MIT's Multimedia Lab's chat circles.[45] In this software environment, users are able to move a coloured dot labelled with their chosen 'handle' (pseudonym) across a featureless plane and produce 'chat' text visible to those in physical proximity to their current position for a brief interval. Much more complex spaces utilizing various virtual-reality tools are possible, but, as they often require the purchase of specialized software, the number of users is limited. While presently somewhat underdeveloped, this is likely to be a growth area in the future as the speed of Internet connections supports more complex dynamic visual displays and the ability to update objects in the form of the personal computer connected to a data telecommunications network. Affinity group convenors in this category seem to have artistic or creative motivations and vary considerably in the provisions that they make for unconstrained inter-user communication.

---

[44] Sites for all of these interests were identified from the HotBot web search engine. With some further searching, virtual communities could be identified for all of the subjects as well. The locations are not reported here simply to conserve space.

[45] See **chatcircles.media.mit.edu** (accessed 3 Oct. 2000).

There are certainly other motives for creating affinity-based virtual communities than the three discussed here. The aim, however, has been to create categories that would capture a large number of the currently available communities. Although of interest, this collection of motives leads back to the common problem of the convenors' motivations to continue activity, given its demands on their time and financial resources. While interest or even obsession may fuel the creation and initial operation of such communities, individuals' interests and lives change over time. In the case where the community that has formed around a particular affinity site dissipates, the loss of interest of the convenor is inconsequential. It is more problematic, however, when the community remains interested but the convenor loses interest. In these cases, it may prove desirable for the convenor to 'pass on' his or her creation to another individual with a keener interest in the project. In this way, it may be possible to pass through several generations of convenors, each of which may make incremental improvements to the site. This kind of exchange also serves to limit the contributions, both in time and finance, which the original convenor makes. In some cases, the convenor may be resistant to giving up his or her creation or have concerns about possible legatees. As experience accumulates with the transfer of web communities, it is likely that further norms, self-imposed rules, and standards will emerge, enriching the institutional structure of the Internet.[46]

Innovations in the mechanisms for sustaining user engagement in both brand-name and affinity-based virtual communities are likely to occur as these communities seek to recruit and retain membership. One example is the possibility of new approaches for users to accumulate 'reputation capital'. An individual gains a reputation in any voluntary association. Harnessing this universal result of social interaction for the purpose of supporting virtual-community engagement is a promising area for innovation. Users that make useful contributions to the community improve their reputation with other members of the community and it is a simple step to provide mechanisms for community organizers or members explicitly to recognize these contributions by conferring a recognizable status on these users (Mansell and Steinmueller 2000).

For example, one might imagine that a user who had given useful technical advice to fellow community members in the Aardvark community might be awarded the designation of 'technical wizard of the Aardvark community'. Communities might choose to exchange information about these kinds of members in a bid to create even higher levels of status for the individual and to achieve the benefits from the free flow of these individuals among communities. The development of ways for users to visit other communities, bringing along their 'reputation', would provide a means to generate network externalities among these smaller communities (Mansell et al. 2000). Such a development would parallel the social institutions of science, which provide rewards for individuals in recognition of their contributions (Dasgupta and David 1994; National Research Council 1993). In virtual scientific communities such as collaboratories, powerful workstations provide a high-quality collaborative environment. These are important testing grounds for the

---

[46] Raymond (1999) argues that passing on the project when interest fades is an important ethic to be promoted in the open-source software movement discussed below.

environments that increasingly will be accessible from ordinary personal-computer desktop environments in the coming years. There are deep problems with the achievement of adequate recognition and funding for these projects because they lie outside ordinary conceptions of how scientific research can or should be conducted. Moreover, the individuals responsible for initiating these projects often fall into dangerous interstitial career positions by having to become deeply involved in information technology, social engineering and the subject matter of their collaboratory. The collaboratory is, therefore, emblematic of the difficulties of achieving a purpose-built virtual community whose site is capital intensive (both in tangible and intangible or human capital).

6.2. The Dynamics of Purpose-Built Communities

Much of the recent attention on purpose-built virtual communities has been directed to the open-source software movement.[47] Here the cases of the development of complex software such as Linux and Apache indicate a substantial and new possibility for voluntary virtual-community association. In the case of open-source software, the explicit intention is to produce a collective public good that anyone can, in theory, use. In practice, a fairly high degree of expertise may be necessary in order to install and configure these products properly, but this outcome can be seen as largely inadvertent rather than as being planned by the community. What motivates highly skilled professional programmers to contribute large amounts of their valuable time? What are the defining features of a successful purpose-built virtual community such as those created to write specific open-source software products?

At this point the reader should anticipate that the answer offered here is based upon an exchange principle. Individuals who participate in such projects do contribute their time and, in the process, are able to gain substantial expertise in a software product that is of inherent interest to them and are able to interact with their peers at a high level of expertise to pursue a common goal. Moreover, these efforts are coordinated by a relatively small group that serves as a procedural authority in attempting to negotiate closure on a particular line of development without the creation of 'forks' or parallel and competing lines of software development.

Raymond (1999) explicitly argues that open-source software development methods are not appropriate for the initial conceptualization of a software product. Instead, they are best organized around the improvement of a relatively complete, if still rudimentary, software design. Having hundreds or even thousands of collaborators to test and refine the software has proved to be of enormous value in achieving reliable software code and offers rewards to the contributors who have unparalleled understanding of the resulting software application. This understanding may then enhance their professional standing and/or remuneration, an example of the accumulation of reputation capital discussed earlier. A particularly interesting feature of the

[47] See Dibona, Stone, and Ockman (1999) and Raymond (1999). A possible exception occurs in scientific environments, which may vary considerably in planned duration. See Finholt and Olson (1997), National Research Council (1993), and the University of Michigan, School of Information web site, **www.si.umich.edu/research** (accessed 3 Oct. 2000).

process noted by Raymond (1999) is the opportunity to receive individual recognition in the comments accompanying source code (which is the basis for 'compiling' the actual computer instructions contained in object code). Programmers, who normally strongly resist documenting their programmes, have produced extensive and high quality documentation for open-source software because of the opportunity it provides for them to receive recognition from their peers.

The open-source model is still relatively new but it offers intriguing possibilities for the collaborative development of products and services. The beginning points of such experimentation are well defined. (1) There needs to be a kernel that is capable of being improved through collaborative effort. (2) A technology for collaboration is required, which, for projects involving computer code, text, and a number of other information products including web-site design, can be achieved through using the standard tools of the Internet. (3) It is helpful if there is a procedural authority that can provide the basis for averting conflict and preventing the fragmentation of the community. It is possible to speculate that a very wide range of collaborative efforts may be organized in this way. If such efforts are successful in overcoming the problems of making themselves known to possible participants, there may be substantial output of goods and services in this form in the years to come. It is important to note, however, that voluntary activity is unlikely to be successful in accomplishing routine or recurrent activities.

7. CONCLUSION

Virtual communities are one of the foundations of the 'new economy' in which interpersonal interactions achieved through computer-mediated communication can meet needs and accomplish goals that would be difficult or impossible without the benefit of the networked information technologies that are creating an ever more populous cyberspace. The prevalence of information goods and services in cyberspace raises new opportunities for the realization of increasing returns in economic activities and new forms of social interaction and exchange that are of substantial value to the users of these technologies. To exploit these opportunities, voluntary associations of individuals using the new computer-mediated communications—that is, virtual communities—are emerging. These communities are governed by many of the same principles as other forms of voluntary association, the establishment and articulation of procedural and institutional authority, which can be steps towards the creation of formal organizations.

This chapter has proposed a minimalist definition of virtual communities in order to encompass the range of related developments stemming from computer-mediated communication and to recognize that important and worthwhile social interactions arise in a variety of different contexts. It establishes that these communities can render services and accomplish purposes that would not be possible with other types of voluntary association. This is the case either because of the role of cyberspace in overcoming barriers of distance or the possibility of forming voluntary associations that would otherwise be impossible to construct due to the difficulties of identifying like-minded individuals.

Despite the unique potential of virtual communities, their operation is governed by the principles underlying other forms of voluntary association. Like other voluntary associations, virtual communities face issues of recruitment, governance, and sustainability. Each of the issues was illustrated in the examples of the USENET, the open-source software movement, and the peer-to-peer information-exchange community. While each of these communities achieves outcomes that would not be possible using other methods of organization, they share the issues of procedural and institutional authority that influence the composition and operation of other forms of voluntary association. In examining other proposals for the uniquely 'intimate' qualities of the virtual community, such as those proposed by Rheingold, it is concluded that they are principally an effort to 'privilege' an anti-commercial or communitarian ideology that is useful for recruiting and engaging users. Exclusively identifying virtual communities with any particular ethos serves to devalue and disenfranchise the wide variety of meaningful social interactions that can occur simply because of the possibility of interpersonal communication.

The virtual community does not provide a fundamentally new approach to the issues of procedural and institutional authority, although it provides many new tools for exercising and articulating these types of authority. The virtual community does, however, provide a unique method for recruiting and maintaining a sustainable membership in a voluntary association. The virtual community faces very similar governance issues to other forms of voluntary association, sharing with them the problem of regulating instances of graffiti attacks or other anti-social behaviours.

The minimalist definition of virtual community—that is, locations in cyberspace where individuals can communicate with one another—unites a vast array of Internet activities. These include brand-name virtual communities such as AOL with large memberships, a vast number of smaller affinity-based communities, and purpose-built communities engaged in the production of clubs (for their members) or public goods. The competition between these communities as well as their internal stability and sustainability are influenced by the problems of congestion (locating a relevant and engaging community in the rapidly expanding cyberspace) and coordination failure (identifying individuals that will take on the costs and responsibilities of convening the virtual community).

Economic, social, and technological factors are influencing the competition between virtual communities for the recruitment and retention of members as well as for the depth of their engagement and involvement. The economic resources for supporting virtual-community development are highly concentrated, with the fifty largest web sites absorbing the preponderance of advertising revenue, a pattern that is likely to continue into the future as these sites make investments in content and promotion of their brand names. Many affinity-based virtual communities are, however, sustained by the interests and passions of their convenors and still others by the collective interests and goals of their participants. In these cases, the problems of congestion may influence recruitment and engagement and the problems of convenor exhaustion and coordination failure in passing on the leadership role may influence sustainability.

Specifically, the types of virtual communities discussed in this chapter offer substantial opportunities for further development. While all of them confront problems of recruitment, governance, and sustainability, each seems likely to produce a widening array of activities. The principal limitations of the largest brand-name virtual communities arise from the difficulty of maintaining user interest. User interest seems likely to fade as the size of these communities makes their diversification similar to the larger community of Internet users. It is also possible that, as users gain experience in navigating in the broader Internet environment, they will discover useful general-interest information services and resources as well as more specialized virtual communities catering to their specialized interests that better serve their needs.

The growth of affinity groups appears to stem from economic problems. Some affinity-group convenors will seek to build large enough communities to provide an economic return and many will be disappointed by the structure of the web advertising market. Alternatively, affinity groups face the problem of convenor exhaustion requiring new institutions for bequeathing active communities to others with the interests and resources to maintain them.

With respect to purpose-built virtual communities, it is important to note that many of these are created and maintained through the provision of government, university, or foundation support, which partially alleviates their economic sustainability problems. It was noted, however, that there might be problems in sustaining support for these virtual communities or in providing adequate career incentives for their convenors. For voluntary associations that construct purpose-built virtual communities such as the open-source software movement, the contribution of effort can be explained as a form of exchange in which the participants receive a number of benefits including an 'insider's' knowledge of the resulting project. Although enormous opportunities appear to exist for extending the open-source approach to purpose-built virtual communities, this model has several important requirements and is unlikely to prove useful for routine or recurrent activities where professional employment rather than voluntary contribution is more likely to prevail. The principal question of whether virtual communities offer more than technological novelty for organizational innovation and change can be answered affirmatively. There is little evidence that virtual communities provide fundamentally new approaches to issues of procedural and institutional authority. Indeed, most of the problems and constraints to the growth of virtual communities involve issues of institutional and procedural authority. The unique contributions of virtual communities as a technology of voluntary association lie in their capacities to develop new methods for recruiting and retaining participants, new tools for cooperative interaction and the achievement of collective goals, and new locations in which to realize existing human aspirations and goals.

3

Cyberspace and Social Distinctions: Two Metaphors and a Theory

DAVID C. NEICE

1. INTRODUCTION

Imagine arriving on another planet where clues to people's social position are either missing or do not matter. Your curiosity is aroused. How do these people assemble their perceptions of each other? How do they decide who is worthy of respect? What social distinctions mark insiders and outsiders? In 'cyberspace' there is an absence of the types of clues about social position that matter in our traditional environments. Let us call our real-life planet RLP and our cyber-life planet CLP. On RLP the heavy baggage of physical traits and material possessions marks most of our social distinctions. Gender, colour, age, education, money, property, and power combine to signify social position and to mark social esteem. Physical traits and material possessions are signs reflecting an underlying code of social prestige.

On RLP social clues are constantly being detected and read like a language and these readings influence the patterns of our interactions. We adjust our interactions when we approach a stranger for information, a boss for a job, a professional for advice, a sales clerk for a service, or a family member for affection. Social distinctions mark our place in numerous hierarchies; they enable us to select and to discriminate. Social clues in cyberspace permeate two processes—getting there and being there. Getting there is a process that mimics the immigrant's experience. A newcomer to any foreign land must find the means to arrive and then must cope with new languages and customs. For newcomers (or 'newbies') to CLP there is some trepidation, and much learning and adjustment.

After arrival the newcomer begins to find out about being there and slowly notices that the prominent physical and material traits of RLP are often missing. Life on CLP is largely disembodied and immaterial. Disembodied life means that few physical presence clues are available, such as gender, age, or dress codes. Immaterial refers here to the subordination of the social significance of material objects, such as possessions, to the significance of weightless bits that are transmitted at the speed of light.

CLP lacks many of the visual clues that define social position and esteem on RLP. The same clues cannot be used on CLP to tell you whether a correspondent is alert or

The author wishes to thank SPRU for financial support during this research and Professor Robin Mansell for editorial guidance. Dr Sally Wyatt also provided helpful comments on an early draft.

asleep, black or white, living in a mansion or a tent, wearing a tuxedo or naked. Sherlock Holmes would find it difficult to intuit a social profile from the scattered clues offered on CLP, clues that can easily be altered, camouflaged, or concealed.

Despite vivid writings by Internet enthusiasts such as Negroponte (1995) and Rheingold (1993, 1995), and others such as Gilster (1997), social science understanding of the characteristics of interactions in cyberspace is limited. Scholars are only beginning to probe its radically different dimensions. Leading sociologists are seeking to comprehend the structure of the information or network society (Castells 1980; Tremblay 1995; Tremblay and Lacroix 1997; Webster 1995). Economic geographers are considering the significance of transactions in markets that are increasingly characterized by network relationships (Saxenian 1994) and economists have long pondered the value of information (Eliasson 1987; Lamberton 1983). Communication and legal specialists are examining electronic network design and governance issues (Lessig 1999; Mansell and Silverstone 1996). Turkle (1996) and others, such as Boisot (1995), are developing a social psychology of computer-mediated communication (CMC). While advances in understanding in these areas are significant and while research on communities in cyberspace is expanding (Smith and Kollock 1999), serious study of social processes in cyberspace is just beginning.

This chapter presents results from a qualitative empirical study designed to investigate the subjective experiences of intensive Internet-users. It examines their perceptions and interpretations of the social distinctions that matter to them in their interactions in cyberspace.

The members of all societies construct a social order based on identifiable categories of social status and esteem. Once shelter is found, hunger satiated, and libido sublimated, human beings turn to defining and reinforcing rich layers of social distinctions. This propensity appears to be universal (Scott 1996). The social order of cyberspace is far from becoming stabilized, but certain patterns or emergent social distinctions are beginning to appear and they are available for inspection.

In order to develop a greater understanding of the network economy and society, we can look at features of a significant part. In this instance, the way that social esteem is assessed and acquired in cyberspace is examined. The rules for acquiring social esteem are usually located precisely at the point where the economy and the wider society intersect. For instance, in a pastoral society these rules may involve complex tribal negotiations over cattle and daughters. On RLP the principal signs of social status are linked to the acquisition of property and material consumption—essentially, people are defined by what they own. Focusing on this point of intersection between the economy and society on our cyber-life planet can help to illustrate what is socially and economically valued in that environment.

2. TWO METAPHORS FOR CYBER-EXPERIENCE

It is helpful to distinguish between two different accounts of types of experiences and interactions within cyberspace. The first type of account is generally presented in the form of social metaphors about cyberspace and the second is presented in the form of social distinctions within cyberspace. The analysis of social metaphors *about* cyberspace often focuses on folk theories and 'common-sense' explanations and the objects of study are the metaphors that people use to describe Internet participation. The analysis of social distinctions *within* cyberspace falls within the domain of social theory; the objects of study are the signs and rules that shape specific interaction processes. In this chapter attention is on both types of accounts.

Wyatt (1998, 2000) notes that when people sense that something unusual is happening metaphors are often used to supply a sense of familiarity through rhetorical figures of speech and this is especially so when the underlying process of change is opaque. Metaphors condition us to think that objects or processes are like something else, something that is more familiar. Examples are the expressions 'time flies' or one feels 'a bit rusty'. In a similar fashion, Wyatt notes that people employ metaphors to think about the Internet. She argues that metaphors of revolution, salvation, progress, and the 'American dream' pervade many popular accounts of the Internet's development.

Unfamiliar processes are often described by social metaphors that may suggest a strong sense of social distinction. For example, the rhetorical power of social metaphors is illustrated by the use of phrases such as 'congealed labour' and the 'capitalist class' in the Marxian literature. Two widely used social metaphors about the Internet, which suggest social distinctions, are captured by the assertions that 'the net is a frontier' and that 'the net is a digital divide'. These metaphors draw attention to the potential benefits of early use of the Internet and they urge people to join with others in using it. They also hint at the potentially dark and exclusionary implications of being left out. Media articles nourish these metaphors by using labels such as 'cyber-pioneers', 'the wired', 'the cyber-elite', 'digerati', and the 'digital haves'. Those who affiliate themselves with the categories designated by these labels are often compared invidiously with the cyber-excluded or the 'have nots'.

These two metaphors may or may not accurately portray underlying social processes. This is an empirical question that is discussed below. Social metaphors offer interesting entry points for developing interpretations of the nature of interactions in cyberspace and, as folk accounts, they offer people familiar signposts and a sense of direction that is in line with evolving social norms.

The second type of account of cyberspace concerns distinctions within it that are more elusive. These accounts primarily are about the experience of 'being there', of interacting there, and of making a life there. In order to understand these accounts, it is necessary to develop a conceptual model. In this chapter the concept of *network peer reciprocity* is used to depict a particular form of social interaction that is facilitated with unusual ease and rapidity within cyberspace. Based upon the new rules of network peer reciprocity, it is expected that experienced Internet-users will make increasing use of network signs of social reciprocity to assess social esteem. By the end of the 1990s, these signs and markers were detectable and the main purpose of the empirical study that is discussed in this chapter was to isolate and identify them.

During the nineteenth century Veblen's notion (1899/1994) of 'conspicuous consumption' aptly described the material basis for community perceptions of social status and prestige. Social status in market societies is signalled by visible displays of property, the accumulation of material objects, and status-group choices regarding leisure pursuits. In highly materialistic societies, this intensely competitive activity supports a vertical social hierarchy that is replete with complex status distinctions based on an individual's (or families') consumption powers and patterns.

Within market societies, however, other important bases for community social esteem coexist, albeit in a context in which conspicuous consumption is predominant. Philanthropy flourishes, community giving occurs, selflessness marks instances of heroism, voluntarism pervades many behaviours during peacetime and in times of crisis, and social esteem in kinship networks is often rooted in acts of gift giving and benevolence. Gifts, social reciprocity, and community contributions, based on a calculus of social exchange, flourish despite the absence, in most cases, of an explicit market calculation, or of obvious benefits of status that may be awarded as a result of their conspicuous display.

As discussed in the following sections of this chapter, analysis of the views held by intensive users of the Internet suggests that the attribution of social esteem in these communities occurs within a system of rules and values that *amplifies* peer reciprocity and peer social exchange. This system diminishes the importance of market-related forms of conspicuous display. Social esteem derived through interaction in cyberspace appears to favour the markers of peer social reciprocity and to de-emphasize consumption markers. This change in emphasis appears to be associated in significant part with the disembodied and immaterial character of Internet transactions. Within cyberspace, visible consumption display is rendered subordinate to building relationships and the basis for allocating social respect and esteem shifts towards valuing reciprocity in lateral or horizontally defined networks of relationships.

A close parallel is found in peer science communities. For example, researchers found previously that social credit (Latour and Woolgar 1986) and social esteem (Hagstrom 1965) in science networks are cumulative processes derived from shared perceptions among peers about researchers' contributions to science. In science networks conspicuous consumption is inverted such that social esteem is accorded to those who can offer *conspicuous contributions*. It appears that the 'taken-for-granted' signs of conspicuous consumption that mark social position and esteem in market societies are being challenged by the prevalence of another model in cyberspace based upon network peer reciprocity and the visibility of conspicuous contributions.

3. NEW COMMUNITY DATA

To assess the implications of the two metaphors discussed above and the insights yielded by the conceptual model of network peer reciprocity, a study was designed to generate information about the views of intensive Internet-users. The participants in the study were drawn from the population of a medium-sized community in the south-east of England in 1998–9.[1] A qualitative research technique called focused interviewing was updated and applied to probe the perceptions of four defined

---

[1] The study was conducted in Brighton and Hove in East Sussex, an area with a modest reputation for being one of the more 'wired' parts of the United Kingdom.

segments of experienced Internet-users: university postgraduate students; new media business developers; Internet community developers, and Internauts (or techies).

The original technique was developed by Robert Merton and his colleagues between 1946 and 1956.[2] A focused interview is a guided or facilitated discussion that follows a series of pre-planned themes. Interview probes are inserted frequently during a discussion session to elaborate upon meanings that may otherwise go undetected. The method shares some of the features of in-depth interviewing, a technique adapted to computer and Internet research by Turkle (1996: 321–4).

Focused interviewing can be applied in either individual or small-group discussion (Merton, Fiske, and Kendall 1956).[3] In the present study, the group method was chosen as a means of stimulating debate among participants and to encourage diverse Internet-user interpretations. Small groups of experienced users were selected, convened, and interviewed to elicit their experiences of being online.[4]

Sample surveys prevail in much of the recent research on the development of Internet use and this has ensured that the results provide profiles of an 'average' Internet-user (Dickenson and Sciadas 1996; Katz 1997; *Guardian* 1999). Surveys suggest that early Internet-adopters and recurrent users are usually young urban males with above-average incomes and education. The same pattern appears for computer ownership, which is not surprising insofar as the majority of Internet access is via computers. However, when 'average-user' profiles are compared with the variety of individual experiences of using this network and its services, they begin to pale. Even casual Internet-users know the 'Net' is an evolving medium that is rich in interactive complexity, full of quirky surprises, and subject to diverse interpretations and perceptions. User experiences (both the epiphanies and the horrors) are often so unique that they rarely make sense to non-users.

Survey-based research needs to be complemented by methodologies that yield qualitative detail and by studies that privilege the perceptions of different types of users. The approach in this study was not to describe the typical or average Internet-user but rather to examine the perceptions of specific types of experienced Internet-users. This approach was based on a common-sense proposition. Intensive users were expected to have the richest base of experience and, therefore, could be expected to offer the greatest direct insight into the complexity of the new medium.

The main objective was to identify experienced types of users to whom reasonable access could be gained. This strategy was in line with the requirements of the grounded theory tradition established by Glaser and Strauss (1967) and extended by Glaser and Corbin (1990). Young people (often males) involved in information retrieval and new

---

[2] Merton *et al.* (1956: 3) used focused interviewing as an information elicitation tool to probe perceptions of subjects after exposure to specific media and their content. Focused interviewing requires that subjects have direct experience with the medium under review.

[3] The focused interview method is not identical to focus group research (see Barbour and Kitzinger 1999). It is prudent to note the similarities and differences in the two methods so that their strengths can be aggregated and weaknesses avoided (Morrison 1998). In later work Merton (1987) compared focus groups and focused interviewing.

[4] Group interviewing is a powerful tool providing the discussion moderator is aware of group effects. Strong individual views are expressed in groups as well as consensus building (see Krueger 1994 and Stewart and Shamdasani 1990).

media work are most likely to be intensive users. Therefore, Internet-users aged 21–35 with these backgrounds were recruited. It was also expected that perceptions and characterizations of the Internet would be very active and intense among younger people. Four segments of experienced users were selected for participation in the study based on an identification of organizations that could be used as contact points for subject recruitment and on a set of hypotheses derived from the literature about Internet experiences.

University postgraduate students were chosen as knowledge workers in training. Graduate schools include individuals trained in the retrieval and use of research information for professional purposes. Today's graduate students spend long hours using the Internet, which, in the United Kingdom, is accessed using high-speed connections (often via Ethernet) and specialized computing resources (such as R&D networks) at their institutions.

New media business developers are frequently interested in exploiting new technology for commercial and personal gain. They are often entrepreneurs who either own or work in small and medium-sized enterprises (SMEs) or in constantly changing micro-business teams. They usually have a strong commercial stake in the Internet's evolution. Unlike postgraduate students, they generally rely on private means and strong personal interest to obtain their generally superior access to digital technologies and high-speed networks.

Community developers use and advocate the Internet for purposes of democratic empowerment. They are often affiliated with voluntary associations and non-governmental organizations with strong interests in the environment, social development, and emancipation movements. They are often enthusiastic about Internet technology as a tool to support greater empowerment and they use it in support of their pursuit of various ideals. However, their access to exotic tools and the latest technology is often constrained by money. People who regard the Internet as a creative forum and are attracted to supplying it with creative content are referred to here as Internauts. They are generally people who have detailed technical knowledge of communication protocols and operating systems or have artistic skills for using digital media. They have significantly enhanced access to computer technology either at home, at work, or at school. While they share many high-level skills with new media developers, their primary interest is not commercial although it may be vocational.

This chapter focuses upon only a few of the many themes that emerged from the group interviews.[5] The participants' perceptions of the two social metaphors are highlighted and considered in the light of the characteristics of the emerging Internet-user base in the United Kingdom. Several hypotheses about likely variations in the perceptions of experienced users are examined and the results are discussed in the form of clusters of observations made by and about intensive Internet-users.

Focused interviewing is an elicitation technique that seeks to excavate perceptions among defined groups of participants to examine research hypotheses. The price paid for the interpretative power of the technique is a reduced ability to generalize and, as a result, inferences must be drawn with care. Fortunately, however, the qualitative research

---

[5] See Neice (2000) for complete details of the research design and research methods.

tradition has been applied in ways that generate insights that can often substitute rather effectively for those produced by statistical inference from larger-scale, structured datasets (Alasuutari 1995) (see the Appendix to this chapter for further details).

4. THE NET AS A FRONTIER

Characterizing the Internet as a frontier has become a publishing cliché (see Ainley 1998; Clemente 1998; Moody 1995; Sterling 1992). The theme surfaces time and again in discussions about Internet communities (Miller 1995; Rheingold 1993). Enthusiast publications, such as *Wired* magazine, are liberally sprinkled with frontier descriptors, as are thematic issues on digital technology of *Time*, *The Economist*, or *Fortune*. The Internet's founding designers, such as Vinton Cerf, use frontier metaphors (Gilster 1993), as do its defenders and advocates in the United States, such as the Electronic Frontier Foundation (EFF) and its affiliates.

Specific social images and distinctions are stirred up by the frontier metaphor. It evokes dreams of new territories that are wide open to entrepreneurs. It offers the promise of participation in an egalitarian community; of being among equals, but as rugged individualists. It suggests that cyberspace is tolerant of unusual customs and idiosyncratic characters and that it is a place where personal freedom of thought and expression abide. It projects the sense of a loose meritocratic federalism, realized through the daily actions and choices of frontiers 'men'. It implies a rejection of conformity, blind obedience, and control. In short, the frontier metaphor harnesses the possibility of a return to life as it was historically in the sparsely settled west of the United States.

The ubiquity of the metaphor invites interpretation.[6] Two such interpretations are offered by Healy (1997) and Turner (1999). Healy characterizes the space available to users of the Internet as a 'middle landscape' where contradictory desires for independence and involvement seek expression. For Healy (1997: 60) the new medium offers a means for 'ontological individualism' and, at the same time, 'an almost limitless potential for an associational life'.

In sharp contrast to Healy's 'Walden Pond' reading, Turner focuses on the relentless and aggressive promotion of the frontier metaphor. He regards this metaphor as one that gives ideological expression to the interests of an emerging digital elite. Thus, this group has relentlessly promoted a vision of computer-mediated communication as frontier exploration. Turner (1999: 2) suggests that they have done so partly in order to gain social and economic advantages for their class.

4.1. *Experienced Internet-Users and the Frontier Metaphor*

What do experienced Internet-users in the United Kingdom make of this metaphor? Analysis of their comments generated by the study discussed here suggests that the frontier metaphor is regarded as being mono-cultural and nationally specific; it is neither much admired nor embraced.

---

[6] The word 'cyberspace' and its vivid depiction as a wild frontier can be traced to Gibson's novel *Neuromancer* (Gibson 1984).

Internauts

INTERVIEWER. Some say cyberspace is a new frontier ... what do you think?

RESPONDENT. It is American.

RESPONDENT. It is a very American driven metaphor ... it is more submerged here ... there is a different European view of what goes on the Internet and the way Europeans look at it ...

INTERVIEWER. You mean the new frontier idea is American ...

RESPONDENT. ... yes it uses an American metaphor ... the domination of the west, the social attitudes that come along with it.

RESPONDENT. But I think it does have a truth. I do think it is like pioneers moving into the wild west. Looking for something on the Internet ... until recently it was like walking through a small town in the wild west and yes there might be a bar and yes there might be a few restaurants ... you had no way of telling which was the good one.

RESPONDENT. Exactly. And there is also a sense of people staking out huge amounts of territory that is not currently productive, but they hope to make it productive later. All these companies are trying to establish themselves on the Internet, not because it makes money now, but because it will one day.

INTERVIEWER. Staking claims?

RESPONDENT. Yes, hopefully.

RESPONDENT. It is a particularly American obsession, pushing back the frontier, and presumably the metaphor itself is sort of bound up with their own history.

RESPONDENT. There is a frontier, but it is not particularly in technology, it is more in the uses and in the way people interact with that technology ... and with each other through that technology rather than the actual technology itself ...

RESPONDENT. There is a great story about email. When they first started linking computers together they thought it was all going to be computers talking to computers ... sending instructions and so on ... and it wasn't ... it was email; it was people talking to each other. This gets a bit buried under territorial metaphors but it is social interaction that drives stuff ... and not just territorial viewpoints or technological viewpoints.

INTERVIEWER. Well ... when I tabled this topic, the frontier, you said it is an American perception but there is a different European attitude ...

RESPONDENT. ... the Europeans seem to have much more of a communal tradition. European politics has a much stronger left emphasis than there is in American politics and that can be very true on things like USENET ...

Community developers

INTERVIEWER. Some say that cyberspace is the new frontier. What do you think of that description?

RESPONDENT. Nonsense.

INTERVIEWER. OK, explain ...

RESPONDENT. ... that is a very American-centric phrase anyway. You think about what the frontier actually meant; it was nicking land and killing Indians and that is not very attractive. That presumes that a society has a common goal and a common aim and something they want to conquer.

RESPONDENT. Do you disagree with the word frontier or do you disagree with the possible changes it could make? Do you disagree with the vocabulary or with actual changes ...

RESPONDENT. That's a particular concept, the concept of cyberspace as the new frontier ... it doesn't make sense to me ... that is not to deny that cyberspace has changed the way I live or can change the way other people live, but I don't think it is meaningful to think of it in terms of frontier.

INTERVIEWER. Do you think though that other people may think it is appropriate?
RESPONDENT. I think it does have potential for being a new frontier, but again probably I would disagree with the language a little bit. I think it will make quite a big change to how things are done.

Graduate students
RESPONDENT. . . . also about the frontier . . . it is in a way like a frontier because there are many new aspects you can't control from any one country . . . you can't control the contents . . . and some governments have had to find that out the hard way. They don't really like their people to know what's going on outside. Also everyone can put whatever he or she likes on the Internet.

New media developers
INTERVIEWER. Since the Internet is basically a United States developed technology, do you think it is perceived differently here?
RESPONDENT. I think it is.
RESPONDENT. It's content as well . . . I mean all our computers are run by American software, there is Apple and Windows basically, but that is just an interface. With the Internet you have got so much new content *so suddenly the content is continent specific*. So it is not the fact that it comes from America, that the technology comes from America, it is the fact that the *content* comes from America . . .
RESPONDENT. There is another problem that is sociological as well, the fact that the English are not Americans. Americans do things very differently to the English and it's not so much a matter of bandwidth and computers and the prices that are a problem in this country, it is the English themselves. You know, we are still used to going down the street, going to Joe Bloggs on the corner, and buying our paper and bread. There are a lot of people who just like going and getting the shopping at the end of the day, they don't do their monthly shop in a place out of town. In England shopping is a social thing, so the whole idea of trading on the Internet, shopping on the Internet and all the rest of it . . . I mean shopping channels are just not going to happen here.

4.2. *The Frontier Metaphor is Monocultural*

Many things can be described as a frontier and often are (Ainley 1998; Guthke 1990); however, actual frontiers exhibit certain common social and cultural features. In the late 1970s Murray Melbin (1978) wanted to know whether the 'night' hours in the United States had come to be regarded as a frontier. Melbin hypothesized that when uncharted geography in the United States declined as settlement spread, the frontier would be transformed. He suggested that time had become the frontier as a result of an increase in wakeful activity during the evening hours. He identified ten features common to both geographic and temporal frontiers and concluded that frontier-like expansion into the night was superseding the geographic frontier. As geographic space had become full, the frontier had moved into time.

The list of common features of frontiers identified by Melbin is displayed in Box 3.1 with several extensions into cyberspace (Neice 1998a). The ten features isolated by Melbin appear on first inspection to apply to the Internet. This suggests that the description of cyberspace as a frontier has some validity. However, this parallel with other frontiers is self-referential. Almost anything fresh, territorial, and exploitable is

Box 3.1. Some common features of frontiers

1. Advance is in stages, involving successive steps in the colonization of regions
Old West. Development was marked by successive stages, as fur traders were followed by
  cattlemen, ranchers, and farmers.
Night. Successive steps involved isolated wanderers, then night and graveyard shift workers
  and consumer activities.
Net. Many technical and design steps preceded the World Wide Web and global reach.
2. The population is sparse and also more homogeneous
Old West. The line between settlements and wilderness was unclear; settlers were mostly
  vigorous young males, sometimes looking for scarce females.
Night. The colonization of night also fits this pattern; males first.
Net. The Internet initially involved mostly male academics and researchers; even with
  emergence of the .com domain names, it is still populated by many young males.
3. There are welcome solitude, fewer social constraints, and less persecution
Old West. Escape from duty, class or religious oppression, and social complexity was a common motivation.
Night. The possession of streets at night is more easily accomplished; there are less crush,
  less surveillance, and more direct expression of identity by those who are different.
Net. Computer-mediated communication (CMC) invites seclusion, anonymity, aliases,
  and constructed identities.
4. Settlements are isolated
Old West. Settlements were small and scattered with few links to either each other or the east.
Night. Night involves separate pockets of wakefulness, with little communication between
  pockets.
Net. Communication nodes are often very distant from each other; communities are
  specialized.
5. Governance is initially decentralized
Old West. The interpretation of law was selective and idiosyncratic or locally constructed
  and amended.
Night. Different internal rules apply at night even in institutions such as hospitals; the same is
  the case for street activity.
Net. Initially the Internet was not governed or 'owned' by any formal legal apparatus; even
  now with Internet Society (ISOC) and Internet Corporation for Assigned Names and
  Numbers (ICANN), governance is nebulous and decentralized.
6. New behaviours and styles emerge
Old West. Rugged individualism applied and the frontiersmen left 'proper' society behind
  for something different; isolation often meant creating deviance.
Night. The night world includes jazz musicians (and audiences), transvestites, and schizo-
  phrenics, and is thought by day people to be 'the haunt of weirdos and strange characters'.
Net. Read the Internet .alt newsgroups; all fetishes and interests are catered to in abundance.
7. There are more lawlessness and violence
Old West. Guns, shoot-outs, thievery, gambling, bawdy houses, and lynching were, if less
  common than popular imagination might suggest, still a fact of life.
Night. Night is more crime ridden and fraught with dangers and nefarious activity than
  day, even through this activity is concentrated in certain places.

Box 3.1. *Continued*

Net. Viruses, spamming, and flaming continue to abound; many believe the net is 'out of control'.

8. There are more helpfulness and friendliness

Old West. Accounts of the geographic Old West are replete with numerous examples of community activity and warmth.

Night. Tests conducted that compare activities requiring the cooperation of strangers found that night produces more instances of volunteer help than the daytime hours.

Net. 'Newbies' can usually get help from those with more experience provided they read the widely posted frequently asked questions (FAQs).

9. Exploitation of the basic resource becomes national policy

Old West. There was a long delay between settlement and formal political recognition.

Night. The development of labour policies for shift work as well as standards to facilitate round-the-clock activity came long after time was colonized.

Net. The National Information Infrastructure Agency (NIIA) and the Canadian Network for the Advancement of Research, Industry, and Education (CANARIE) were introduced thirty years after the Advanced Research Projects Agency Network (ARPANET) of the US Department of Defense, the global Internet's progenitor in the USA.

10. Interest groups emerge

Old West. Agricultural blocs led to the formation of political protest parties.

Night. Some night-leaning people formed alliances such as the Gay Liberation Front and Call Off Your Old Tired Ethics (COYOTE).

Net. The Internet Society (ISOC) and the Electronic Frontier Foundation (EFF) formed as well as other international advocates and bodies.

Source: Based on Melbin's original depiction of the 'the night' (1978: 6–18) with major adaptations to address the Internet developments by the author of this chapter.

metaphorically characterized in the United States as a new frontier.[7] Nuclear power, space travel, genetic engineering, and bio-medicine are examples. Furthermore, the dominant social and economic model for expressing new aspirations in the United States is based on examples drawn from the history of Western settlement patterns. Given this 'one-note' repertoire, it is not surprising that, in the face of new opportunities, frontiers are defined metaphorically in this way or that behaviours are encouraged that resonate with this basic pattern.

Images of frontier life in all likelihood trigger specific mythic meanings in the United States, and particularly in the south-western states that are home to many aspects of the digital revolution. It may be that thinking about the Internet as a new frontier of space and time serves to update and reinvigorate old myths while reasserting American exceptionalism (Hofstader and Lipset 1968; Wrobel 1993). More generally, the frontier

---

[7] Perceptions of economic growth in the United States are heavily influenced by Frederick Jackson Turner's frontier thesis where settlement and exploitation of the West are posited as the explanation for American development (Simonson 1963).

metaphor may help to heighten corporate interest and drama in the commercially exploitable aspects of the Internet. These aspects may be intensified by the possibility that Internet-based transactions may offer an escape from conventional rules of commerce.

In summary, while the social metaphor of the Internet as a frontier appears to apply with some interpretative force to a reading of the dynamics of Internet participation in the United States, the social distinctions it implies are not synchronized with those outside the United States. The chauvinist and monocultural roots of the metaphor do not resonate particularly well in the United Kingdom, at least not with the participants in the study that is reported here. This is not surprising, given the fact that the Internet is, after all, a global communication system with a vast number of uses, applications, and cultural meanings.

5. THE NET AS A DIGITAL DIVIDE

Discussions about an emerging digital social divide and about whether action should be taken to reduce this divide have existed for some time. Descriptions of this divide vary considerably. Some descriptions refer to the cleavage as the digital 'haves' and 'have nots', others refer to the included and excluded, and still others to digital citizens, or 'the wired', and the rest of the population. Concerns about the implications of a growing divide are most frequently articulated by government bodies (European Commission 1997; Department of Commerce 1998), although, when sales of digital appliances begin to slow down and markets begin to saturate, these concerns flourish even within industry.

Evidence of the existence of this divide comes mainly from the results of statistical monitoring of the public's access to the Internet.[8] In market research, statistics on the penetration into people's homes of devices such as televisions and video cassette recorders are commonplace. The novelty of Internet adoption in the mid-1990s, stimulated by numerous surveys based on relatively large-scale samples, produced indicators of differences in the usage characteristics of Internet-users and the reasons for non-use and a host of explanations for the variations in usage and non-usage patterns.[9] In some countries detailed comparisons of variations between user groups have been published based upon population segmentation and typologies of media-users. These types of indicators of differences in Internet use are continually being developed and refined.[10]

Table 3.1 provides figures on Internet penetration.[11] Generally in these surveys the North American and Nordic countries rank highest for Internet use. The lead established by the countries in these two regions creates an incentive to introduce 'catch-up'

[8] In some countries the national statistical agencies or other branches of national governments collect access statistics (see Department of Trade and Industry 1996, 1998*a*; Dickenson and Sciadas 1996; and Industry Canada 1997). Data on Internet access and use are also being collected through public opinion surveys, sometimes paid for by rival media companies such as newspapers (see the *Guardian* 1999).

[9] Most surveys measure only the presence or absence of an Internet appliance in the lives of respondents (Neice 1998*b*). Actual patterns of Internet use and user perceptions of it are mainly uncharted.

[10] The panel survey *Home On-Line*, being carried out by the Institute for Social and Economic Research at the University of Essex, is an example.

[11] Caution should be applied to international comparisons as these data sources vary in sampling technique and in the questions asked of respondents. However, some countries do tend to have particularly high levels of connectivity per capita (*Globe and Mail* 1997).

Table 3.1. *Percentage of Internet-users for selected countries*

Location	Percentage	User estimate (000)	Period	Source
North America:				
Canada	42.0	12,700	June 1999	Angus Reid
USA	39.0	106,300	July 1999	Nielson/Net Ratings
Western Europe:				
Belgium	16.0	1,400	Feb. 1999	Initiative Media, Brussels
Denmark	34.0	1,700	May 1999	Business Arena, Stockholm
Finland	32.0	1,600	May 1999	Business Arena, Stockholm
France	12.9	6,200	July 1999	Mediangle
Germany	10.0	8,400	Mar. 1999	GfK
Ireland	13.5	38	June 1999	Amarach Consulting
Italy	7.9	5,000	June 1999	Osservatoria Internet Italia
Netherlands	13.7	2,300	Mar. 1999	ProActive
Norway	35.3	1,600	May 1999	Business Arena, Stockholm
Portugal	2.0	188	Jan. 1998	IDC Research
Spain	8.7	3,100	June 1999	AIMC
Sweden	41.0	3,600	May 1999	Business Arena, Stockholm
Switzerland	12.0	870	Sept. 1998	Swisscom
UK	18.0	10,600	Dec. 1998	NOP Research Group

Source: Compiled from NUA Internet surveys. All figures are estimates available December 1999 at www.nua.ie/surveys/how_many_online/index.html (accessed 15 Mar. 2000).

policies when comparisons are made by agencies and task forces with a mandate to understand the determinants of variations in the explosive growth of online communication in different countries.

By the mid-1990s, the demographic distinctions between 'the wired' and the rest of the population had become a media mantra. The 'wired' were characterized as savvy, white, young urban males with superior education, upper middle incomes (or aspirations), and a lust for new technical devices; in other words, 'geek chic' had arrived. Once these demographic features became well known, debates about the privileges of the 'haves' as compared to the 'have nots' became more apparent in policy discussion.

Concerns about social inequality hover around the digital-divide metaphor. This focuses attention on corresponding social advantages and disadvantages that stem from relatively intensive use of the Internet as compared to low levels of use or non-use.[12] In general, the objective of policy in this area has been to ensure that all members of society are on a level playing field as far as Internet access is concerned. Survey instruments rarely elicit information about whether respondents are concerned about something called a 'digital divide' or whether the Internet 'haves' have something worth having. It is simply presumed by those advocating the elimination of the 'digital divide' that having Internet access is always better than lacking it.

[12]   Some of the author's earlier work employed this line of argument (Neice 1996).

5.1. *Experienced Internet-Users and the Digital Divide Metaphor*

In the United Kingdom, participants in the group interviews were generally sensitive to the 'digital-divide' metaphor and regarded it as being more useful and constructive than the frontier metaphor. However, they were also somewhat sceptical of its accuracy.

New media developers

INTERVIEWER. It's possible that digital technologies are creating new divisions in society between the wired and those who aren't—between the digital 'haves' and the 'have nots'. Do you see this happening?

RESPONDENT. I guess it's the same division, it's just another kind of expression of it . . .

INTERVIEWER. Another expression?

RESPONDENT. It was colour TVs at one stage, wasn't it, going back to when I was a very small boy?

RESPONDENT. Yeah, but it's the distinction here . . . if you've got a TV set you have access to information and news and all that kind of stuff. If you've got Internet access and there's a war and the TV is out you can find out the real information from other people from other countries through the Internet . . . which has happened in revolutions.

RESPONDENT. But would you be disadvantaged if you didn't have it?

RESPONDENT. Well, in that context, yes, you're right.

RESPONDENT. Would you as an individual feel disadvantaged?

RESPONDENT. . . . well, that's the whole idea of cultural capital. I mean . . . there are certain things that certain people don't have access to . . . whether it be books in their houses or . . . it doesn't have to be technology! You know, some people grow up without books in their house. I think that in terms of people who don't have television, they have a different sort of cultural relation to the world they live in. I just think they must. I think it's the same and more with the Internet . . . that if you don't have access to that information, whether there's a war or not, there are certain frames of reference that you're going to miss out on . . . in terms of communication and information and the rest of it.

RESPONDENT. I think you're right but . . . at the moment the Internet is being driven by what's happening in the real world. I think that we will soon pass a line where things happen first on it and then the real world gets driven by the information that's on the Net. That's going to be the big difference . . . it's when it happens in the digital media first . . .

RESPONDENT. . . . like (digital) music distribution methods . . .

RESPONDENT. Yes. Music distribution is a really good one.

Community developers

INTERVIEWER. Some people think society is becoming digitally divided. What do you think?

RESPONDENT. Yes.

RESPONDENT. Very much so.

RESPONDENT. And it is a worry.

RESPONDENT. We are an elite.

RESPONDENT. In terms of the global population . . .

RESPONDENT. The majority of people on this planet have less priority in having a computer . . . they need somewhere to live. We are a very privileged minority actually.

RESPONDENT. I think we are all in danger of thinking it is more important than it is.

RESPONDENT. . . . possibly, but I would be prepared to argue those aspects myself, but I think for me quite a useful illustration is this. I get into environments . . . and in those environments one starts talking to people and developing contexts and things like that—'are you on email, can I email you?'—and if they are on email typically things get happening, and if they are not on email then forget it. It is . . . the 'haves' and the 'have nots' that are identified in those terms and it's not arrogant or anything on my part. I have got so much to do and it's an order of magnitude easier for me to do it electronically than it is in other modes. If it's electronic, it gets done, and if it's not it doesn't.

Graduate students

RESPONDENT. My favourite statistic is that there are 500 million people on the Internet but two-thirds of the world's population have never used a telephone . . . in England or the States it seems like lots of people are wired but vast chunks of Africa and Asia and South America have no comprehension of what the Internet is . . .

INTERVIEWER. What about within an advanced society like the United Kingdom? Do students for instance have any advantage because they have better Internet access than other people?

RESPONDENT. Yes, because they come out of university with some skills . . . there is some exposure at least . . . an advanced step up. I mean computer skills in almost any job now is definitely wise . . . just being able to use Microsoft Word or something.

RESPONDENT. I kind of get the feeling that we are quite a transitional generation, sort of people in their 20s now. We have gone from people in their 40s never having done it, never having (computing) skills and not really having to use them at university . . . to now.

Internauts

INTERVIEWER. Some people think there is an emerging digital divide? Do you see that happening?

RESPONDENT. Absolutely, it already has [happened].

RESPONDENT. It is just another language. There is a split between literacy levels in the [use of this] language and it's powerful . . .

RESPONDENT. I think that will become more and more diffuse as time goes by. It is like when the television was invented and then one little boy would be dead popular because his parents could afford a telly and everyone in the street watched that big black and white telly to see whatever the programme was . . .

RESPONDENT. . . . . Oh there are plenty of divisions and I think we can either let digital technology reinforce those or we can actively do something to try and use it to break them down.

RESPONDENT. I think we need to actively do that.

RESPONDENT. I think the divisions are there already. I think the Internet is the latest thing to highlight that it is already there, and I think in a lot of ways the Internet can actually break it down, especially among the younger people that might not even be as aware of the class issues as perhaps people are when they are older . . .

RESPONDENT. I think it is the people again though, the Internet can't break anything down, it is the people who use it and in some ways the Internet can mask divisions and in some ways it can highlight them—it depends what you draw attention to . . .

RESPONDENT. The divisions will be in what will happen in so-called third world countries when we are totally computer literate . . . and they are not . . . that concerns me.

5.2. *The Digital Divide as a Policy Construct*

These extracts from the interviews suggest that in the United Kingdom the digital-divide metaphor is more favourably perceived than the Net-as-a-frontier metaphor. It appears to have greater transnational salience although it is also less descriptively accurate. In one sense, the digital-divide metaphor is a survey-derived statistical artefact. The profile of an average Internet-user emerges from large-scale surveys because the empirical method predetermines this result. Before the survey data are collected, an outcome suggesting a social divide is a foregone conclusion because only some survey respondents have Internet access. If the words Internet 'wants' and 'want nots' are substituted for Internet 'haves' and 'have nots', the effect is altered.[13]

In most surveys Internet access is treated as a consumer object—a thing—as if it were interchangeable with toasters, tomatoes, or toothpaste. By defining Internet access in this way, information on user experiences and perceptions is reduced to the formulaic indications that can be derived from checking boxes on a telephone survey instrument. Demographic profiles provide useful information, but such baseline data must be supplemented by more in-depth information if the factors motivating Internet use are to be better understood. The results of Internet surveys, and particularly market surveys, do not yield insights into the complex paradoxes that pervade Internet-user perceptions. In particular, they overlook the interest of intensive users in applications that favour social reciprocity and exchange, and non-consumptive interaction.

Large-scale surveys also tend to be insensitive to very small incremental breakthroughs that can produce dramatic changes in any country's access and participation statistics. Examples are the opening of the America Online (AOL) portal to the World Wide Web in 1996 and the explosion of 'free serve' Internet service-providers in the United Kingdom, a development triggered in 1999 by Dixons, the electronics retail merchandiser. These service-provider solutions are the real drivers of the rapid increases in Internet-users.

What is the attraction of using dividing lines when metaphorically thinking about the Internet? Why do people talk about the presence or absence of Internet access using the vocabulary of social inequality? After all, access to other appliances and utilities, such as dishwashers and cable television, does not produce such concern. The heightened attention that is being given to the digital divide metaphor may be a reflection of its cultivation by governments and state-appointed bodies. The 'divide' metaphor updates eighteenth-century debates about citizen equality and highlights issues around the evolving rights and duties of citizenship in an uncertain digital era. Since one of the most important duties of citizens in the twenty-first century is to consume and use goods and services (including information), it has become important for governments to ensure that any new means for their consumption and use are universally available.

A further attractive feature of the metaphor for policy purposes is that digital divides can be found everywhere. Such divides can emerge from within societies, they may exist between industrialized countries, and they may exist between the industrialized world and the developing world. As in the case of global warming, everyone can either suffer or benefit from the effects of digital dividing.

---

[13] This semantic trick was pointed out in a Canadian Broadcasting Commission (CBC) research paper (Kiefl 1996).

Does the 'digital-divide' concept offer an accurate depiction of social distinctions in cyberspace? In many ways the 'divide' is a modern example of a pervasive statistical mythology, born out of the availability of technology and service penetration statistics and cast as an updated form of an eighteenth-century political debate rooted in class distinctions. While this may serve the aims of policy reformers, statistical and political myths are not very durable when they are confronted with empirical reality. The Internet 'haves' and 'have nots' are a statistical artefact produced by off-the-shelf data and hasty interpretation. The statistical classes are representations of neither political interest communities nor a political movement. The digital 'have nots' are storming neither the Bastille nor Westminster. In fact, the concept of a digital divide is far removed from the main arena in which social distinctions are emerging in cyberspace.

6. NETWORK PEER RECIPROCITY

The social metaphors discussed in the preceding sections allude to the attractions of going into cyberspace and participating in it. In this section, the social distinctions that appear to be emerging with direct experience and immersion in Internet-related activities are considered. The signs and markers of network peer reciprocity in cyberspace are contrasted with the physical and material markers within a market society. Understanding the main features of these new signs and markers requires that we draw upon social theory.

Social distinctions provide reference points for comparison and for social standards. These standards indicate to the members of a society the way people 'should live' and what they may aspire to; the social order is a kind of moral or normative order. The social distinctions found in traditional market societies, for example, involve categories of employment (or unemployment), education credentials, professional accreditation, the acquisition and display of property, the consumption of commodities, and social class distinctions that are referred to as 'tastes' and that may include aspects of dress and speech.[14]

In cyberspace, social distinctions are under construction but certain identifiable patterns are appearing that differ from those in the conventional marketplace. The new signs and distinctions are emerging out of disembodied and immaterial characteristics of cyberspace interaction. To grasp this idea, a thought experiment is useful, beginning first with what is familiar: life in the market society or RLP. In the material world there is consensus within certain culturally bounded communities about how social status is valued and expressed. People seek fortune and fame! This often involves the development of a business, media, or political personality, lavish displays of consumption, unflagging pursuit of public honour, and the exercise of influence through appointments at the apex of organizational hierarchies. In short, in today's worshipful celebrity culture (*Guardian Weekly* 1999), it is possible to aspire to become Geri Halliwell or Tony Blair.

---

[14] The literature on this topic, formally called social stratification, is voluminous. See Crompton (1993) and Scott (1996) for excellent overviews. Bourdieu (1984) has also written extensively on social stratification and taste.

When a social space is occupied in which most, if not all, of the status markers of the material world are either absent or subordinate to other markers, what factors mediate social interaction? There are no physical bodies in cyberspace that can provide visual physical presence cues (Lombard and Ditton 1997). There is no sense during online interactions of property, commodities, or other material clues that mark consumption. Communication on the Internet is largely stripped of formality, hierarchy, and authority. Corporate or institutional positions can be flagged in electronic signature files, but these are disembodied and seem to lack the physical trappings of offices, office workers, and letterhead. Geographical location is irrelevant and nationality is only an incidental addressing feature of some domain names. Nearly all of the clues that signal market-derived social status and position and that are so crucial to face-to-face conversations and interactions are absent or only dimly present. What the Internet imposes is a dematerialization of communication and, in many of its aspects, a transformation of the subject position of the individual who engages within it (Poster 1997: 205).

Dematerialization of economic and social processes involves a 'lightening-up' of industrial culture and this has major implications for economic behaviour (Coyle 1998; Quah 1993, 1996) and for the social processes that are of central concern in this chapter. Dematerialization is a process whereby perceptions of economic value appear to be shifting from matter to binary code or from physical objects to bit-streams. Changes in the location of perceived value are expected to induce changes in consumer preferences and in what is produced in the market in response to demand. Over time an immaterial realm of entities and experiences is constructed through the valorization of immaterial outputs, a process that seemed to gain further momentum with the emergence of microelectronics and the thirty-year trajectory of silicon chips (Freeman and Perez 1988; Perez 1983, 1985). Fundamental changes in perceptions of value, the main component, require citizens to be convinced that the immaterial realm is better. This is occurring at all levels of the market as consumers are stimulated by the burgeoning media sector to believe that they want or need better digital games, software, entertainment, and virtual experiences. Industrial production is being affected too. Compare a 1955 Rolls Royce with a year 2000 Lexus or Jaguar. The former was massive and highly mechanical. The latter are smaller, lighter, bursting with design, and a full of smart chips. The past decade's swell of the NASDAQ exchange and the decline of global primary commodity prices, particularly gold, would suggest that processes of dematerialization are likely to continue to unfold for some time.

The changes are not only economic; changes in social processes are intimately bound up with the evolving immaterial realm. Computer-mediated communication involves a massive experiment with immaterial social forms. The attractiveness to young people of virtual worlds and virtual reality is another expression of this trend. The goal of media convergence that is so enticing for big media businesses involves the 'digital rendering' of most forms of artistic expression and efforts to control our access to them. The rising primacy of bit-streams facilitates or amplifies certain forms of social relationships and enhances certain patterns of interaction. The Internet is a working prototype of a highly disembodied and dematerialized medium in which increasing numbers of humans regularly immerse themselves. Are status perceptions and distinctions in this environment affected by the Internet's process characteristics? How does a person become somebody within this type of social space? What are the consequences of immersion in this dematerialized medium of interaction? Partial answers to these questions can be derived by considering the activities of the Internet's 'elders'—people defined as 'somebodies' within Internet culture. The status and prestige of highly esteemed participants in the Internet culture appear to be gained in several ways:

1. by contributing openly and visibly to the Internet's tool base (examples include Vinton Cerf (the TCP/IP model), Tim Berners-Lee (http and www), Marc Andreeson (Mosaic and Navigator browsers), and, most recently, Linus Torvalds (Linux));
2. by moderating, editing, and contributing to important network discussions (examples are Jon Postel and writers of more than 5,000 request for comment (RFC) technical postings crafted by members and affiliates of the Internet Society (ISOC), and the Internet Engineering Task Forces);
3. by free posting of exceptional insights, particularly those that offer guidance or are prescient and oriented towards the future;
4. by the clever use of words through erudite online interaction (examples are the selection and use of moderators for discussion groups such as those found on the San Francisco-based Whole Earth 'Lectronic Link (WELL) (Rheingold 1995; Seabrook 1997));
5. by providing answers to, specialized knowledge about, and help with requests (examples are the myriad of technical postings to USENET newsgroups and to discussion groups set up by web-site managers; the volume of posting and replying indicates that this activity provides its own rewards);
6. by acting as information forwarding agents and by managing specialized private mailing lists and Listserves (for example, the Red Rock Eaters newsletter);
7. by the visibility of a good 'hack'—or by other feats of virtuoso programming wizardry (examples are the Obfuscated C contests that evolved on the USENET system and even more recently the appearance of a small, elegant, open-source operating system V2_OS).[15]

The activities in this list are similar to the types of communication activities that are often present in scientific research activity. In this case, however, they appear as properties and functions within Internet social space. These activities signal that some of the most highly esteemed agents in cyberspace are accumulating esteem by making 'network contributions' and that these contributions are realized through interaction under the peer reciprocity rules of Internet space.

---

[15] These exercises may appear meaningless but they counter the question 'why would you want to do that' with the assertion 'because you can'. In hacker circles this is an important part of encouraging inventiveness. For USENET, see **www.ioccc.org**; for V2_OS see **www.v2os.cx** (accessed 20 June 2000).

Social credit in science networks is cumulative and derives from shared perceptions of a scientist's contributions (Latour and Woolgar 1986).[16] These are marked by acknowledged puzzle-solving prowess and displays of codified knowledge gained through publications. In science networks Thorstein Veblen's notion of conspicuous consumption is inverted; social esteem is given to those who offer 'conspicuous contributions'. Since the cumulative character of science activity depends upon the benefits of peer contributions, this feature is socially reinforced through strong positive sanctions.

Social esteem in market societies is normally hierarchical and centred in consumption. The lack of a material base within cyberspace for assembling social clues means another system of signs of esteem is required. The social model that underpins the esteem and respect accorded to the Internet's elders is centred in a perception of their contributions to the larger network. Levelling, facilitated by the remarkably flat network architecture of the Internet, produces a type of space—cyberspace—that fosters a lateral social system of peers, and pursuit of esteem within this system appears to operate much like a science network. Network dynamics precipitate network externalities; users reap benefits from the relative size and activity level of a network and these externalities influence both economic and social relationships. The benefits of network externalities are maximized when it is transparent to agents participating in a network that they will gain through mutual sharing and reciprocity. Marcel Mauss (1969) examined the role of gift exchange within the economy and society and found gift exchange is a process that helps to establish group ties of social reciprocity and to facilitate moral and behavioural reciprocity. Gifts invoke a system of balances and obligations based on sentiments that flourish outside the framework of market transactions. Processes of gift exchange often tend to flatten social relationships. People seek to give and receive as peers and, if social barriers are too high, exchange may occur between momentary quasi-peers. Within a well-functioning peer network, recognition is a lateral process because the more one gives, the more one receives in social honour. This is very reminiscent of the way peer science operates and this parallel offers a starting point for defining other properties of network peer reciprocity.

The main properties found in peer networks and the corresponding signs of social approval can be derived from analysis of the views expressed by participants in the focused interviews. This analysis suggests ten properties that are grouped here within five broad themes: peer interaction, persona, information, skill, and freedom and creativity. Each of these is highlighted by extracts drawn from the observations.

6.1. *Peer-Interaction Themes*

Two clusters of properties appear to embody the norm of peer reciprocity. The free exchange of information and knowledge is the primary basis for the conduct of basic science culminating in refereed journal publications, but there are many exchanges

---

[16] Latour and Woolgar use the concept of 'cycles of credit' in analysing science careers. Another reading of their evidence based on the pursuit of social esteem is also possible (Neice 1998a).

that precede formal publication and that support trade among peers. This process has parallels in Internet space that are very striking.

Graduate students

RESPONDENT. I think it actually gives me the freedom to speak to a load of people whenever I feel like it and quickly and easily. I can certainly send someone a document who I couldn't actually reach ... I'm part of academic groups that are like a clever network so that I can actually see what other people have published and the way they are working. This is all coming through the email address.

RESPONDENT. It's just a different sense because you know there are other people for you ... so they are like 'other' personal contacts.

INTERVIEWER. Right.

RESPONDENT. It creates a sense of community when you are discussing issues with people that have a vested interest in your topic or your project or whatever. But then again of course you can say that you miss out on the richness of the interaction, the face-to-face interaction. The interaction and the verbal cues that go with talking to people on a personal level when people are in a room ...

New media developers

RESPONDENT. ... well what we have not touched upon is the subculture that exists on the Internet, which I think is fantastic.

INTERVIEWER. ... tell me something about that.

RESPONDENT. Piracy, for example, I think that is superb (*laughter*).

INTERVIEWER. You mean as a concept?

RESPONDENT. As a concept, as an actuality, as a reality ... it is about time.

RESPONDENT. I think it has got massive implications for the music business.

RESPONDENT. I mean right now ...

RESPONDENT. It has got a lot of implications for a lot of businesses, anybody that supplies information or entertainment, anything, it has got huge implications ...

RESPONDENT. I mean right now when a Sony Play Station game comes out there are huge kudos on the Net, to have that cracked, coded, and downloaded within three hours. I think this is the fastest they have ever done it.

In other cases, there is evidence of fostering reciprocity and community attachment.

New media developers

RESPONDENT. One of the things I really like is feeling that I'm a part of a community that I might have felt intellectually a part of. That's the thing you were saying about the Zapatistas isn't it! I would have felt very intellectually drawn to it and emotionally drawn to it, but having that kind of practical connection of being able to talk to people and finding a community of people with the same political interests, I find that on a humanitarian level very moving, very involving and very positive. Actually I have to say that I find communities of people like that which are ... I mean I might feel differently if it was voyeuristic ... but I feel it is a very positive thing. I like being part of those communities. I like being sort part of the virtual Zapatista community or whatever ...

RESPONDENT. Yeah. And also I would feel part of a community just by logging on and by requesting information ... and things like that.

Community developers

RESPONDENT. Digital technology also opens up infinite possibilities that would not be there without it ...

RESPONDENT. They are different qualities, different proclivities, but the thing in the analogue world you know where you are, it is riveted to the spot. With digital information what you can access, what it can open up, how it can change your life isn't so fixed ... you can't say these are the limits, this is where it stops. At the moment I have got somebody in Hawaii doing spiritual healing on me. Now whether I believe that is having an effect, whether you want to believe that it potentially could ... to me it doesn't matter; the possibilities are there and they are infinite.

RESPONDENT. But that is almost trying to put a limitation on spiritual healing isn't it? I mean the fact is all you did was contact them to say okay we can work out something about spiritual healing and then at that point you leave the Internet behind and it ceases to be relevant.

RESPONDENT. Yes, but without the Internet ... it couldn't happen.

RESPONDENT. I mean how many people actually do remote healing anyway across space?

RESPONDENT. Yes, but I am just saying the way it is affecting my life is that a lot of things are happening that couldn't have happened ...

Graduate students

INTERVIEWER. What about email?

RESPONDENT. It helps to start communicating with someone who you might not otherwise start a conversation with, someone who might be quite superior to you, like I am not afraid to send this professor at university an email saying, 'Can you tell me how to do this extraction?' Whereas I am pretty sure I wouldn't phone him up because I feel it is totally encroaching on his time ... and I am just sort of doing it from somewhere else.

INTERVIEWER. So it sort of cuts across authority relations?

RESPONDENT. Yes.

6.2. *Persona Themes*

Two patterns emerge under this theme that have become stereotypical of the science personality type. The characteristics in this case include the primacy of reflection, thought, and enquiry over body and matter and the self-effacing humility of scientists who speak of 'standing on the shoulders of giants'. These characteristics are also noticeable in Internaut interactions. There was evidence of a preference for the disembodied, for example.

Internauts

RESPONDENT. I am a bit worried about everyone, because it is almost like everyone ends up being somewhere else, nobody is actually ever where they actually are.

RESPONDENT. It is like the old comedy sketch isn't it where one person is on a mobile phone and they are talking to someone else on a mobile phone, who is standing right behind them just round the corner. We think that is facilitating communication but ...

RESPONDENT. It is making you aware of those things.

RESPONDENT. It is like an externalization of the mind ... in some kind of way ... and what you learn when you come off is how you want your life to be different.

RESPONDENT. Telecommunication is always easier than communication, it is always easy to just communicate with somebody who is a long way away and who is not there than it is to look someone up . . . because you are not looking someone in the eye and you are not faced with their physical presence, you are not faced with all the things that irritate you about them . . . all of that is not there . . . so you feel connected, but you are not. In some ways it is quite good because it means we do get to connect without all the surface stuff getting in the way, but I worry about how well equipped we are then going to be to actually deal with each other.

RESPONDENT. Really communicating . . .

RESPONDENT. . . . I wonder whether people reconstructing themselves continually in chat rooms will finally lead them to realize their identities are a construction and perhaps somehow that may change . . . in the mainstream, not just in deconstruction academic seminars . . . with mainstream people . . . the penny might start to drop, you know, because of the nature of this thing, because it is not real, it is a virtual space, because everything we put into it is our own construction.

Community developers

RESPONDENT. When people get on line first they go over the top.

INTERVIEWER. They go over the top?

RESPONDENT. Yes, often, if they are going to take to it, then they are likely to overdo it. But I think that is a honeymoon period, you adjust sooner or later.

RESPONDENT. You spend less time on it now than you did?

RESPONDENT. It is decreasing all the time but the quality of the time I will spend is increasing. I have definitely moved from quantity to quality, although my initial period I thought was terrific. I spend less time but I use it better.

RESPONDENT. It is so absorbing . . . there is just so much on there . . . you get absorbed in it . . . you get drawn into cyberspace and lose touch with your immediate . . .

INTERVIEWER. You lose touch with your immediate reality?

RESPONDENT. Yes, which can't be good can it? I don't know but it is happening to me.

There was also evidence of self-effacing humility.

New media developers

RESPONDENT. I consider myself nothing more than a glorified taxi-driver. Somebody comes to me, they are at point A and they want to get to B. They are a company that want to reach a person and talk about their particular product. I just give them the tools or the car and drive them there to it. That is all it comes down to at the end of the day. Anybody can go and get a web-site design package, anybody can go get a colour printer and develop their own leaflets but they have limitations because they miss the whole fundamental point . . . 'what is marketing, what is that' . . . how do I communicate effectively? It is more than putting pictures on a page or putting information on there.

Graduate students

RESPONDENT. I think I fool myself by it some of the time.

INTERVIEWER. Sorry?

RESPONDENT. I think I fool myself by it some of the time. Because you often think, 'Oh, I'll just go and get something off the web' or, 'I'll just look here or there' and then like two hours later you're still sitting there going, 'What am I doing? Where am I going really?' Because it wasn't that simple. You needed to do various searches or you went down several dead-ends going through . . . from one place to the sub-pages and so on and you find . . . at the end of the day after reading or going through several gates or whatever, then you realize there's nothing there and you have to start again. But you know, as time goes past you get more realistic about that. I don't have a level of expectation from other people, expecting me to move fast because I'm using a PC or the Internet. It's more sort of the way that I think that it can help me. I deceive myself really more than anything else.

6.3. *Information themes*

In this case, the Internet participants express themes reminiscent of the scepticism of scientists towards evidence and information and their continuous sifting and examination of evidence to establish its validity. Scientists are also generally relatively open to novelty and spontaneous connections between entities and ideas. Similar characteristics appear to permeate Internet perceptions. Participants in the study referred to issues of screening and controlling information.

Graduate students

RESPONDENT. I also find myself relying more and more heavily on [Internet] catalogues which have been achieved by real people behind them . . . which sort out the sites and maybe produce a quick descriptive text for each site and sort of filter it down so that you are only presented with a limited list of sites so that you have sites where you can you get real information and not just the rubbish.

RESPONDENT. Yes. [With search engines] . . . you start taking out words or adding more words and then before you know it, you've actually lost important stuff in the same way that you can gain other stuff, you know.

RESPONDENT. . . . and sometimes they're not there are they?

RESPONDENT. Yeah. They're around, but it's been taken off, but it's still around somewhere.

INTERVIEWER. Sorry? I didn't . . .

RESPONDENT. . . . . missing links.

INTERVIEWER. Missing links.

RESPONDENT. You think, 'All right, that article looks good', and you click on there and it says, 'error—file not found' and that means the host has obviously taken that file off, you know, because it's a couple of years old or whatever but there's obviously some references somewhere else and they have been picked up by the search engine.

Internauts

RESPONDENT. I suppose I think that you could argue that the use of a computer, communication by computer, gives you an illusion of control you don't get over the phone . . . and [it avoids] that frustration of talking to an individual because you compromise what they are saying and adjust to their agenda. You don't get that on a computer because you have all these options, you can switch around and change what you are doing . . . and it has to a degree the same allure as playing a computer game when you can find yourself up all night under the illusion that you are controlling it, that you are building this future world, whereas it is hopelessly out of control if you are staying up all night and going to be late to work the next morning. There is perhaps an element of that; it is a very dictated form of communication.

RESPONDENT. And there is the other type of free too . . . conceptually, you know, free information, you don't know where it is coming from, you don't know who is targeting that information, what lies between the lines. If you have got a piece of text, by its very presentation etc. through your door, you know it's just junk mail that is coming through your door. But there is a kind of ambiguity there . . . you feel disabled, you can't kind of read between the lines about what the ulterior motives might be . . .

Some participants also appeared to value serendipity.

New media developers

INTERVIEWER. Any other 'most' likes?

RESPONDENT. Tangents. I really like the fact that you can just go off on the weirdest tangents and find yourself on weird web sites.

INTERVIEWER. Tangents. OK.

RESPONDENT. She's on some bad links.

RESPONDENT. Or just odd links as well and the fact that . . . I mean that's one of the benefits, I mean, [perhaps] it's nothing . . .

RESPONDENT. Yes, I mean, in my industry every week I simply go on every single ticketing web site just to check what everybody else is doing and sometimes, you know, you get a great idea and there's one or two that are run really well and they have some really good ideas and you think 'Oh well, we'll incorporate those kind of strategies into what we're doing', and then it sparks off an idea and you think, 'We'll take that a bit further' or 'that's an interesting idea they've done', but they've done this with a bit of Java or whatever and we could do something much more interesting and much more powerful or . . .

Internauts

RESPONDENT. It depends though, I can get distracted really easily. Say I am trying to find something, and I am using the search engine to find other sorts of things . . . and then I think, 'Oh I will just have a look at that' and you go on all different tangents . . . and I am gone.

Community developers

INTERVIEWER. When you say its presence, can you give me more details about that . . .?

RESPONDENT. It is hard to doubt that those lines of connection exist. It is hard to say particularly whether one particular technology or experience can mark it out but there is a sense of kind of connectedness . . . being connected to a lot of other things simultaneously, or potentially connected as well . . .

6.4. *Skill Themes*

Scientists generally respect technical expertise and 'how-to-do-it knowledge' as well as methodological rigour—characteristics that may set them apart from the common-sense views of the lay public. Similar characteristics appear within cyberspace. There was evidence, for example, of respect for technical expertise.

Internauts

RESPONDENT. I think not wanting to be seen as some kind of nerd puts people off as well because I quite enjoy the Internet, and I was the kind of kid who liked looking things up in encyclopaedias and cross referencing things, so I quite like it, but I know that people just find that really boring and nerdy. So sometimes I don't enthuse about it as much as I would like to.

Community developers

RESPONDENT. I think the media put a lot of people off because as you were saying it implies that you are a nerd if you spend time in front of the computer, it implies that you are strange and you need to spend hours talking to other people on email. There has been a lot of bad press about that recently, that I found quite irritating in the *Guardian* and the *Telegraph*, putting across very bad publicity. There is nothing wrong with it, if you want to . . . it is your life, you choose how to spend your time . . . and therefore that would put off some people I think . . . and the cost, but there are ways of getting round the cost, you can access it for the time that you need to . . .

Participants also indicated that they valued 'geeks'.

Internauts

RESPONDENT. It is a sense of a digital leap . . . the phenomenon of it being young people. The digital elite are 15 years old or the 16 year olds who are capable of running businesses . . . and I think I get the sense that to be in the digital elite involves a huge amount of investment of time, it involves all day all night, it is not something that you get by doing it nine to five. It is something you get by being 15 or 16 and spending your nights and days and getting just so up to speedy levels . . .

New media developers

RESPONDENT. For me one of the things that I really respect are some friends of mine who are actually digital artists, and people who are . . . just . . . like my brother who is trying to build robots at the moment. He's successful with robots and he's trying to get into some money building robots. People who are just going in completely different directions; people who are thinking way beyond the screen and way beyond the mouse and way beyond the clipboard. People who are thinking of just the craziest, the most expensive, the most stupid things . . . just all that kind of stuff. People who are the real visionaries aren't thinking at all about whether they can show (something) at the end of the year.

6.5. *Freedom and Creativity Themes*

There is widespread recognition that scientific endeavour is a creative pursuit not unlike artistic creation and that both activities offer the allure of relatively unconstrained freedom of action or thought. Both these values appear to pervade Internet communities. In some instances, participants advocated 'following the free'.

New media developers

RESPONDENT. (What) . . . we're having to deal with is hundreds and hundreds of small companies that we are trying to link up.

INTERVIEWER. Right.

RESPONDENT. And the only way for us to move them into the twentieth century is to give it away for free. So we're having to . . .

RESPONDENT. Virtually for free?

RESPONDENT. No. Free. Completely free. So we're having to . . .

INTERVIEWER. You mean to get them in the loop?

RESPONDENT. Yes, absolutely. So what we're having to do is to come up with business models whereby . . . we will put one kiosk in for every five new businesses in the town . . . we'll connect the hub with ISDN . . . we'll put ISDN lines into every single venue in that town and we'll put all the software gateways into all those venues, and we'll do the whole lot for free—and that's the only way to do it.

RESPONDENT. And what? Charge for the transactions?

RESPONDENT. Right . . . charge for the transactions.

Community developers

RESPONDENT. I think Hot Mail has been a very useful device for a lot of people because it gives people access who don't have their own terminal and anywhere they are in the world if they have got a Hot Mail address they can go there and they can speak to their friends. They don't have to pay for it and it is free.

RESPONDENT. It's free . . . that's why it is popular.

RESPONDENT. I have a friend in New Zealand and wherever she is she just needs to find an access point and we are always in touch.

RESPONDENT. Because it is free see . . . it is the whole thing . . . the whole system of the Internet was supposed to be a universal device . . .

RESPONDENT. I think the culture is very different on the Internet too. I don't know, I am trying to think of a parallel but if you look at it . . . certainly from my experience with developers . . . we have got programmers at our place and they see the Internet as a means of meeting other people who are interested in the same area and they will post up bits of code for free, and it is just a way of sharing information and they see that as like some kind of forum, it is like almost their friends in that particular area. They don't see it as a way to make money at all.

RESPONDENT. That is the major subculture of the Internet. It is not a money-making venture, it is the kudos and shafting the big companies.

Participants also indicated their interest in pursuing creativity.

Graduate students

RESPONDENT. I don't see it like that so much. I think computer programming is a very creative process, just as creative as writing a novel or composing a piece of music, but it's a different type of creativity, definitely, but it's still a creative process. I don't have the same feelings like, 'You should go out more or I can't really talk to you because you use the computer all day.'

New media developers

RESPONDENT. They have totally revolutionized their (job) caption because I can't function now as a designer without my technical boy. He is a creative person, he is a creative person that designs something creatively in a way that blows my mind. He shows me something that technology can do which enhances my creative mind and I think I can sell that. I can't do that. As a designer I can come up with pretty pictures which can sell a product but he adds functionality to something on a web site which means for a new media person that stretches both of us. I will say I want to do this and he will say 'No, you can't, because have you thought about this' and my mind suddenly goes off into the next stage and together as a team we come up with a product which satisfies our client in the way the client wasn't even dreaming of doing.

7. JUGGLING SYSTEMS OF VALUES AND SOCIAL ESTEEM

These extracts from the focused interviews, as well as many other quotations not presented here, also indicate that experienced users are juggling two systems of values and social esteem. The biases of the Internet stress reciprocity and contribution. This appears to facilitate peer interactions that are similar to those that occur in the context of science activity. These values, however, are sometimes at odds with the taken-for-granted values that facilitate both traditional physical forms of market exchange and emerging forms of electronic commerce where the emphasis is on consumption, hierarchy, and formal organization.

This divergence produces paradoxes and discontinuities for Internet-users, developers, and entrepreneurs. Their lives are rooted in the traditions of consumption hierarchies as well as in a different set of values, mostly experienced while being online. One set of values is governed by rules of conspicuous consumption and the other by the rules of conspicuous contribution. In the face of the tensions between these sets of values, accommodations are required. This is noticeable in the paradoxical Internet business exhortation to 'follow the free' (Kelly 1997), where profits by *.com* and *.co* start-up companies are expected to follow on from the distribution of gifts.

Examples of the requirement to 'follow the free' are legion. For instance, Netscape gave away its browser in a bid to secure the server trade and Adobe gave away its Acrobat Reader to promote the .pdf file format. Electronic mail programs, such as Eudora and Pegasus Mail, have been posted free for downloading and Corel ported WordPerfect 8 to Linux, posted its free availability, and gained a million users in two months. Sun Microsystems purchased StarOffice (a Linux office application suite) from a German firm, Star Division Corp., and then gave it away.[17]

Although these strategies could be regarded as promotional activities, their recurrence does not fit traditional business models of rational profit-maximizing activity. Of course, these companies have expectations for revenue streams and profits derived from related products and services. Internet-user expectations of free access and usage have been visible in the pressures placed on Internet Service Providers (ISPs) in the United Kingdom in 1999 to offer 'free Internet servers'.[18] Free servers are expected to reduce a major disincentive for Internet use in the United Kingdom (and Europe) by reducing the charges for network usage and/or subscription fees.

Interaction and exchange between participants in the Internet medium seem to give rise to a disposition in the user community to seek out free goods and services. In a medium that privileges 'gifting and contributions', these types of expectations are not surprising. For companies seeking commercial gain from Internet-based interactions, the lesson is that, if they are to generate substantial revenues, they will need actively to engage in processes that produce conspicuous contributions.

8. CONCLUSION

This chapter has illustrated some of the changes in social distinctions that appear to be induced by processes of network peer reciprocity and has demonstrated that Internet processes are reshaping patterns of social signification for Internet-users. The question as to whether shifts in the basis of social esteem reflect a deeper transformational shift

---

[17] Sources for these examples include: for Netscape, **www.netscape.com/company/press/index.html**; for Adobe, **www.adobe.com/aboutadobe/pressroom/main.html**; for Eudora, **www.eudora.com/**; for Corel, see Corel news archives 19 May 1999, **www3.corel.com**; and for Sun Microsystems, **www.sun.com/products/StarOffice/news.html** (accessed 20 June 2000).

[18] The client pays local charges via 045 services. All other ISP service charges, except telephone-based technical support, are provided at no cost to the user.

away from the predominance of market values and towards social reciprocity within network environments is very complex.[19]

The proliferation of *.com* and *.co* addresses on the World Wide Web may suggest to the casual observer that commercial forms of market exchange are capturing the Internet. However, the evidence from the empirical study highlighted in this chapter suggests that the inverse may also be occurring. The rules of social reciprocity embodied and fostered by the Internet may come to envelop market forms of exchange. To the extent that something like this is occurring, companies may benefit from radical adjustment of their strategies for generating revenues and profits.

Phil Agre (1999) argues that the economic and social tendencies that a particular technology *amplifies* are essential to any useful analysis of its wider significance. Insofar as the Internet is a medium that amplifies network peer reciprocity and social reciprocity, the nature of the dynamic interaction of this form of exchange with traditional forms of market exchange will provide a focus for research and analysis well into the twenty-first century.

To further the understanding of these important processes more interdisciplinary work will be needed that crosses the economics and sociology of cyber-exchange, both of which rest on transformations in perceptions, values, and beliefs. The research community will benefit by drawing upon the guidance provided by classic works that sought insights into the integration of economic and social processes — for example, the works of Adam Smith, Karl Marx, and Max Weber. They and others treated industrial capitalism and the market system as whole entities and they focused on interacting systems of values and beliefs about economy and society, market and community. Today, the challenge is to investigate the properties and processes involved in emerging forms of network interaction and to examine the transformations in belief systems that govern and limit the possibilities for human agency.

Appendix. Note on Recruitment of Sample, Incentives and Analysis

The selection of focus-group subjects employed a sampling approach developed within the grounded-theory tradition. The goal of recruitment was purposively to stream subjects into four broadly defined types (or segments) of experienced Internet-users without being excessively rigid. In the implementation some overlap between segments occurred.

Representative organizations were identified that had lists of active members through which access to subjects could be gained.[20] Heads of these organizations were approached to inform them of the purpose of the study and to obtain their consent for further action. For each

---

[19] Some interesting work on the notion of an Internet gift economy includes that by Eric Raymond, **www.tuxedo.org/~esr/writings/cathedral-bazaar/**, and by Richard Barbrook, **www.hrc.wmin.ac.uk/**, but the idea of a gift economy is not without its critics (accessed 20 June 2000).

[20] By example, for new media developers an affiliation with 'Wired Sussex' was developed, an organization representing a very large proportion of all new media developers in the study area. All members were canvassed using email and flyers.

organization a collaboration and a contact strategy was arranged to canvass all possible cases fitting the user type.[21]

Potential subjects could signal their interest only by email. This strategy was an essential screening step. Attentive use of email, including the required proximity to Internet appliances, was taken to be a good unobtrusive corroboration that subjects were experienced Internet-users.[22] All contact until the subjects entered the discussion room was exclusively by email; no phone calls, letters, or other forms of contact were employed.[23]

Three incentives were offered to participants: (1) an assurance that the discussion would be interesting; (2) light refreshments following the sessions; and (3) a modest cash stipend for 'booking their time'. Six people on average attended each focus-group session. Each of the four segments was repeated once to ensure that responses were not idiosyncratic. In total fifty-six people participated in eight discussion groups and a separate pre-test.

Audio tapes were made of the sessions and then transcribed, edited, and indexed. The transcriptions were submitted to two types of detailed coding, thematic and heuristic. *The Ethnograph*, a qualitative coding and analysis computer package, was used as an aid in the heuristic data analysis.

---

[21] The intention was to let experienced users become volunteer subjects through auto-identification with characteristics listed on flyers and postings.

[22] The fact that 'email contact only' worked demonstrates that regular use of email and the commitments made through this medium are good behavioural indicators of Internet use.

[23] The advantage of this approach is that most participants become fully engaged and this enhanced the quality of group discussions.

4

Knowledge Management Meets the Virtual Organization in the Newspaper Industry

JENNIFER J. GRISTOCK

1. INTRODUCTION

If the word virtual is 'a huge vessel of semantic vacuity waiting to have meaning poured into it' (Woolley 1993: 58), then the virtual organization is a bottomless (Alasuutari 1995: 22) vessel. Despite all the attempts that have been made, none has been sufficient to fill these words with meaning. Indeed, since the publication of Davidow and Malone's *The Virtual Corporation* in 1992, the continuing search for definition of 'the virtual organization' has done little but muddy the management field of research with conflicting definitions and slippery references to the unexplained enabling powers of new technologies.

Based on an analysis of the organization of work in the UK newspaper publishing industry, the theme of this chapter is that the management researchers who continue to wrestle over the definition of the virtual organization are, in effect, arguing about whether this conceptual vessel is half-empty, or half-full. The research summarized in this chapter was undertaken to illustrate that in order to develop an improved understanding of virtuality in organizations—and of the ways in which it is supported—it is necessary to investigate the nature of virtuality in organizations with reference, not to the structure of the glass, but to the ways in which this vessel may be filled; that is, with reference to the ways meanings are created.

The virtual organization rose to prominence in the 1990s, at the same time as the field of knowledge management, a field of study associated with management's realization that 'what an organization and its employees know is at the heart of how the organization functions' (Davenport and Prusak 1998: p. x). However, although this decade saw an increasing recognition of knowledge as 'the most fundamental resource in the modern economy' (Lundvall 1992a: 1), the link between knowledge and competitive advantage was made long before knowledge management became a familiar concept in the literature. For example, as far back as the 1960s, Drucker (1964: 147) noted that, 'Economic results are the results of differentiation. The source of this specific differentiation, and with it of business survival and growth, is a specific, distinct knowledge possessed by a group of people in the business.'

Over time, understanding of knowledge and innovation processes has come to be regarded as the necessary focal points for those wishing to stimulate economic growth and development. Those who first studied innovation saw it as a linear sequence of events, which either channelled research and development (R&D) results to the marketplace (the technology-push model), or 'pulled' technology out of the laboratory in response to a market need (the market-pull model). In many ways, the concept of the virtual organization is to the technology-push model of innovation what knowledge management is to the market-pull model (Edquist 1997; Freeman and Soete 1997). Just as the proponents of the technology-push model of innovation focus on the power of R&D to push technologies to the marketplace, virtual-organization enthusiasts point to the power of new technologies to allow access to information beyond the reach of traditional organizations that are not virtual. Similarly, just as the market-pull model of innovation focuses on the market environment's ability to draw innovations out of R&D groups, the supporters of knowledge management hold that their techniques allow them to tap into sources of knowledge that are beyond the reach of traditional organizations. Both knowledge management and the virtual organization, therefore, are ideas that have risen out of the process of readjustment that is taking place as managers and economists struggle with the management/organizational implications of doing business in the 'new economy'.

While management theories are proliferating in the literature, empirical research into the factors associated with successful innovation has established that 'understanding user needs, good communication, and effective collaboration' tend to be strongly associated with successful innovation. This work has taken the market-pull/technology-push debate 'to a qualitatively different level' (Coombs, Saviotti, and Walsh 1987: 1012). Innovation has come to be seen, not as a sequence of events, but as a process (Lundvall 1992*a*: 9), 'where interaction is the critical element' (Tidd, Bessant, and Pavitt 1997: 29). At the heart of the question of how to support innovation, therefore, is an understanding of how these interactions take place—that is, how they are mediated.

The possibilities for the mediation of such interactions are many and varied. The development of information and communication technologies (ICTs), such as email and the Internet, have opened up new possibilities for the ways in which data can be shared, equipment may be accessed, and links with colleagues can be made and maintained. However, where knowledge management meets the virtual organization, there are two fundamental misunderstandings that are making it difficult for businesses to develop strategies to support the interactions necessary for innovation to flourish. This chapter draws attention to the nature of these misunderstandings. It shows how a more accurate understanding of the nature of virtuality in organizations, and the means by which it is supported, can help managers to develop business strategies that are likely to create and support communication processes that are the keys to successful innovation in the current economic environment (Gristock 2001).

This chapter draws on the results of a survey of new media developers and editorial staff of regional daily newspapers in the United Kingdom, and follow-up interviews with those involved in the creation of newspaper web sites, including those contributing to the Internet portal 'This is Britain'. The concept of 'the virtual organization' is shown to be a misleading and inaccurate model that does not fit the way in which these organizations do business. The evidence from this study suggests that what has been mistakenly described as the arrival of a singular new organizational form is, in fact, a subtle, and yet pervasive, extension of the spatial, temporal, and community boundaries that have traditionally limited the way in which work is organized. This chapter suggests how extensions to these boundaries are experienced by new media developers in the UK newspaper publishing industry and it identifies the possible limits of organizational virtuality in the context of the variety of organizational transformations that appear to be occurring in this industry.

The arguments presented here have implications for the support of what has been called virtual work, but they also have major implications for our understanding of the implications of the virtual organization and knowledge management more generally. This chapter argues that the perspectives on knowledge that are at the heart of the virtual organization, knowledge management, and the innovation process need to be re-examined. There is a need for a move away from the view that sees the outcome of knowledge exchange as directly and irrevocably related to the essential characteristics of knowledge—that is, the knowledge 'type' (see Gristock 2001). Most theories of knowledge management draw a veil over the mediation processes that encourage or discourage innovation—that is, the processes involved in the creation and attribution of meaning. The concept of intermediation is understood here as a process that directs the search for meaning, allowing communities of interest to create knowledge. In this context, therefore, intermediation is the missing link between knowledge and innovation. Innovation is defined here as the mediated process of creating and exchanging meaningful knowledge over time, across space, and/or in one or more communities, with results that are perceived to be positive in a given context.[1]

To understand the nature of virtuality in organizations, it is necessary to identify and consider the foundations upon which the concepts of both knowledge management and the virtual organization are built. Section 2 of this chapter opens with a brief discussion of research on the virtual organization and knowledge management that is concerned with the role of ICTs and the nature of knowledge. After identifying areas of weakness in the virtual-organization literature, the chapter continues with an analysis of the differences in the views as to what the virtual organization might be (Section 3). In Section 4 three themes in the literature are identified and then integrated into a framework for investigating organizational virtuality that is referred to as a three-dimensional space or organization space. Section 5 introduces the activity chosen to examine the organization of work in the organization-space—new media product development in the UK regional daily press. This focus facilitated an investigation of the activities and communication patterns of a group of product developers separated

---

[1] *Mediation* is the process of *making sense* of the world through forms of expression and interaction. An *intermediary* is someone or something that changes the ways in which knowledge may be made sense of and used in particular forms of expression and interaction.

by space (geographical space), time (availability), and organizational community boundaries (company allegiance, capability). Section 6 presents the results of the investigation with respect to the spatial, temporal, and organizational separation of those involved in creating these products, and with respect to the state of development of this industry in January 1999. Section 7 discusses these results in relation to several of the activities associated with two newspapers in the 'This is Britain' network, 'a unique portal linking city and community sites across the country [the UK] with compelling national content' (Associated New Media 1999). The chapter concludes with a summary of the key issues that arise from the analysis and a discussion of their implications for policy-makers and those concerned with corporate strategy who wish to support and respond to virtuality and innovation.

2. INFORMATION, TECHNOLOGIES, AND ORGANIZATION

In much of the management literature, the rise of the virtual organization is attributed to the development of ICTs. For some authors, the association between these technologies and virtuality is one of straightforward 'enablement'. For example, virtual organizations are said to be 'reliant on the medium of cyberspace' and 'enabled by new computing and communications developments' (Barnatt 1995: 79). By the same token, Makridakis (1995: 799) writes about the 'impact of information technology' as 'resulting in . . . flat, horizontal organizations'.

On the other hand, some authors seem to eschew terms like 'enabling' and prefer, instead, to describe the relationship between virtual organizations and ICTs as facilitatory. For example, Davidow and Malone (1992: 95) write that 'information technology is already facilitating the emergence of new, co-ordination intensive structures'. At first sight, it would seem that, by using the term 'facilitating' instead of 'enabling', these authors avoid the problems associated with a view that sees a set of technologies, however used, as triggering organizational change. However, on deeper investigation, this is found not to be so. For Malone and his colleagues, the emergence of virtual organizations can be attributed to the technologies' innate ability to 'enable people to co-ordinate more effectively, to do much more co-ordination', and so to 'form new, co-ordination intensive business structures' (Malone and Rockart 1991: 92). It appears that facilitation, in this context, is enablement in another guise.

In his article 'The Coming of the New Organization', Drucker (1988: 45) argues that the driving forces for new organizational forms are: (1) a demographic shift in employment from manual and clerical work to 'knowledge work'—work that requires what Gibbons *et al.* (1994: 111) refer to as 'skills and resources which are not easy to imitate', (2) an increasing pressure to innovate in the face of global competition, and (3) 'above all, information technology'. Drucker argues that, although new organizational forms may be created without advanced data-processing technologies, the use of ICTs will result in new organizational structures and companies that 'bear little resemblance' to textbook company models. Drucker (1988: 45) proposes that information technology is used to 'endow data with relevance and purpose', and so, using his definition, converting it to information can transform decision processes, management structures, and the way 'work gets done' in organizations. He therefore draws a distinction between, on the one hand, using ICTs for data processing for simple number-crunching, and, on the other, using ICTs for 'analysis and diagnosis'. His view is that, once ICTs become involved in these processes, the function of management is changed from a decision-making to an intermediating role that is needed in the new organizational form, a role that is described as 'boost[ing] communication signals' (Drucker 1988: 46).

Where such technologies are prevalent, Drucker (1988: 45–6) says, 'we have to engage in analysis and diagnosis—that is, in "information"—even more intensively, or risk being swamped by data'. But it is uncertain how these analytical and diagnostic processes, which, in his words, 'endow data with relevance and purpose', can be thought of as 'information'. It is not clear whether the meaning in Drucker's statement is that the sheer volume of data produced by ICTs increasingly will require the use of the results of analysis and diagnosis, that is, information, or an engagement with analysis and diagnosis in knowledge creation processes. Such ambiguity is not uncommon in the literature. Just as Drucker appears to equate information with knowledge creation, DeLisi confuses the ability to create knowledge with information availability:

With information available to all individuals in an organization regardless of status, and across time and space, the company no longer needs multiple levels of management to pass information up and down the hierarchy. Also, since executive decision makers are able to access information directly, they no longer need staff personnel to prepare and filter the information. (DeLisi 1990: 86)

With the principal emphasis on information availability, information processing, and information technologies, it is hardly surprizing that Upton and McAfee (1995: 126) write that 'Managers . . . do not understand why their heavy investments in IT have not radically changed the way their companies work'.

The source of this confusion lies in the fact that data cannot be packaged to guarantee their status as information. Whether or not something informs depends on many factors, including the knowledge and experiences of the audience. The same event, set of data, or message may inform one person, may be obvious to some, and be useless noise to others. Hence, work requires, not that information is available, but that knowledge, which is 'by definition, specialized' (Drucker 1988: 46), is used to create meaning from—that is, to make sense of—the world.

3. REALITY AND THE VIRTUAL ORGANIZATION

The virtual organization has been defined in many ways. For authors such as Handy, this type of organization's defining characteristic is the fact that it has no one physical location. Thus, 'Virtual organizations . . . do not need to have all of the people, or sometimes any of the people, in one place in order to deliver their services. The organization exists but you can't see it. It is a network, not an office' (Handy 1996: 212). For other writers, the virtual organization's essential element is that its members do not share the same time frame. For example, Dess *et al.* (1996: 22) comment that the virtual organization is 'a continually evolving network of independent companies—suppliers, customers, even competitors—linked together to share skills, costs and access to one another's markets'. In a similar vein, Byrne *et al.* (1993: 37) describe the virtual organization as being located in a specific and relatively narrow window in time, as 'a temporary network of independent companies linked by information technologies'.

Particularly with regard to technical or R&D activities, other authors define the virtual organization as one that supplies outsourced services or products. For example, Chesborough and Teece (1996) and Harris *et al.* (1996) fall into this category. This outsourcing perspective sees the virtual organization as one that uses capabilities that are not its own. 'Using high-powered, market-based incentives such as stock options and attractive bonuses, a virtual company can quickly access the technical resources it needs, if those resources are available' (Chesborough and Teece 1996: 66). The authors point to a trade-off between incentives to take risks (which are said to be higher in the outsourcing virtual organization) and the 'ability to settle conflicts and co-ordinate activities'. Following this model, organizational strategies are said to require an understanding of the 'information flow essential for innovation' (Chesborough and Teece 1996: 67). And yet some differences of opinion about what the virtual organization might be exist even here. While Chesborough and Teece (1996: 66–7) perceive 'virtual organizations and integrated companies' to be 'at opposite ends of the spectrum', they also suggest that 'alliances occupy a kind of organizational middle ground'. Harris *et al.* (1996: 32) appear to include alliances in their definition of the virtual organization, which is 'one in which non-critical/non-core functions are resourced externally via collaborations, partnerships and/or straight outsourcing'. Much time and effort have been spent in the search for an accurate definition of 'the virtual organization' (Bultje and van Wijk 1998; Strausak 1998). Although no consensus has yet been reached about what 'the virtual organization' might be, many scholars still sing its praises because of its assumed superior efficiency, flexibility, access to markets, and potential for giving rise to productivity gains.

In terms of structure at least, each description of 'the virtual organization' tends to be different. The definitions that have been put forward include any number and combination of the following features: (1) an organization having no one physical location; (2) an organization that is temporally distributed; and (3) an organization that accesses capabilities externally. Generally, the virtual organization is perceived to be ICT enabled, such that activities are organized in a control structure that crosses some combination of space, time, and organizational boundaries.

4. EXPLORING THE LIMITS OF VIRTUALITY

Although reference to activities that are carried out by highly fluid groups of personnel, in multiple locations or using the capabilities of different organizational communities, does not help to define the virtual organization, it does help to demonstrate that virtuality in organizations is multidimensional. Organizations may be regarded as having different degrees of virtuality in three dimensions, each dimension corresponding to the way that activities are organized over space, time, and community

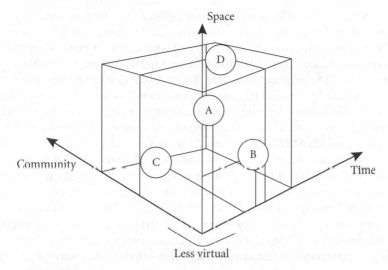

Figure 4.1. *Four hypothetical organizations in the organization-space*

boundaries.[2] This framework is illustrated in Fig. 4.1, which shows hypothetical examples of several different organizational forms.

The letters A, B, C, and D on the figure indicate where the following hypothetical organizational forms might be found in the organization-space.[3]

- A: a group of collaborating high-energy physicists, accessing equipment and communicating online on a long-term project; geographically disparate, sharing the same time frame, shared organizational ties, and experimental capabilities common to all.
- B: a group of collaborating industrial scientists, accessing equipment and communicating online; high turnover of personnel within firm; duty rotation.
- C: a group of collaborating journalists, based in one newspaper, bringing in a signif icant proportion of 'outside' knowledge (from freelance specialists, etc.) to work on a long-term project.
- D: a continually evolving group of scientists (suppliers, customers, etc.) accessing equipment at many sites and communicating online; members based at many sites.

---

[2] The dimensions of organizational virtuality are the shared contexts that assist the creation of meanings. Many such shared contexts might be considered. For example, physical location is one aspect of the dimension of space. Availability, shiftwork/timezone, turnover, and past/future history might be considered to be potential aspects of the time dimension. Language, capability (discipline/skill), gender, and culture (national, organizational, age) might be considered to be aspects of the community dimension.

[3] The organization-space is a three-dimensional representation that can be used to compare different organizational forms. Separation in the organization-space indicates that some context(s) is/are not shared, and that interactions are mediated to overcome the limitations that this absence of shared context imposes on meaningful knowledge exchange.

In effect, those who wish to exploit organizational virtuality are seeking to escape from the confines of the physical environment, the moment, and the community to which they belong. This could be expressed as a wish to overcome the limitations that separation by distance, time, and community impose on the senses. This framework is an extended multidimensional version of Biocca *et al.*'s view of telepresence in virtual reality: 'The desire for physical transcendence is cast not as just some general desire to overcome the limitations imposed upon humans by the physical environment, but to overcome the limitations of the senses, to augment the senses through electrical means' (Biocca, Kim, and Levy 1995: 10). In fact, many parallels can be drawn between virtual reality and organizational virtuality. Just as organizational virtuality has been described in terms of the arrival of a single organizational form, the virtual organization, virtual reality was once described in terms of the development of a particular technology—for example, a head-mounted display. Steuer (1995: 33) notes that, while this was useful for the manufacturers of virtual-reality-related hardware, for communication researchers, policy-makers, software developers, or media consumers, a device-driven definition of virtuality is unacceptable; it fails to provide a conceptual framework within which to make regulatory decisions. This is because it does not provide an aesthetic from which to create media products, and it fails to provide a method for consumers to rely on their experiences with other media to understand the nature of virtual reality (Steuer 1995).

In response to these problems, Steuer defines virtual reality without reference to any particular hardware. The key to this definition of virtual reality, which is made in terms of human experience, is the concept of presence. Durlach and Slater (1998) describe presence simply as 'the sense of being there'. Presence, however, refers 'not to one's surroundings as they exist in the physical world, but to the perception of those surroundings as mediated by both automatic and controlled mental processes' (Steuer 1995: 35). A virtual reality, as defined by Steuer, is therefore a real or simulated environment in which a perceiver experiences telepresence—that is, the experience of presence in an environment by means of a communication medium. A book can, therefore, be a virtual-reality medium.

Virtual-reality media, like a novel or the virtual-reality headset, are successful in creating a 'virtual reality' when their users experience a 'sense of being there' (Durlach and Slater 1998), if 'the perception of those surroundings as mediated by both automatic and controlled mental processes' (Steuer 1995: 35) is meaningful. This meaningful perception is presence, and the relationship between this particular mediated world and the wider world is called virtual reality.

Similarly, it is suggested in this chapter that communications technologies and communications media such as interviews, lunch meetings, email, and telephone calls will be successful in supporting organizational virtuality if they are able to help support meaningful exchanges of knowledge across some combination of spatial, temporal, and community boundaries. The relationship between organizational forms, which involve such boundary-crossing interactions and more traditional ways of organizing work, is called organizational virtuality.

To explore organizational virtuality, and the means by which it is supported, it is necessary to recognize that all technologies mediate and that all technologies are mediated. Technologies that we acknowledge to have emerged from, or to send ripples through, a sea of human activity are called media. Where this influence is not so well acknowledged, the word technology is used. Technologies are the result of mediation because they 'emerge from complex processes of design and development that themselves are embedded in the activities of institutions and individuals constrained and enabled by society and history' (Silverstone 1999: 20).

Technologies mediate, changing 'the scale or pace or pattern' that they 'introduce into human affairs' (McLuhan 1964: 8). Managers and technology strategists often refer to this as the 'impact' of new technologies. However, users do not passively adopt technologies for homogeneous activities; rather technologies are adapted to everyday lives and vice versa (Silverstone and Hirsch 1992: 1–2). If one wishes to design processes to support and respond to innovation, it is not enough to say that technologies impact on society. Rather, what needs to be recognized is that socio-economic and political systems both give rise to, and are influenced by, the patterns of use and adaptation of technologies by different communities. Systems of innovation *are* systems of mediation (see Gristock 2001). To be successful in supporting organizational virtuality, technologies and other media are used and adapted to overcome the limitations that separation by distance, time, and community impose on the senses. In this way, these 'extensions of man' (McLuhan 1964: 8) are used to support meaningful exchanges of knowledge across some combination of spatial, temporal, and community boundaries.

The key to defining both virtual reality and organizational virtuality is the concept of presence. If the concept of presence that is found in the virtual-reality literature is extended, it can be argued that organizations display some combination of *presence, temporal presence*, and *virtual community* through the mediation of knowledge-exchange interactions across space, time, and communities.[4] Presence is created where an exchange of knowledge is mediated to make it meaningful[5] to those who are separated by a distance in space. Temporal presence is created where a meaningful exchange of knowledge is mediated between those who are separated by a distance in time. Virtual communities are created where a meaningful exchange of knowledge is mediated between those who are separated by organizational or other community boundaries, such as gender and capability.[6] In this sense, given the diversity of activities and knowledge systems that exist, 'all communities are virtual communities' (Silverstone 1999: 104).

5. EXPLORING THE CONCEPT OF VIRTUALITY

An ideal subject of study to test the concept of organizational virtuality began to take shape in the mid-1990s with the development of Internet products in the regional

---

[4] Following Durlach and Slater (1998), presence might be defined as 'a sense of sharing the same physical environment, at a distance'; temporal presence might be defined as 'a sense of sharing the same timeframe, at a distance in time'; and virtual community might be defined as 'a sense of belonging to a community to which one does not belong'. However, many instances where mediated interactions might produce meaningful, but very different results, might be imagined. The definitions in the text account for these differences.

[5] 'To make sense' and 'to be meaningful' are seen as equivalents.

[6] For example, organizational departments, language, capability (discipline/skill), gender, and culture (national, organizational, age).

newspaper publishing industry in the United Kingdom. The first regularly updated local newspaper web site in England (Beamish 1998: 141) was the *Wycombe & South Bucks Star*, owned by Newsquest. This site, which went live in April 1995, contained twelve pages of news, features, and sports, with a clickable image on the front page representing each section. The creators of the site, editor Alan Cleaver and systems manager Robert Whittlesea, also produced a newsletter that was emailed to those who requested it (Cleaver 2000). Other newspapers soon followed the lead of this paper, and of a national newspaper, the *Daily Telegraph*, producing sites that included advertising, images, search facilities, and links to other local web sites, supplementing their trusted paper products with their less familiar electronic counterparts. By January 1999, 82 per cent of the ninety-five daily regional newspapers in the United Kingdom had a presence on the Internet, bringing audio, video, web forums and shopping opportunities to their readers.[7] By 2000, there were over 3,500 newspapers online around the world (*NewsCentral* 2000).

The many regional newspapers are owned by a variety of companies, but their numbers continue to decrease through acquisitions and mergers. In 1999, for example, 1,400 local newspapers were owned by 126 regional newspaper publishers, sixty-five of which owned just one title (Peak and Fisher 1999: 36). By July 2001, this had declined to ninety-six regional newspaper publishers, forty-three of which owned just one newspaper title (Newspaper Society 2001*a*: screen 1). Ownership changes are fast moving: 77 per cent of the regional press is said to have changed hands between 1996 and 1999 (Peak and Fisher 1999: 36). This rate of change resulted in a regional press acquisitions-and-mergers spend of nearly £6 billion between 1995 and 2001 (Newspaper Society 2001*b*: screen 1). Regional newspaper organizations, therefore, are characterized both by interrelationships and by local identity. The relationships that exist between newspapers, their local companies, and the regional press publishing groups such as Northcliffe, Trinity, Newsquest, and Johnston Press, and the distribution of these newspapers across the United Kingdom, represent a geographically dispersed network of companies involved in the development of both paper-based and Internet products. In 1999, therefore, the opportunity existed to study the organizational virtuality of such product development. Those involved in these activities are based in different geographical locations and frequently are unavailable when needed because of the high-pressure deadlines associated with the daily newspaper business. They also belong to different companies—that is, newspaper companies, local groups, and/or regional publishing groups. New media development by the UK regional daily press thus provides an opportunity to explore and compare the spatial, temporal, and organizational community relationships between those involved in the development of the different, yet similar, products.[8]

This exploration took place through a variety of mechanisms between January and August 1999. First, to confine the investigation, the study was limited to the regional daily newspapers in the United Kingdom with web sites in January 1999, a total of seventy-six out of ninety-five regional daily newspapers. Secondly, in January 1999 a type of content analysis was performed on various aspects of seventy-three of these newspapers' web sites.

---

[7] Results of a telephone/Internet survey of all ninety-five regional dailies carried out by the author during January 1999; for further details, see Gristock (2001).

[8] This chapter focuses on communities as defined by organizational boundaries.

The results of this analysis were combined with trade data (circulation figures, ownership information) from the United Kingdom's Newspaper Society, and distance/travel-time data to assess the geographical separation of towns and cities in the United Kingdom from a Royal Automobile Club (RAC) database.[9] Thirdly, a survey questionnaire was developed and sent to the new media developers of these newspaper web sites that asked questions about the organization of their work and their use of meetings, technologies, and other media.[10] The questionnaires were returned by thirty-five of the seventy-six potential respondents in the sample, representing a response rate of 41 per cent. Finally, five follow-up interviews with new media developers and managers were used to gather more detailed information about the different kinds of activities carried out by those involved in regional daily newspaper web-site development. Case-study subjects were chosen with reference to the spatial, temporal, and community data gathered in the questionnaire to ensure that the use of different forms of mediation-supporting activities organized very differently across space, time, and organizational communities could be investigated.

The principal aims of this investigation were as follows. First, the theories of the virtual organization in the management literature suggest that such an enquiry would uncover evidence of a new organizational form. The study aimed to examine the organization of newspaper web-site development in the United Kingdom to assess whether there is evidence of the arrival of a new organizational form or of the expansion of the spatial, temporal, and community boundaries that may limit the way in which work is organized. Secondly, organizational changes were investigated in an effort to discover whether they were closely associated with the introduction of new technology; the pursuit of the development of new knowledge; or with changing systems of mediation. The third aim was to determine whether organizational virtuality can be meaningfully described as the organization of activities, and hence the mediated exchange of knowledge across space, time, and communities. The expectation was that, if this perspective could be validated by the results of the empirical study, this would provide a more useful foundation for further investigations of the changing environment in which firms operate. The virtual organization is believed to ignore the effects of space, time, and community on our ability to create knowledge. However, organizational virtuality, while offering no easy answers, may give policy- and strategy-makers a conceptual setting that can be used to consider, design, and maintain the systems of mediation that are so necessary to innovation.

5.1 Analysis of Regional Daily Newspaper Web Sites

In January 1999 four aspects of the content of all seventy-six regional daily newspaper web sites were analysed to make preliminary judgements about the nature of innovations

---

[9] See the RAC website, **www.rac.org.uk** (accessed Aug. 1999).

[10] The survey was sent in the form of an interactive web page via email to identified new media developers in the ninety-five regional daily newspapers. Questions asked included, 'What organizations are involved in the development of this news site?', 'How frequently are you helped to obtain a useful face-to-face meeting by using the following? (Secretary, phone call, email message, etc.)'. The survey was implemented and analysed during the period January–August 1999.

in the development of such sites. Previous studies had discussed the importance of embracing features of the new medium (the Internet) not normally found in the printed version of a newspaper, such as search facilities, round-the-clock updating, and Internet shopping (Molina 1997; Neuberger *et al.* 1998; van Dusseldorp 1998). In the light of these expectations, the content analysis of the web sites was limited to four key areas—namely, whether or not the newspaper had a searchable news archive with a clear date in the news story; a different layout or hierarchy for the front page as compared with the printed product; updates before the printed version; and opportunities for readers to buy products or services online.

Of the regional dailies that were online in January 1999, seven out of ten updated their Internet newspapers before 6.00 p.m. on the day of publication. Only one in three allowed readers to add context to news archive search results by providing a clear date. One in four newspapers was selling goods or services online. Fewer than 10 per cent had chosen a new hierarchy for the front page of the new medium. Each newspaper was allocated a score based on the above indicators,[11] and the average of these four scores was taken as a proxy for the relative innovativeness of the newspaper web site. The results are shown in Table 4.1.

Different publishing groups appeared to have varying ideas regarding the features of an innovative newspaper web site. For example, the few newspapers that provided a context to search archive results belonged to Associated Newspapers, Eastern Counties Newspapers Group Ltd, Newscom, Newsquest plc, Northcliffe Newspapers Group Ltd, Portsmouth & Sunderland Newspapers plc, Scotsman Publications Ltd, Scottish Media Ltd, and Trinity plc. At the time of the analysis, Trinity and Newscom joined the set of relatively independent publishers[12] in creating a completely new hierarchy for the front page of the newspaper web sites. Major publishers[13] such as the Eastern Counties Network were involved in selling goods online. Some sites, such as the joint site of the *East Anglian Daily Times,* the *Eastern Daily Press,* and the *Evening News,* advertised services to help readers go online. At the *Belfast Telegraph*'s site, the Cyberfood section allowed readers to order pizzas and ice cream. This newspaper also allowed users to order books from the virtual bookshop run in association with a local bookshop. Some of the smaller publishers[14] appeared to be adopting a 'wait-and-see' attitude to regional daily news site development. Some of these groups had weeklies or

---

[11]   Each newspaper web site was allocated a mark out of four for each of the following: (i) *searchable news archive*: no archive (0), archive with no date with story text (2), archive with date with story text (4); (ii) *new layout*: front page similar hierarchy to printed version (0), front page, new hierarchy (4); (iii) *update time*: update after 6 p.m. (0), before 6 p.m. (4); (iv) *shopping*: online shopping opportunity (4), shopping opportunity via telephone or postal service (2), no shopping opportunities (0). The final mark for the newspaper was calculated using an average of the four scores.

[12]   Hirst Kidd & Rennie Ltd, Newscom, Portsmouth & Sunderland Newspapers plc, Regional Independent Media, Trinity plc.

[13]   Associated Newspapers, D.C. Thomson & Co. Ltd, Eastern Counties Newspapers Group Ltd, Guardian Media Group plc, Newscom, Newsquest plc, Northcliffe Newspapers Group Ltd, The Midland News Association Ltd, Trinity plc.

[14]   C. N. Group Ltd, Clyde & Forth Press Ltd, Johnston Press plc, Kent Messenger Group, Nuneaton & District Newspapers.

Table 4.1. *Average scores of regional daily newspapers*

Regional newspaper group	Average score of group per daily (max score = 4)
Associated Newspapers	3.75
Eastern Counties Newspapers Group Ltd	3.31
Scotsman Publications Ltd	3.25
Scottish Media Ltd	3.25
Northcliffe Newspapers Group Ltd	3.21
Trinity plc	3.06
Newsquest plc	2.98
Irish News Ltd	2.75
Mirror Regional Newspapers	2.69
The Midland News Association Ltd	2.68
D. C. Thomson & Co. Ltd	2.50
Hirst Kidd & Rennie Ltd	2.50
Regional Independent Media	2.30
Jersey Evening Post	2.25
Newscom	2.13
Portsmouth & Sunderland Newspapers plc	2.00
North Wales Newspapers Ltd	1.75
Guardian Media Group plc	1.63
Bristol United Group plc	1.41
Yattendon Investment Trust plc	1.13
C. N. Group Ltd	0.00
Clyde & Forth Press Ltd	0.00
Johnston Press plc	0.00
Kent Messenger Group	0.00
Nuneaton & District Newspapers	0.00

Note: See n. 11 for method. Scores have been rounded to the nearest one-hundredth.

free papers available on the Internet, but most had yet to launch their first electronic publication. A number of the newspapers that were on the Internet had put together unusual features that catered to the local and global communities with a potential interest in their circulation area, but many sites provided little context to help readers understand the news stories. Two-thirds of the sites surveyed either provided no news archive, or provided readers with dateless news stories. Considerable differences in the approaches taken by new media developers around the country were apparent. Some of the characteristics, such as the provision of bare text without context-setting dates, suggested that many of the web sites had been created by those with excellent computer skills, but weak journalistic expertise.

The next stage of the investigation examined the organization of the product development, and how the creators of these web sites were interacting with others in the regional daily newspaper business.

5.2. *Survey of New Media Developers in Regional Daily Newspapers*

The 1999 survey of new media developers generated considerable data regarding the organization of web-site development, and about the ways in which meetings, technologies, and other media such as newspapers, books, and magazines were being used by website developments to communicate with those involved.[15] Multidimensional scales were constructed to investigate the organization of newspaper web-site development across the three dimensions of space, time, and organizational community. Spatial separation was assessed in relation to the time taken to make a journey. This was important to avoid overemphasis on those aspects of space that distance as a continuous variable ignores (see R.J. Johnson 1997: 326). The aim was to take account of the barriers to travel that are not directly related to actual distance (for example, stairs, hills, poor transport opportunities). The approximate measure of the spatial distribution of those involved in web-site development used in the survey was the average time to achieve co-location.[16]

Deadlines and short timescales are features of the daily newspaper industry. It was therefore decided that availability would be an appropriate indicator of temporal separation in this context.[17] Respondents' perspectives on the company allegiance of those involved were taken as indicators of the extent to which those involved were separated by organizational boundaries.[18] This approach is similar to that used by Clarke and Roome (1995). They differentiate between organizational groups that are 'intra' (those belonging to the same or different departments in the same company), 'inter' (those belonging to different companies), or 'supra' (those belonging to a mixture of organizational types—academic, industrial, government, etc.).

5.3. *Building the Organization-Space*

To analyse and compare the different newspaper web-site development organizations, the spatial, temporal, and community separation data from the questionnaires were then sorted into 'zones of virtuality' in the organization-space. As a first step, the space axis was divided into six zones: zone 1 (the least virtual) corresponding to a near

[15] The words 'those involved' were used instead of 'the team', so that the questionnaire was not limited by structure-oriented terminology.

[16] Under the heading, 'geography', respondents were asked: 'to get together when the product concept and strategy were being developed, would/did most of those involved have to travel a journey of . . .'. Respondents indicated whether this average separation was under 1 minute, over 1 minute to 2 minutes, 3–4 minutes, 5–9 minutes, 10–15 minutes, 16–20 minutes, over 20 minutes to 1 hour, over 1 hour to 2 hours, over 2 hours to 4 hours, a journey that required an overnight stay, a journey that required several nights' stay. The data were simplified by arranging into zones: 4 minutes and under (Zone 1), 5 to 20 minutes (Zone 2), over 20 minutes to 1 hour (Zone 3), over 1 hour to 4 hours (Zone 4), over 4 hours to 24 hours (Zone 5), over 24 hours (Zone 6).

[17] Respondents were asked, 'Of those involved in developing the product strategy and concept, what percentage were: (1) frequently unavailable for face-to-face contact?, (2) frequently unavailable for real-time contact?' Respondents indicated whether this percentage was 0–30, 31–50, 51–80, 81–99, or 100 per cent. The data obtained from the second question were used in this study.

[18] Respondents were asked, 'Of those involved in developing the news site strategy and concept, did they belong to . . .'. The questionnaire was structured to allow respondents to indicate whether or not they were from: the same department, different departments, the same firm or organization, different firms, or a mix of organizational types.

co-location separation (an average time to co-location of four minutes and under), zone 2 corresponding to an average time to co-location of 5–20 minutes, and so on. The time and space axes were divided in a similar fashion.[19] Table 4.2 gives a more detailed account of the boundaries between zones. These thresholds of virtuality were used to indicate degrees of virtuality in the three dimensions. This approach was deemed to be more appropriate than attempting to distinguish between the 'virtual' and the 'real' or 'traditional'.[20]

This step allocated three numbers to each organization, which represented the average separation in space, time, and organizational community—that is, a matrix of the form s,t,c. For example, a newspaper web site created by a group of co-located people from the same newspaper company, only 10 per cent of whom were frequently unavailable, was represented by the matrix 1,1,1. Each organization could therefore be represented as occupying a volume in the three-dimensional organization-space defined by its matrix (see Fig. 4.2).

After the spatial, temporal, and community data from questionnaires had been sorted into virtuality zones and matrices (see Table 4.3),[21] the matrix of each organization was plotted in the organization-space.

Table 4.2. *Division of the organization-space into virtuality zones*

Dimension	Separation		Zone
Space	Average time to co-location (minutes):	4 and under	1
		5–20	2
		21–60	3
		61–240	4
		241–1,140	5
		over 1,140	6
Time	Percentage of those involved who were frequently unavailable for real-time contact:	0–30	1
		31–50	2
		51–80	3
		81–100	4
Organizational community	Those involved belonged to:	same or different departments 'intra'	1
		different firms: 'inter'	2
		mix of organizational types: 'supra'	3

Note: These thresholds of virtuality were used to indicate degrees of virtuality in three dimensions. Thresholds were estimated and set to reflect the distinguishing features of the virtual organization in the management literature.

[19] For time, zone 1 (the least virtual) relates to 0–30% of those involved being frequently unavailable. For organizational community, zone 1 (the least virtual) relates to those involved belonging to the same department in a firm.

[20] For instance, Silverstone (1999: 104) notes that 'all communities are virtual communities'. Thresholds were set by the author to correspond to the distinguishing features of the virtual organization in the management literature.

[21] Three of the twenty-six questionnaires could not be used for this purpose, as the data were incomplete.

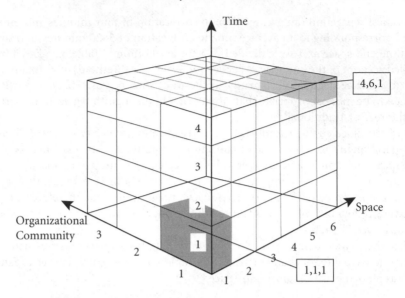

Figure 4.2. *Division of the organization-space into virtuality 'zones'*

The results of this analysis (see Table 4.3) failed to locate each regional newspaper new media organization in one part of the organization-space (as might be expected by the theory of the virtual organization and the associated perspective of the arrival of a single new organizational form). Neither were the regional newspaper new media organizations located in every part of the organization-space. They fell into particular parts of the organization-space, as indicated by Fig. 4.3. The results indicate that, in the context of Internet product development, there is no one organizational form—that is, the virtual organization. Neither is organizational virtuality completely unlimited by the shared contexts of space, time, and organizational community. There are limits to organizational virtuality in Internet product development, and these limits appear to relate to the minimum shared contexts or what might be referred to as the tacit contexts of space, time, and community that help those involved create the meanings that are necessary to create knowledge.

Examination of the data from the survey revealed that around 35 per cent of the web-site development activities were organized in the least-virtual sense—that is, 35 per cent (eight newspapers) of the organizations studied consisted of a group of co-located people from the same newspaper company, less than 30 per cent of whom were frequently unavailable. Therefore, 65 per cent of the web-site development activities were virtually organized beyond these thresholds (this accounts for fifteen out of the twenty-three newspapers).

Closer inspection of the data using the separation zone specifications outlined in Table 4.2 revealed that 73 per cent of the product development activities that were organized virtually crossed these thresholds of virtuality in one dimension only[22]—that

---

[22] Strictly speaking, more virtually—that is, beyond the virtual threshold of zone 1,1,1 in Table 4.3.

Table 4.3. *Virtuality matrices of Internet newspapers: distance from London and average score of newspaper web sites*

Regional daily newspaper[a]	Average score of web site (ii) (max=4)[b]	Distance from London (length of journey in hours and minutes)[c]	Virtuality zone matrix		
			Space	Time	Community
n1	2.5	01.31	1	1	1
i1	3	01.53	1	1	1
i2	2.75	02.14	4	1	1
i3	2.75	02.36	1	1	3
n2	2.5	02.52	1	1	1
i4	3.5	02.53	2	1	2
n3	3.75	03.01	4	2	1
n4	2.75	03.20	1	4	1
n5	3	03.24	4	1	1
n6	3	03.25	1	1	1
i5	3.25	03.41	3	4	1
n7	3	03.53	1	1	1
n8	3	03.53	1	1	1
n9	3.25	04.02	1	1	3
n10	3.75	04.06	4	3	1
n11	3.75	04.27	4	1	1
n12	3.5	04.50	1	4	1
n13	3.25	04.55	1	2	1
i6	3	05.14	5	1	1
i7	2.25	10.01	1	1	1
i8	2.75	11.42	1	3	1
n14	3.75	11.49	1	4	1
i9	2.25	12.13	1	1	1

[a] To maintain data confidentiality, each newspaper has been allocated a code that indicates whether it was part of the 'This is Britain' network in January 1999 (code ≡ n) or whether it was independent of the 'This is Britain' network in January 1999 (code ≡ i).
[b] The score gives an indication of four aspects of the relative innovativeness of the web site, taking into account updating time, search facilities, front page layout, and Internet shopping facilities (see Section 5.1). Data gathered and compiled by the author, Jan. 1999.
[c] See www.rac.org.uk (accessed 10 July 1999).

is, they were described by either the matrix location *x*,1,1, 1,*x*,1, or 1,1,*x*. While the Internet product development organizations studied here took many different forms, the organization of this activity appears to have limits, as shown in Fig. 4.3. Eighty-three per cent of the organizations studied were either located in the least virtual region near the origin of the organization-space, or were to be found in the regions of the organization-space that followed the dimensions of virtuality. The organizational virtuality of this industry appears to be limited to organizational forms in which those involved are separated by space, time, and organizational community, but not by all three dimensions at once. For the most part, those involved in this activity are separated

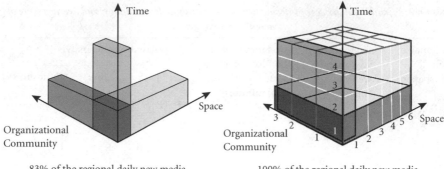

83% of the regional daily new media   100% of the regional daily new media
organizations fall within this shaded space organizations fall within this shaded space

Figure 4.3. *Distribution of Internet newspaper organizations in the organization-space*

by one of these three dimensions only. If the boundaries that limit the way we work are unravelling across space, time, and organizational communities, they appear to be doing so in one dimension only.

6. THE LIMITS OF ORGANIZATIONAL VIRTUALITY IN THE REGIONAL NEWSPAPER INDUSTRY

One of the aims of this study was to address questions about whether the influence of new technologies can meaningfully be described as the arrival of a singular new organizational form—that is, the virtual organization. None of the models described in the literature appears to be adequate to describe the virtuality of Internet product development that seems to characterize the regional newspaper publishing industry in the United Kingdom. The survey results reported in Section 5 indicate that, while some 35 per cent of the Internet newspaper development activities involved only co-located members of intra-organizational communities who shared the same time frame,[23] the 65 per cent that did not cannot be described as representing a single organizational form. This result calls into question the validity of any argument that proposes that a new organizational form is being enabled by new technologies. Other factors are at work.

The analysis of the data collected for this study indicates that these factors are related to the ways in which knowledge exchanges are mediated across space, time, and organizational communities. The results also shed light on whether organizational virtuality can be described meaningfully as the organization of activities, and hence the mediated exchange of knowledge, across these barriers. Analysis of the survey data showed that, of those product development activities that were organized virtually,[24] 73 per cent of regional daily Internet product development activities in the United Kingdom were virtual in one dimension only—that is, they were described by either the matrix $x,1,1$, $1,x,1$, or $1,1,x$ (see Section 5 and Tables 4.2 and 4.3).

---

[23] Belonging to the same department or newspaper company.
[24] Strictly speaking, organized more virtually, i.e., beyond the virtual threshold of zone 1,1,1 in Table 4.3.

Where such activities are organized more virtually across either space, time, or organizational community boundaries, they are organized less virtually in the remaining two dimensions. It appears that there may be a minimum shared context for knowledge exchange. Space, time, and community are the tacit contexts that endow communication, however mediated by meetings, technologies, or other media, with the meaning that is necessary for the creation of knowledge. Where one of these contexts is diminished, there appears to be a need for other contexts, or dedicated processes of mediation, to compensate.

The survey results suggest that presence—that is, an exchange of knowledge that is mediated to be made meaningful between those who are separated by a distance in space—is encouraged if those involved are not separated by time or community. They suggest that, if knowledge exchanges are to be mediated across time, space, or community, they are more likely to be made meaningful if mechanisms can be created to bring those involved together in the remaining two dimensions. Since the creation of meaning necessary for the exchange of knowledge in very different contexts is the key to innovation, the design and implementation of such dedicated systems of mediation are potential sources of innovation. They are also a means by which innovation may be responded to.

6.1. *Organizational Virtuality as the Extension of the Organization-Space*

Systems of mediation involve the use and adaptation of meetings, technologies, and other media across space, time, and communities. They are deeply rooted in the activities of organizations and individuals, and are constrained and strengthened by different communities that may have different interests in society. They also have a history. Case studies of the development of such systems suggest how the processes of innovation and mediation were interwoven in three distinct, and yet intimately related, newspaper products: 'South Wales Online', 'This is South Wales', and 'This is Britain'.[25]

Case study one: South Wales Online and the transition to This is South Wales
South West Wales Publications is a local publishing company that owns five newspapers, the *South Wales Evening Post*, the *Carmarthen Journal*, the *Llanelli Star*, the *Carmarthen Herald*, and the *Swansea Herald*. Of these, the *South Wales Evening Post* is the only daily paid-for newspaper, the others being free papers or weeklies. South West Wales Publications is owned by Northcliffe Newspapers, the regional publishing division of *Daily Mail* and General Trust. The Internet version of the *South Wales Evening Post* was designed initially by the editorial staff from the newspaper who had an interest

---

[25] Case studies were chosen with reference to the spatial, temporal, and community data from the questionnaire to investigate different parts of the organization-space. The material for these case studies was drawn from interviews that took place between January and May 1999 with the new media developers of the *South Wales Evening Post*, the *Carmarthen Journal*, and 'This is Britain'. Additional information was gathered from the portal 'This is Britain' on the Internet: **www.thisisbritain.co.uk** (accessed 31 July 2000).

in the World Wide Web. 'South Wales Online', therefore, was first and foremost a product of journalists on the Swansea newspaper.[26]

The relatively limited computer programming skills of those involved at this early stage made the content of this site faithful to its audience and the newspaper and was strongly tailored to local interests, but the overall system was impractical to manage on the timescale required by daily newspapers. 'South Wales Online' gave news, sports, and local politics the greatest prominence on the home page, with special sections for the young, features, and readers' letters. Its structure was very like the printed paper. Successful *Evening Post* special supplements, such as the pull-out section on Swansea poet and writer Dylan Thomas, also featured as links from the home page. Although the site included many sections that were specific to place and time (for example, the Y2K bug), it was in essence a text-based site with very few pictures and very simple graphics. More significantly for the new Internet product developers, the site had been designed to accept updates only by the laborious pasting into the text of each story using the Quark software used to produce the paper newspaper and each update took six to seven hours.

When the *South Wales Evening Post* decided to expand its Internet activities, it pooled resources with its sister papers in Carmarthen, moving the nucleus of the web development activities to offices above the editorial floor of the *Carmarthen Journal*. These offices are an hour's journey away from the *South Wales Evening Post* buildings by car or train. Over time, a growing web development group began to produce web sites not only for the *South Wales Evening Post*, but also for the *Carmarthen Journal*. The new media developers automated the updating process so that each update took half an hour instead of six or seven hours. This left those involved free to concentrate on the features of the web site and on how it might change in the future.

In the meantime, the parent group, Northcliffe, developed its own web strategy, and began to take more control of the Internet activities of its newspapers, which included, not only the *South Wales Evening Post* and the *Carmarthen Journal*, but also titles such as the *Sentinel* (Stoke), the *Hull Daily Mail*, and the *Plymouth Evening Herald*. Under the direction of the parent group, and using a template system designed by Northcliffe Electronic Publishing in London, those developing 'South Wales Online' concentrated their efforts on taking the newspaper web site through the transition from an automated and relatively independent Internet news product, to a branded Internet presence, 'This is South Wales'.[27]

During this time, developers, managers, and editorial staff had face-to-face discussions about changes to the web site, keeping in touch and exchanging documents using the telephone, email, and postal mail. The senior web developer visited the *South Wales Evening Post* in Swansea two or three times a week in this transition period. Frequent emails and telephone calls linked the bases at Carmarthen, Swansea, and London. Forums, workshops, and conferences also brought those involved together with web developers from other Northcliffe newspapers, such as the *Leicester Mercury*. To aid the

---

[26] See www.swep.co.uk (accessed 6th Apr. 1999).
[27] See www.thisissouthwales.co.uk (accessed 31 July 2000).

exchange of knowledge between newspaper developers, Northcliffe also employed five 'roving idea distributors' to visit the individual web developers every two weeks. The sole task of these representatives was to help pass ideas on from one newspaper to another. In the summer of 1999, the new web site, 'This is South Wales', was launched.

The activities of, and the relationships between, the new media developers in 'South Wales Online', the editorial staff of two print-based publications (one in the same building and the other one hour's drive away), and the management team in the publisher's offices in London, illustrate how communication is mediated and made meaningful across space, time, and organizations.

When separated by space, knowledge exchanges need to be mediated to be made meaningful if innovation processes are to give rise to, and emerge out of, the relationships that exist between different communities of interest[28] across geographical barriers. This is illustrated by the evolution of 'South Wales Online' into 'This is South Wales'. If news is 'first and foremost an organizational product' (Berkowitz 1997: 40), then 'This is South Wales' is an organizational product that is, spatially and organizationally speaking, organized more virtually than 'South Wales Online', which was, in turn, more spatially and organizationally virtual than the paper product, the *South Wales Evening Post*. Although those involved in the creation of the printed newspaper did not all share the same time frame, they were all based at one location—that is, Adelaide Street in Swansea, and belonged to the same company—that is, the *South Wales Evening Post*. In contrast, the organization of 'South Wales Online' was distributed over two sites separated by an hour's journey. Furthermore, while 'This is South Wales' is similarly distributed in space, those involved in the organization of this news product belonged to other newspaper companies as well.

The virtuality of the news product, and the delivery of news text via email to Carmarthen, can be associated with the extension of the product's organization space. But this extension was also associated with the creation of a completely new product, 'This is South Wales'. This product not only extended the organization space still further, but also paved the way for much more significant changes in the role and function of the regional newspaper as mediated by the partnerships, communication, and shared template/databases associated with 'This is Britain'.

Case Study Two: 'This is Britain' and 'Fish4News'
The first Northcliffe newspaper to undergo the transition from local newspaper web site to the branded 'This is . . .' product was the *Sentinel*, a regional daily paper in Stoke-on-Trent. In 1998, a partnership was formed between Northcliffe and another major regional publisher group, Newsquest, to produce 'This is Essex'. By March 1999, ten Northcliffe newspapers had launched web sites allied to this brand, and, in April 1999, Newsquest and Trinity and Northcliffe, through Associated New Media, established what the companies called 'a unique portal linking city and community sites across the country with compelling national content'. The portal was effectively a gateway to

---

[28] Here, the term 'communities of interest' is used to describe communities that may not share the same space or time, but that have some common interest.

regional news from all over the United Kingdom accessed via the 'This is . . .' web sites of the Northcliffe, Newsquest, and Trinity newspapers in England, Scotland, Wales, and Northern Ireland.

The evolution of 'This is Britain' also illustrates how systems of mediation, and the communication media that support them, give rise to, and emerge out of, the mediated relationships that exist between different communities of interest over time. Although the portal initially took the form of a gateway, by October 1999 the newspapers drew on the success of Trinity's 'Adhunter', an electronic classified database that allowed users to search for jobs and cars using the Internet, to reframe 'This is Britain' under a new brand called 'Fish4'. 'Fish4cars', 'Fish4jobs', and 'Fish4homes' were integrated with the 'This is Britain' regional news platform to create an Internet-based classified database supported by news from around the country—that is, 'Fish4news'. However, if the faceless portal of 'This is Britain' was a door to a journey down a lonely corridor, 'Fish4news' attempted to provide a guided tour to the news in the regional communities around the UK. The portal model that had guided the design of the 'This is Britain' product was based on the perception of Internet newspapers as providers of news. In contrast, 'Fish4news' was developed in a way that acknowledged that the characteristic that distinguishes one newspaper from another is the way in which it adds context to news.

The Internet product 'This is Britain' is itself a mediating technology, a communication media that co-locates news and events from many communities in the UK. 'Fish4news' is, in effect, a type of virtual regional newspaper that relies on the organization of work that is distributed across the entire country. Although the United Kingdom Newspaper Society defines a regional newspaper as 'any newspaper not available throughout the UK', integrated news delivery mechanisms and communication between regional newspapers, their local publishers, and a number of different parent group companies have created a regional newspaper that is available to those online in the United Kingdom and beyond.

In 2000, 'Fish4news' retained a link to a map of the United Kingdom to allow users to search for news by region. At this stage, the front page organized its content by region, time, and community of interest. Stories were grouped under headings such as hot topics, in the courts, would you believe it, animal crackers, health, and that's showbiz. For example, in June 2000, the first headline under health was 'Cardiff: Meningitis alert at hospital creche'.[29]

Designing and Implementing Process of Mediation

These two cases studies offer a basis to consider the design and implementation of the processes of mediation that are necessary to exchange knowledge and support innovation in the developing Internet newspaper industry in the United Kingdom. The organizational virtuality of Internet product development in the UK newspaper publishing industry—that is, the extension and redistribution of activities over space, time, and organizational community boundaries—does not appear to be simply enabled or

---

[29] See www.thisisbritain.co.uk (accessed 30 June 2000).

facilitated by the use of ICTs. Face-to-face meetings were identified by those participating in the survey as critical to the web development process, particularly during times of great change, or when communicating with those from different backgrounds (editorial) and different companies. The 'roving idea distributors' were also valued as they offered a means of helping those involved in the creation of 'This is Britain' to exchange ideas in a face-to-face setting, without ever meeting face to face. The intermediaries that supported the exchange of knowledge across space, time, and communities, in some instances, were appropriately used technologies. However, in other cases, they were trusted people who interacted directly with other people or traditional forms of media that were used by Internet site developers.

While the use of email and the telephone was important, the mere possession of such tools does not appear to have guaranteed their efficient or effective use for a particular purpose. For example, despite the fact that the text of the *South Wales Evening Post* could be emailed to Carmarthen, this did not automatically enable the newspaper web site to be maintained and developed at a distance. When text from the *Carmarthen Journal* and the *South Wales Evening Post* had to be pasted into the respective web sites, the availability of the printed Carmarthen paper, with hand-written annotations made by the editor, had a significant impact not only on the ease of extraction, but also on the types of stories that were included in the online version. Because of the hour's journey time between Swansea and Carmarthen, the same method could not be used to update the Swansea newspaper, as the printed newspaper would invariably arrive too late in the afternoon. Without the meaning given to the newspaper text by its location in the printed newspaper, updates were much more difficult to carry out. Web developers commented that 'it helps to know where the stories are on the page', 'It is a different kind of reading', 'The paper gives me an idea —I can see the photograph, and then look at the Quark document and see it straight away'. These experiences suggest that access to highly codified knowledge and the use of ICTs do not necessarily enable virtual activities.

Nevertheless, those working in the Carmarthen offices developed the capability to automate the text retrieval process. The results of these labours, and the use of the web templates supplied by Northcliffe Electronic Publishing in London, freed up more time to focus on the development of 'This is South Wales' and, indeed, to work on other projects. For example, in 1998, the new media developers created a web site for the car company Vauxhall. Although many such web sites were created for external clients, they may be regarded as 'newspaper' activities. This is because one of the main selling points of a web site created by a regional newspaper company is the Internet traffic that can be directed from the newspaper to the client's web site. Regional newspaper new media departments have the added advantage that their clients often advertise in the printed paper and know and trust their ability to do their job.

For those developing new media products in the newspaper publishing industry in the United Kingdom, there were no simple answers to the question: 'For whom do I work?' A group of web developers might produce web sites for a number of different newspapers and clients on a freelance basis. From one newspaper's perspective, the developers are the in-house, co-located web development team. From another's, they are the in-house, non-co-located web team. From the point of view of the clients who commission web sites, the developers are carrying out work that is both outsourced and non-co-located.

Organizational virtuality does appear to have its limits. In the context of the creation of Internet newspapers by web developers in the regional press, all of the organizations fell within three perpendicular planes, which intersect the origin of the organization-space (see Fig. 4.3). There are limits to organizational virtuality in Internet product development, and these limits appear to relate to the minimum shared contexts—that is, the tacit contexts of space, time, and community—that help those involved create the meaning that is necessary to create knowledge. If successful knowledge exchange, and hence innovation, is to be mediated across time, space, or community, they are more likely to be meaningful if mechanisms can be created to bring those involved together in the remaining two dimensions.

Virtuality is not centrally a matter of the role of hierarchies or the management structure for team working. Organizational virtuality can be described as being multi-dimensional across space, time, and communities, but it is more than that. People who work on virtually organized newspaper web-site developments are *simultaneously* part of organizations that occupy very different parts of the organization-space. They are working both in-house and externally; they are both available and unavailable; they are both co-located and separate from the people with whom they work. The organization is a network whose constellation depends on the perspectives of all those within it and on the identity and requirements of each individual project in a portfolio of activities. This is the reality and the virtuality of new media product development in the newspaper publishing industry in the United Kingdom.

7. CONCLUSION

While the advantages of the virtual organization have been sung from the rooftops, the limitations of the concept of 'the virtual organization' often seem to have been disregarded. The reasons for this are threefold. First, most definitions of the virtual organization imply that they are enabled or facilitated by the power, or more modestly the use, of ICTs. The extent to which these technologies can or cannot support virtual activities has not been considered in the management field as an issue for empirical enquiry. Despite the existence of research on computer-mediated communication that highlights the diversity of the mediation processes that are necessary to create knowledge, there is little if any acknowledgement that the mere possession of such tools does not guarantee their efficient or effective use for a particular purpose.

Secondly, as long as the contributors to the management field continue their search for a single definition of *the* virtual organization, it will remain difficult to compare different kinds of organizational forms in terms of their implications for the spatial, temporal, and community boundaries of work and how they are changing over time. Without a means of conceptualizing the relationships between these different manifestations of virtuality in organizations, the different processes that support the various working experiences involved cannot be identified or compared.

Thirdly, the prevailing concept of the virtual organization, with the mystical enabling powers of ICTs, fails to acknowledge the need for a conceptual model to help design the mediation processes that can direct the search for meaning in diverse communities of interest in particular places, time frames, and communities.

The word 'virtual' may have been defined as a 'huge vessel of semantic vacuity waiting to have meaning poured into it' (Woolley 1993: 58), but one cannot pour meaning into the vessel of virtuality. Meanings have to be created if they are to mean anything at all. One thing is transparent, however. Those who see the changes that are currently taking place in the so-called knowledge economy simply as the result of the enabling powers of ICTs, such as the Internet, are likely to be rather powerless to respond to the new environment.

It is the processes of mediation that help make knowledge exchanges meaningful. These processes of mediation help people in very different contexts to make sense of the world. It is only through these processes that exchanges of knowledge can contribute in support of and in response to innovations that are so critical to economic and social development. Systems of innovation *are* systems of mediation. The makers of strategy and policy who recognize them as such will be much better placed to create, maintain, and direct dedicated processes of mediation. This will help those in different communities of interest to reach out across space and time to rediscover the sense of togetherness they lost on the way to the global village.

5

Mind the Gap: Digital Certificates, Trust, and the Electronic Marketplace

INGRID SCHENK

1. INTRODUCTION

I am a dog. I use the Internet. So what? A now infamous 1993 *New Yorker* cartoon depicts two dogs in front of a computer discussing one of the perceived virtues of the Internet—anonymity.[1] User anonymity in an open networked environment, however, is seen as particularly problematic by those actors seeking to promote commercial transactions in this medium. Concerns at this level focus on how network-users identify those with whom they are transacting and whether the exchanging parties will do what they say they will. Public key infrastructures (PKIs) are emerging as one solution to the problem of authentication and commercial transaction issues that arise as open networks proliferate. A PKI refers to a system that governs the provision of trust services, such as digital certificates, by trust service providers (TSPs) that will allow users to present certified credentials to support secure electronic exchanges (Winn 1998*b*).[2]

Since secure electronic commerce requires cryptography and is linked to the taxation authority, many industrialized countries, such as the United Kingdom, the United States, Canada, Japan, Australia, and member states of the European Union, are taking an active role in attempting to shape market developments.[3] In general, national governments are

The author acknowledges the financial support of the NCR Financial Services Knowledge Lab, London, UK, for the research reported in this chapter. The views and opinions expressed here are the sole responsibility of the author and do not necessarily represent those of any institution or organization.

[1] Cartoonist, Peter Steiner, caption: 'On the Internet, nobody knows you're a dog.'

[2] There is no accepted definition of the term 'public key infrastructure'. The term is often associated with specific designs for distributing and managing digital certificates on open networks. 'Trusted third party' and 'certification service provider' are also terms used to describe this role. PKIs based on peer introduction or a 'web-of-trust' model are also an option for securing online exchanges (for a detailed description, see Garfinkel and Spafford 1997).

[3] The controversies surrounding governments' visions of a common infrastructure and TSPs involve the perceived need of governments to ensure key escrow, or key recovery. Key recovery legislation seeks to give governments access to the keys of businesses and individuals for law enforcement, taxation, issues of national interests, and other purposes. See the Electronic Frontier Foundation (EFF) in the USA, www.eff.org, and, in the UK, the Cyber-Rights and Cyber-Liberties Organization www.cyber-rights.org, and the Foundation for Information Policy Research, www.fipr.org (all sites accessed 8 July 2000).

seeking to establish a predictable legal, technical, and regulatory environment,[4] or common infrastructure model, for open network electronic commerce. Some of these government proposals include voluntary licensing—that is, accreditation schemes for TSPs as a standard for trust services provision. Opponents of the endorsement of a common infrastructure model envisaged by national governments argue that these initiatives introduce market distortions, encourage the externalization of costs to users, and lead to the promotion of an inappropriate business model in the trust services industry that is only beginning to emerge (Biddle 1997; Feigenbaum 1998; Greenwood 1998; Winn 1998a,b). Many of these debates are highly speculative, since few TSPs operate in either the public or the private realm, and most PKIs have not advanced beyond the level of pilot projects. Consequently, there is a lack of empirical evidence to support controversial claims and assertions regarding the use of trust services and the role of the TSP for the uptake of secure electronic commerce.

The principal question this chapter addresses is whether newly emerging trust services offer a solution in terms of commercial viability for TSPs and responsiveness to user concerns about authentication. The chapter examines how key contenders, the TSPs, for entry into the trust services marketplace have been developing strategies for market entry and development. Particular emphasis is placed on how TSPs are proposing to use digital certificates to mediate online exchanges. Based on an analysis of thirty-four in-depth interviews with parties engaged in the development of PKIs, the evidence suggests that TSPs are not relying on the presence of a common PKI to support their strategies. Rather, TSPs are proposing to enter and develop the trust services marketplace on the basis of their existing sources of trust—that is, by a process of sticky trust. The discussion suggests that TSPs regard the certification process as critical for establishing TSP legitimacy as well as for creating the presumption of trust in the use and reliance on digital certificates by the parties who engage in electronic exchanges. It is these certification practices that both mediate user expectations and enable users to enter into and negotiate electronic exchanges using an electronic handshake.

The changing role of the TSP is important because these providers are well positioned to influence new patterns of social interaction within the electronic realm. The analysis in this chapter highlights the potential gap that exists between certification practices that are being established by TSPs and how these practices may be used to mediate direct exchanges in open environments for secure electronic commerce. This gap reflects the different social dynamics associated with the various types, or levels, of trust that appear to facilitate online interactions and encourage the development of long-term electronic commerce relationships.

---

[4] Competing electronic signature legislation is being proposed and introduced by many national governments and, in the United States, by different state governments. Some legislation appears merely to shift the risks of online exchanges to consumers and merchants by building in safe harbours for TSPs and technology providers, while other legislation, such as that proposed by the European Commission, attempts to define minimum liability requirements for all parties to an exchange. For a review and summary of electronic signature legislation worldwide, see **www.mbc.org/ecommerce.html** (accessed 8 July 2000).

2. TRUST SERVICES: AN EMPIRICAL INVESTIGATION

A TSP can be treated as an institution for mediating the interests of online parties to an exchange to provide each with a reliable means of making commercial exchanges in an electronic environment. TSPs are expected to meet security and transactional require-ments for online exchanges by ensuring the authentication of transacting parties, access to authorized users, the data integrity of online exchanges, and confidentiality as well as non-repudiation of transacting parties (Summers 1997). This may be achieved through the provision of trust services such as digital certificates.

Digital certificates are the data or characteristics of an entity that enable users to cre-ate a surrogate presence for secure exchanges in the electronic realm.[5] Digital certifi-cates are based on public key technology and supported by cryptographic techniques and organizational processes. TSPs are expected to establish the veracity of specific authentication criteria—for example, identity, attribute, and creditworthiness—of a certificate subscriber, which are then certified by binding the public key of a user to a digital certificate (Feghhi, Feghhi, and Williams 1999). The provision of trust services may require the creation of new institutions or the establishment of new trust rela-tionships with existing social institutions (Froomkin 1996).

By January 1999 there had been relatively little commercial activity in the trust serv-ices market in the United Kingdom, continental Europe, or North America. One excep-tion at the time the study was undertaken was a new entrant, VeriSign, of the United States, which held the largest share of the digital certificate market globally, estimated at approximately 90 per cent.[6] The main potential market entrants were financial insti-tutions including banks and credit-card companies, telecommunications operators, and post-office organizations. Additional contenders included, but were not limited to, Chambers of Commerce, major retail and media companies, major consultancy firms, and ISPs, as well as computer and software companies. This disparate mix of institutions with a potential interest in the provision of commercial trust services indi-cates the diverse ways in which the TSP role may be interpreted. It also indicates that the potential size of the market is deemed to be significant by many types of firms and organizations.

Given the technical, legal, and regulatory uncertainties associated with trust services provision, thirty-four in-depth interviews were conducted between February 1998 and January 1999 to investigate how the trust services marketplace was likely to evolve. Particular emphasis was placed on the market entry and development strategies for TSP contenders in the United Kingdom and North America. Included among the interviewees were financial institutions and postal authorities from the United States,

---

[5] Digital certificates include data elements such as a copy of the public key of the subscriber, a reliance limit, an expiry date, and/or a reference to a URL where the TSP's certification practice statement (CPS) resides. For a detailed description of certificates, see Feghhi, Feghhi, and Williams (1999) and Ford and Baum (1997). For a description of public key technologies, see Garfinkel and Spafford (1997). For an excellent primer on modern cryptographic techniques, see Schneier (1996).

[6] There was tacit recognition of the market share held by VeriSign since formal statistics on the penetra-tion rate of digital certificates (excluding server certificates) were not available.

Canada, and the United Kingdom, British Telecom (an International Affiliate of VeriSign), new players, including VeriSign (United States), InterClear and Entrust (United Kingdom), a major UK consultancy firm, and the British Chambers of Commerce (UK head office). Additional interviews were conducted with hardware and software suppliers, United Kingdom-based industry representatives, supporting organizations, and corporate trust service users in the United Kingdom, as well as with representatives of the Department of Trade and Industry (DTI) and the Office of Telecommunications (OFTEL) in the United Kingdom, and representatives of Industry Canada and of the European Commission.[7] The information in the following section is based on an analysis of the main perspectives offered by the interviewees.[8]

2.1. *Trust Services and Market Entry Strategies*

There is a variety of trust services that a TSP can choose to offer to support secure exchanges between transacting parties. The employees of firms who were interviewed all suggested that they would not enter the market to provide trust services without first becoming a Certification Authority (CA) for issuing and managing certificates. Aside from certification, there was considerable uncertainty about the kinds of trust services that might be provided. Some interviewees from firms expressed the view that, although becoming a TSP in the public realm was not a particularly high priority, it did represent an extremely important commercial opportunity.

The primary motivating factors for market entry and involvement in the design of some kind of PKI were a perceived need to demonstrate an image of competence and sophistication as well as the capacity to take a leadership role in the development of the electronic commerce marketplace. Despite the high level of perceived risk associated with becoming a TSP in an uncertain market environment, all the firm-based interviewees noted that it was 'riskier not to do anything than to do something and get it marginally wrong'. Since many commercial applications for trust services have yet to be defined, the majority of interviewees tended to perceive risk in terms of the potential lost opportunities in an environment where the growth of electronic commerce markets seemed inevitable.

Most of the interviewees from firms appeared to be modelling their future service offerings on the experience of the trust services market leader, VeriSign. Like VeriSign,

---

[7] The names of those interviewed by the author and their company affiliations are protected because of the commercial sensitivity of this emerging market.

[8] Interview data for potential and existing TSPs were used as the basis for several case studies. Data obtained from each interview were aggregated into thematic categories including the background for public key infrastructure initiatives, technical and infrastructure standards, legal and regulatory issues both nationally and internationally, and market entry, market development, and competitive strategies. The empirical results were derived from the analysis of data obtained within and across the case studies as a means to establish similarities and/or differences in the views expressed. Additional interviews with representatives of organizations involved in the development of trust infrastructures were used to achieve triangulation and to verify the results. Secondary data sources, including published and unpublished documents, press releases, attendance at related conferences, special interest group sources, and a review of the DTI Trusted Third Party and Electronic Communications Bill consultation process, supplemented the data collection and analysis.

existing and potential market entrants were expecting to segment the trust services market by offering different certificate policy classifications, for example, Class 1, 2, 3, or 4; high, medium or low; or Bronze, Silver, Gold, and Platinum, indicating the levels of permission and assurances provided by a certificate. These initial market segmentation strategies were to be followed up by efforts to introduce attribute or other transaction-specific certificates because of their ability to customize transaction requirements. These certificates were perceived to offer greater value added as new commercial applications and markets for trust services are developed. In line with this view, one interviewee stated that 'at this stage the British Telecom/VeriSign relationship is no more than a "rebranding exercise" . . . market development is limited to the classes of certificates available. There is no capacity to extend how the certificates can be used.'

In establishing levels of assurance for certificate-based exchanges, the emphasis of TSPs appeared to be on improving, rather than replicating, the processes and practices for commercial exchanges in the physical realm. This was reflected in one interviewee's comment that 'doing a PKI and identity certificates would not be worth it if they just replicated the same physical world processes'. Another interviewee commented that his firm's objective was 'to develop the capability to issue an "indisputable identity" ' for the electronic realm. Establishing authentication procedures in the electronic realm was deemed to require not only the transference of existing commercial practices, but also the development of new processes that are capable of maintaining business continuity and the service quality already established in the physical realm. These strategies depended significantly upon the way the interviewees expected to manage the process of certification.

Registration is a key part of the certification process that influences how digital certificates will be issued, verified, distributed, and validated in order to support online transactions. When a user registers with a TSP, the identity or the attribute of a user must be authenticated before a level of authorization for electronic trading is granted. From a TSP's perspective, registration is the first point of contact with a potential client, which provides an opportunity to communicate the levels of security and assurances available for certificate use and reliance as well as to establish a relationship.

Interviewees from organizations that already had significant market reach and an extensive physical infrastructure believed they had a competitive advantage over new entrants who could not reach out directly to users. For example, banks have a branch infrastructure, post offices have local outlets, the British Chambers of Commerce have local business networks, and the major consultancy firms have office locations in city centres. British Telecom lacks physical outlets but has extensive customer links through its network services. In the United Kingdom, it is likely that most TSPs initially will use physical presence and documentation to facilitate registration procedures.

A demonstrated capacity to verify a transacting party's documentation was also considered an important prerequisite for market entry. For example, High Risk Certification generally requires physical presence and documentation such as a passport or other notarized piece of identification. Post offices, banks, Chambers of Commerce, and major consultancy firms have established track records for the collection and verification of personal or business documentation and details, while British

Telecom has not needed to establish competency in this area. At the time this research was conducted in 1998–9, British Telecom could not offer Medium Risk Certification. Unlike its American counterpart, VeriSign, British Telecom did not have access to a public database that would enable it to match and verify the details of the documentation provided by potential subscribers to its trust services.

Establishing the credibility of the registration system was regarded as a challenge for TSPs. Commenting on VeriSign's web-based registration process for Class 1 and 2 Certificates (low and medium risk), one interviewee explained that: 'the VeriSign model is one that is easy and quick. Click of a button. The problem for X with the VeriSign model is one of accountability and credibility at the technical level. That is, if you can acquire a certificate at the click of a button then there is a problem ... if you can talk to it, you can crack it.' All the representatives of TSPs felt that this was an area where practices and procedures needed to be built up in order to establish 'fit-for-purpose' certificate use. Fit-for-purpose certification refers to how the level of assurance of a certificate policy matches the related business risks and security requirements to ensure that the association between a user with a specified public key is 'tamper evident'.

One of the most significant perceived barriers to entry into the trust services market is related to organizational issues surrounding the process of externalizing information about a firm's internal processes, practices, and procedures. For example, all interviewees regarded internal company politics, organizational inertia, and the assignment of responsibility or ownership of the certification system as hurdles to be overcome. The critical importance not only of what was communicated, but also of how it was communicated, both across organizational boundaries and between the public users of their services and other traders, was emphasized. The introduction of authentication procedures into the electronic realm could pose risks to both reputation and business competence because many internal certification practices of TSPs have gone largely unchallenged in the physical realm. At the time of the interviews, only British Telecom and VeriSign had publicized details of their policies, practices, and procedures for certificate use and reliance.

The design and deployment of trust services for authentication appear to be greatly influenced by how the procedures of registration, verification, certification, and certificate management are combined in the context of certificate policies. The combination of these procedures for establishing standards in certificate use involves not only decisions about appropriate security mechanisms, but also decisions regarding the legal and business requirements that must be met in accordance with industry practices. The following section indicates how TSPs were seeking to establish a competitive basis for trust service provision in the context of the regulatory and legislative proposals by national governments.

2.2. *Competing on Business Policies, Practices, and Procedures*

TSP market entry strategies are being developed in an environment where most national governments are moving to reach agreement on specific regulatory and legislative requirements for the provision of trust services. Most national governments endorse a common trust infrastructure that includes voluntary accreditation (licensing) schemes to mediate interoperability between TSPs through standardization as well as legislation for the legal recognition of electronic (digital) signatures.

For many of the British interviewees, the critical factor for business success was the way accreditation or the voluntary licensing of TSPs would be handled by the government.[9] All potential entrants to the market in the United Kingdom agreed, in principle, that an accreditation regime and legislation were important for enhancing their credibility and for building confidence and trust in electronic commerce transactions. However, the interviewees representing firms also felt that the voluntary licensing of TSPs, which would be supervised by OFTEL, was a premature attempt by government to establish standards before sufficient experience in trust service provision had been accumulated by TSPs. In establishing a level playing field for the trust services industry, a common view held by prospective TSPs was that 'everyone is greedy, imagine trying to set standards with a telco (British Telecom) within OFTEL?'

In response to government proposals, TSPs in the United Kingdom felt that they had been racing against time to establish their markets in the hope that legislation would adapt to their initiatives. While the government had been moving towards establishing a public licensing framework, some TSP organizations were seeking to establish standards and an accreditation framework through self-regulatory initiatives.[10] There was also something of a 'race' between TSPs as they tried to establish the criteria that would then become benchmarks for other services providers to measure their products and services against.

Most interviewees from firms and public-sector organizations were of the opinion that the issuance of certificates would rapidly become a relatively low-margin commodity business. There was a sense among the market players that value would be created by establishing standards aimed at minimizing incompetent trust service provision. According to one interviewee, 'the key to differentiation is the policies, practices, and procedures . . . it is easy to copy the technology'. Most firms believed that the establishment of *de facto* standards in the area of certification policies and certification practice statements (CPS) would provide a basis for establishing market leadership with respect to certificate issuance and use. Commenting on VeriSign's perceived dominant position in the marketplace, one interviewee stated, 'everyone criticises VeriSign's efforts in the marketplace but at least they are doing something, no-one else is doing anything. That

---

[9] The OFTEL representative commented that voluntary licensing was really a term for standards accreditation.

[10] Two self-regulatory initiatives were under development at the time the study was undertaken: the European Certification Authority Forum (ECAF) sponsored by the European Electronic Messaging Association (EEMA) and the UK Alliance for Electronic Business (AEB); and the Emeritus project, which was promoting a voluntary, industry-led global trust services infrastructure under the auspices of the AEB with collaborative partners in Belgium and Spain and supported by the European Commission's Trans-European Network (TEN) Telecom programme. The third initiative, tScheme, was initiated in late 1999 by the AEB. The tScheme is intended to be an industry-led accreditation scheme for the provision of trust services in the UK. The intent of the tScheme was to substitute for Part I (Cryptographic Service Providers) of the proposed Electronic Communications Bill (which was enacted as the Electronic Communications Act, 25 May 2000). See www.fei.org/fei/news/tscheme.html (accessed 8 July 2000).

said, we couldn't have our Certification Practice Statement being perceived as meaning nothing, like theirs is. We would lose our reputation . . .'. This statement is consistent with the opinion of all the interviewees to the effect that, since the CPS embodies the parameters of self-regulation, it would serve as a 'unique selling point' for establishing credibility and attracting a user base.[11]

The TSP strategies that focused on standardizing authentication procedures differed from the technical and business standards proposed by national governments for certificate issuance. This is because there is no regulation that applies to how a decision is made by a TSP to issue a certificate beyond disclosing the procedures that it intends to follow in its CPS. It was thought to be likely that certificate policy standards would raise issues of liability as they become incorporated into CPS. According to an interviewee, the 'problem with most companies is that they throw everything regarding liability and risk into the CPS . . . this raises issues for cross-certification'.

Liability is another key issue facing TSPs in the new trust services marketplace. Many expect that electronic commerce markets will grow rapidly if legal uncertainties can be reduced or removed.[12] Competing electronic signature legislation has been proposed and/or legislated by national governments and by the different states in the United States, regarding the legal recognition of electronic (digital) signatures as a minimum standard for binding contracts and avenues for legal recourse that are to be associated with certificate practices.[13] Although there was significant controversy over proposed legislation in the European Union,[14] the European Commission representative who was interviewed suggested that legal initiatives could also be regarded as a move to ensure that electronic documents are signed and that signatures imply contractual terms and liabilities. This would help achieve interoperability for global practices as these terms and conditions could then be embedded within TSP CPS.

---

[11] As with other PKI components, no standards exist for the role, format, and/or content of a certification policy and a CPS. At the time of the study, interpretations of certification policy and CPS standards were based on *de facto* standards established by VeriSign for use on open networks. Other practices may be equally applicable.

[12] Based on recent and projected estimates from the OECD (2000), business to business and business-to-consumer electronic commerce will grow at the rate of 50% per year in 2000–5. For example, starting from a base point of approximately $200 billion, business-to-business electronic commerce will reach $1,400 billion by 2005, while business-to-consumer commerce, starting from a base point of approximately $25 billion, will reach approximately $160 billion by 2005.

[13] It is assumed that the reliability that is, the intent to be legally bound, of a transacting party is established through the presumption of validity of an electronic signature. For details, see Winn (1998*a*, 1999).

[14] Differences exist in terms of how electronic signature legislation can be expressed in law. As of 30 November 1998, Germany, Italy, and France were proposing to make specific technologies mandatory to ensure full legal protection, thereby deviating from the provision of technological neutrality promoted by the European Electronic Signature Directive (de Bony 1998). For a review of country proposals, see OECD Working Party on Information Security and Privacy document entitled an 'Inventory of Approaches to Authentication and Certification in a Global Networked Society', www.oecd.org (accessed 8 July 2000). For a typology of differing legislative models for addressing the tension between technological neutrality and legal specificity for the use of electronic authentication systems, see 'Survey of International Electronic and Digital Signature Initiatives' published by the Internet Law and Policy Forum (ILPF), 1999. See www.ilpf.org/digsig/survey.htm (accessed 8 July 2000).

One of the outstanding issues regarding the evolution of global practices for transactions using open networks is how contractual liability can be established with a 'relying' party. This includes defining the liability entailed in the relationship between the supplier of the digital certificate and a trading party who relies on this supplier to verify the authenticity and identity of that supplier's subscriber. This distinction is important when it is applied to the electronic exchange process, since independent verification and acceptance of the certification process by the relying party—that is, cross-certification—still needs to be addressed. As one interviewee commented, 'The perception of a certificate is linked to the value it is supposed to hold. Therefore most firms are waiting to get it right. If X proceeds with a certificate that no user can attach a value to, then we have lost out on market share. Currently, the need for an accepting party to appreciate its value is what is lacking in the marketplace.' Interviewees held a variety of views with respect to the role of the certification process. Some regarded it as a means of establishing a contractual relationship with a user. With respect to establishing the limits and boundaries for certificate use, one interviewee indicated that, 'although everyone is making an issue of liability, really everyone who is entering the market is just seeking to do it properly. We want to utilize liability and contract legislation. These are the unresolved issues but also the way to move forward.' Others said that the certification process should be one that enables the personalization and customization of the relationship between the certificate issuers and the user.

These differences in approaches to the certification process suggest that open Internet standards are *not* regarded as being synonymous with interoperable trust infrastructures. The majority of TSPs interviewed recognized that achieving their goal of establishing trust and confidence with users of their systems would require more than simply implementing technical standards established by hardware and software suppliers. Aside from technical protocols, trust service implementation also involves new standards and practices for managing information flows and uses. The interviewees suggested that barriers to interoperability could be built into these areas as a means of product and service differentiation. These barriers may potentially limit how a certificate can be implemented or extend how the certificate may be used for multiple applications. Section 2.3 examines the market entry and development strategies that were being employed by the TSP organizations interviewed.

2.3. *Market-Making Strategies*

Most interviewees from firms saw the emerging trust infrastructure and services as extensions of their current lines of business. Some suggested that these developments were indicative of the evolution of an entirely new industry. However, according to one interviewee, the problem was that 'we can't define what the whole new industry is'. Linking trust services to current lines of business was seen as being necessary for responding to both regulatory and perceived social obligations. It was also seen as a means for overcoming the lack of education on the part of businesses and consumers with respect to electronic trading. One interviewee observed that:

To take a leadership role is important. Users don't know what they want. They need to be led. A firm cannot just jump into an uncertain, immature market with a product. They need to tie it in with current business. Later, they will try to extend market boundaries. Therefore, a firm needs to take the competency of what the firm already has and create around it. Differentiation in the traditional sense of the term will occur later on, in a growth period based on particular features. Differentiation occurs after a firm has gone through a learning process.

Most of the representatives of firms in the interview sample said that tying new services to existing products and/or services would help to frame the 'unknown' of trust service provision in a way that would establish general expectations for users. It was also a way of reinforcing a set of commonly accepted rules and standards of behaviour for all parties to an electronic exchange.

Conveying an appropriate level of trust in a certification system for market positioning was recognized as a major issue both for established TSP firms—that is, banks, post offices, and British Telecom—as well as for new entrants such as VeriSign, InterClear (UK), and Entrust (USA and UK). According to one interviewee, the ability to enter a high-risk, uncertain market 'all comes down to brand recognition'. In line with this view, another interviewee commented that the VeriSign model for market development 'demonstrated that acquiring market share was more important than security'.

Firms seeking to enter the trust services market have a number of options for how they manage the brand extension process in the electronic realm. Each option involves trade-offs between conveying an image of technological competence, associated with a firm's existing products and services, and of quality and the transfer of trust, and trusting relationships, associated with a brand name and a firm's reputation. For example, to position Canada Post in the marketplace, the interviewee observed that 'we don't pretend to have technical expertise, that's why we partnered with Cebra, which has a reputation for technical expertise with which we can combine our commerce and service orientation'.

VeriSign interviewees confirmed that they were seeking to acquire 'default' trust, or trust by association, with established brand names. Two examples of VeriSign partnerships include associations with the American Institute of Chartered Public Accountants (AICPA) and the Canadian Institute of Chartered Accountants (CICA), and with British Telecom in the United Kingdom. In the case of the AICPA/CICA partnership, VeriSign was offering a 'digital mark of trust' (digital seal of approval) with the provision of technologies, infrastructure, and business practices for trust services, while the AICPA was providing the authentication or authorization services.[15] The OFTEL representative also observed that none of the contenders in the United Kingdom was in a position to provide all the components necessary for trust services. While players may enter the market alone, they may also consider partnerships with other upstream players who would be better positioned to provide some components.

The scalability of trust services was also deemed to be an important market entry and development consideration. Interviewees commented that considerable innovation would be necessary to scale services up from 10,000, to 1 million, to 10 million

---

[15] For details of the WebTrust seal of assurance programme, see **www.aicpa.org/webtrust/index.htm** (accessed 8 July 2000).

users. In addition to enhancing a current customer base, all firms saw this as a major reason for the focus of trust services on the business-to-business electronic commerce markets. British Telecom was the exception. There was also a common perception among TSPs that, once the technical infrastructure for business transactions was in place, it would be transferred easily to the mass consumer market.

Participation in industry-wide cross-certification schemes at the national and international levels was also regarded as important for achieving market extension. The summary of these differing schemes presented below illustrates two key issues regarding certificate use and reliance. First, the characteristics of the different schemes appear to confirm the viewpoint of the interviewees to the effect that achieving interoperability was more of a business issue than a technical one. Secondly, the features of these schemes illustrate the need to make a distinction between a certificate as an enabler of electronic exchanges, and the level of permission or value that a certificate entails.

At the time of the interviews, the Universal Postal Union, a number of leading banking institutions through a pilot project, Identrus, and the British Chambers of Commerce (BCC) were all involved in industry-specific schemes that were using digital certificates as an identification for pre-transaction authentication. Pre-transaction authentication compares with the use of letters of credit (banks), certificates of origin (BCC), or guaranteed delivery (post offices) in the physical world. In this context, the certificate may authenticate transacting parties, but the relying party is required to perform additional procedures to fulfil the trading requirements of the transaction. Another proposed scheme, outlined by one of the interviewees, was Scheme Y. This scheme was being developed to harmonize 'certificate policies, trust, and underwrite transactions to enable full interoperability of certificate use'. According to one interviewee, unlike other schemes, certificates issued under Scheme Y were intended to have an implicit value, since they would, in theory, be able to be used and accepted in multiple exchange settings and would eliminate the need for additional business processes and practices to complete the exchange. In this regard, the value of the certificate is derived from underwriting the exchange process and providing the related guarantees for particular exchanges. Since Scheme Y allows certificates to be used in multiple exchange settings, it is similar to the common infrastructure model being proposed by national governments. However, this scheme was being proposed at an industry level. Therefore, certificate valuations were likely to incorporate industry-specific business practices regarding how issues of risk and liability would be assessed and managed by participant TSP firms. These differences are likely to result in multiple certificate frameworks as most interviewees for this study indicated that they were considering participation or had chosen to participate in one or more of the schemes being developed.

Box 5.1 compares the market entry strategies of the main contenders in the United Kingdom market for the provision of trust services at the time the interviews were undertaken. A comparison of these strategies suggests that the offerings made by TSPs for mediating certificate use and reliance between transacting parties are influenced by a combination of technical, social, and business factors. Developments in certification procedures and the use of digital technologies such as certificates suggest that there is a complicated set of interactions and changing relationships in the

Box 5.1. Trust service market entry strategies in the UK

Company	Entry strategy	Core focus	Stimulus	Market reach	Role of certification
British Telecom	Market	Logo/Seal of Trust	Intranets	Global	Coordination Role
The Post Office	Infrastructure	Accreditation	Transitional positioning to expand into new markets	National	National identity
Major 'Big 5' Consultancy	Infrastructure moving to market	Accreditation	Expand value add consultancy	Global	Risk management
Banks	Market	Leadership in cryptography and smart card market	Internal requirements, maintenance of consumer base	Global, local	Risk assessment (underwriting)
British Chambers of Commerce	Market	Seal of quality	Reinforce current lines of business	Local, Europe	Arbitration services

electronic marketplace. The role of the certification process is very important in mediating relationships between online transacting parties as well as for shaping how major contenders will extend into new roles as providers of trust services to meet their strategic objectives.

3. BUILDING TRUST: THE SOURCES

The notion of trust is increasingly being viewed as an important precondition for exploiting the opportunities of the electronic commerce marketplace. Although the notion of trust has received significant attention in the literature, it remains ill-defined and vague. For the purposes of this analysis, trust refers to the expectation that the future behaviour of another party (human, organization, or other) will result in an anticipated outcome when the trustor is unable to control the actions of others.[16] This definition offers a synthesis of important aspects of the definitions offered in the trust literature and provides a basic starting point for assessing the different sources from which trust expectations may be developed as a basis for interaction in the electronic realm.

Many national governments are seeking to endorse an open PKI trust model to establish a basis for trust and confidence in the electronic commerce marketplace. Underpinning this model is the assumption that TSPs will become the standard organizations through which online entities can identify one another for online exchanges

---

[16] See Baier (1995), Hardin (1993), Lewis and Weigert (1985), Mayer, David, and Schoorman (1995), Misztal (1996), Rotter (1967), and S. P. Shapiro (1987).

and that data contained in, or derived from, a digital certificate would serve as the foundation of trust for online exchanges (Feghhi, Feghhi, and Williams 1999; Ford and Baum 1997; Wilson 1999). In other words, trust is treated as a commodity in which user expectations are based on both the technical standards and the policies and practices for digital certificate issuance and maintenance offered by TSPs. TSPs, in turn, are influenced by the effectiveness of government legislation and regulations for mediating and governing the provisions of their services to ensure high levels of system performance or system trust (Johnson and Grayson 1998). This perspective is based on assumptions drawn from experience in computer security management, which suggest that if a TSP system is accredited then trust, or trustworthiness, associated with accreditation transfers to the TSP (Feghhi, Feghhi, and Williams 1999).

The assumptions underlying government-endorsed system trust are based on the development of the trusted systems concept in computer science (Schneider 1999; Stefik 1997, 1999).[17] According to the trusted systems concept, security is implemented by means of rules and techniques that are embedded in hardware and software. In other words, the protocols of trust are built into the architecture to optimize an entire system that will guide participants in their negotiations and meet user needs, as well as reduce the risks of market failure. More specifically, trust systems provide two capabilities to users: the legal recognition of a specified user right and the technical means to protect and control access to the distribution of digital goods and services. This view of trust in technology also appears to underpin proposed legislation for the legal recognition of electronic signatures. By legally recognizing electronic signatures as the equivalent of handwritten signatures, electronic signature legislation seeks to establish the presumption of validity of an electronic signature to be used for contractual liability and recourse. The presumption of validity is based on the design of the technology itself and its associated relationship to the entity that has been assigned a specified property right—for example, electronic signature (Biddle 1997; Winn 1998*b*).[18]

The results of the research reported in this chapter, however, suggest that TSPs do not perceive the presence of a trust infrastructure as a necessary precondition for establishing trust in their commercial structures and processes. Rather, TSPs appear to be relying on their existing sources of trust as the basis for their market entry and development strategies. To establish trust expectations for their user communities, TSPs seemed to be proposing the use of certificate polices and certification practice statements as mechanisms to define a set of rules, or terms and conditions, for the use of and reliance upon certificates for exchanges. TSPs perceived certification policies and CPS as procedures to establish the technical and commercial boundaries of acceptable exchanges as well as to assign liabilities to all the parties in the relationship, including the TSP. TSPs were then proposing to associate the use and reliance of digital certificates with the known brand names of current physical products and services.

---

[17] Trusted systems are also commonly referred to as digital rights management systems.

[18] Although electronic signature legislation addresses only electronic signatures, some opponents to the legislation argue that open PKI electronic signature legislation is an attempt to create an infrastructure that supports key recovery (key escrow in the USA) to facilitate government access to keys (Biddle 1997).

The bundling of these elements into a branded digital certificate package in accordance with legal, regulatory, and social norms that already exist in the physical marketplace can be viewed as an indirect means of communicating a specific level of quality of service—that is, by a process of developing sticky trust.[19] Sticky trust refers to the transfer of existing sources of trust and trusting relationships from one realm to another—that is, from the physical to the electronic realm. Sticky trust was also perceived by the interviewees as important for establishing the credibility of TSP offerings and for managing the expectations of users involved in trading relationships. Furthermore, sticky trust as a core component of TSP market entry strategies refers to processes that emphasize the use of a certificate as an enabler, or medium of exchange, with an intended purpose or value associated with it. Creating differentiation in this underlying value offers a way for TSPs to compete in the provision of their service offerings.

This interpretation provides only a partial answer to questions about how digital certificates may be used to mediate secure electronic exchanges conducted using open networks. Although practices that encourage the transfer of sticky trust may be one way of overcoming the barriers confronted in making the transition from the physical to the electronic realm, it also assumes the existence of norms and governance regimes as well as the existence of prior relationships—for example, through conventional business-to-business commerce or as a result of an existing client base. As Zucker (1986) argues, social norms and legal and regulatory rules are capable of encouraging trusting behaviour in social interactions only once trust exists. Ring and Van de Ven (1994) and Luhmann (1988) take this argument further and suggest that the existence of established relationships and norms might even mitigate the need for trust.

The concept of sticky trust, therefore, does not appear to shed much light on whether trust is, or can be, created by using existing sources of trust to facilitate direct exchanges or stranger-to-stranger electronic commerce. Research on relationship marketing and exchange theory is useful for gaining insight into the processes that are likely to mediate the interaction between online exchange parties. This research offers suggestions as to how trust can be used to encourage interaction in the electronic marketplace to facilitate the transfer and creation of value in exchange relations.

3.1. *Trust as an Interactive Process*

Trust relationships are deemed to be very important for electronic exchanges because of the potential use of electronic commerce services for improving the exchange of information between transacting parties as well as facilitating the exploration of new mutually beneficial arrangements and user interactions (Hart and Saunders 1997; Kim and

---

[19] The concept of sticky trust used in this chapter draws on three bodies of literature. Specifically, the economics literature on sticky wages and sticky prices (Bewley 2000), the concept of sticky information in the technology and innovation literature (von Hippel 1994, 1998), and stickiness in customer relationships as described in the marketing literature (Nemzow 1999). In each of these contexts sticky refers to situations where there have been changes in the market but the *status quo* is maintained because of the opportunity costs associated with adapting to change.

Mauborgne 1998; Muller 1996). An electronic market can be defined as: 'a co-ordination mechanism for the market exchange of goods and services and represents the total—or a certain quantity—of the exchange relationships between potential market partners having equal rights. The interaction processes between market participants are thereby supported by electronic market services' (Schmid and Lindemann 1998: 2). In the context of the trust services marketplace, digital certificates can be viewed as an electronic market service that supports interaction processes by mediating the relationships between partners in exchanges. But how can trust be established in order to fulfil the expectations of users and create a balance in technology-mediated relationships? This is an important question, because, in the context of open networks, TSPs are expected to mediate direct exchanges thereby altering the way that fair or equitable exchanges may be negotiated.

The nature of interactive processes that achieve equitable or fair mutual exchanges between actors is a central issue for social exchange and relationship marketing theories.[20] Both exchange and relationship theories posit that cooperative relations emerge when there is a voluntary transfer of value between exchanging parties.[21] The process of reciprocity is seen as a social interaction that influences the ways in which parties negotiate, commit to, and execute an exchange, as well as the degree to which they judge it to be equitable.[22] In the absence of third-party enforcement, parties to an exchange may be willing to rely on trust as a basis for reciprocity for fair dealing in environments characterized by uncertainty and complexity.[23] In other words, trust is an interactive process.

Theoretically, reciprocity based on trust may be one way of mediating user interactions but it does not answer the question of how parties to an exchange may come together to negotiate an exchange. Rengger (1997) presents a process-oriented view of evolving institutional frameworks for governing interactions for cooperation and compliance between states that is helpful for conceptualizing how transacting parties are drawn together. Rengger suggests that habitual expectations or the presumption of trust for cooperation and compliance can be achieved on the basis of authoritative practices that establish a practical association between interacting parties. As an example, Rengger (1997) proposes that the introduction of the practice of international shaming may be one way to ensure that individuals, states, or corporations keep to the agreements that they make. An important feature of this practice is that it needs to be established, monitored, and organized through recognized public forums in order to establish legitimacy. The purpose of the practice is to serve not only as a mechanism for compliance, but also as a way to institutionalize patterns of interactions so that they become habitual—that is, so that they support the presumption of trust. This idea is based on the observation that to strengthen and extend trusting relationships requires

---

[20] The notion of fair as it is used here is consistent with the view of social exchange theorists—i.e. that all parties receive benefits proportional to their investments (Homans 1961) and not according to the economic notion of equivalence of benefits.

[21] See Houston and Gassenheimer (1987), Kim and Mauborgne (1998), Kollock (1994), and Ramirez (1999).

[22] See Kim and Mauborgne (1998), Nooteboom (1996), Ramirez (1999), and Ring and Van de Ven (1994).

[23] See Bradach and Eccles (1989), Kollock (1999), Macneil (1980), and Rengger (1997).

practices such as 'handshakes, and "the reliance on detached third parties as intermediaries" which are reciprocal and occur between those who have arranged, if need be contrived, some rough equality of power and vulnerability' (Baier 1995: 116).

In the trust services market, the practice of certification may be understood as being central to both the role of the TSP and a means for facilitating an electronic handshake between buyers and sellers. Interviewees from TSPs were proposing to use certification policies and their associated levels of assurance as the means of creating a common value system between transacting parties. The use of certificates as a medium of exchange was intended to facilitate the transfer and/or creation of value between exchanging parties and enable users to negotiate and commit to an exchange with an electronic handshake.

For existing and potential entrants into the trust services marketplace, legitimizing and attaining credibility for the certification process in the public realm are central to the TSP's perception of its role. Most of the interviewees from TSPs thought the biggest barrier to entry into the trust services marketplace was communicating the characteristics of both new and existing internal certification practices so that these could serve as the basis for evaluating the quality and value of the levels of assurances that they sought to support. The establishment of authoritative certification practices by TSPs for the intended use of and reliance upon certificates can be regarded as a means of establishing the presumption of trust in online exchanges. Trust in this case is not based solely on the TSP's reputation but is based also on the stability of the guarantees provided by the TSP.

This analysis of the process of certification sheds new light on how firms are seeking to create trust in exchange relationships and to establish practices, such as electronic handshakes, that will encourage potentially greater degrees of interaction and participation in online exchanges. Moreover, since the certification process requires the voluntary participation of users, it appears not only to resolve many of the problems entailed in authentication of user identities, but also to allow users to establish trust or reputation profiles that can be legitimized through negotiated exchanges with other exchanging parties. This insight is consistent with the premises of theories about trust and exchange relations and the role of money.[24] These theories suggest that, when value systems are reinforced through use, they may result in a greater level of interactivity among users as well as a strengthening of the trust that exists between potential users.

Since certification and cross-certification are core processes for establishing trust in the physical realm, they need to be reconsidered in the context of market-making strategies of the TSPs. At the time of the research on which this chapter is based, TSPs were seeking to make certification, embedded in certification policies and CPS, the key means of differentiating their services in the marketplace. However, these mechanisms may also create barriers for cross-certification and interoperability. This is illustrated by the evidence provided by interviewees, which supports the use of industry-wide cross-certification schemes as a basis for scalability. Implicit in industry-based proposals for cross-certification schemes are specific value systems for coordinating certificate

[24] See e.g. Hirschman (1984), Ramirez (1999), and Reagle (1996).

use. While these arrangements may lead to an increase in the number of users, they do not necessarily facilitate innovative forms of interaction or mutually beneficial arrangements for electronic commerce transactions. Moreover, although the representatives of TSPs thought that they would be able to transfer processes from the business to business or the existing customer base, there is little or no evidence to suggest that this will be a straightforward process.

Most of the TSPs were seeking to segment the marketplace according to specific certificate policy classifications. As one interviewee pointed out, there is no effort to extend the use of the certificate. This is important when considered in the context of the electronic exchange process. It points to a gap between the establishment of certification practices for mediating online exchanges in existing communities of interest and in the way these processes and practices may be used to support direct interactions that are carried out electronically. To fill this gap, new processes may be required to integrate the use of certification practices so that they provide an improved means of mediating direct exchanges in the electronic marketplace. In failing to address this gap, the basis for competition among trust service providers may be limited to how issues of liability, risk sharing, insurance, guarantees, and regulatory arrangements are defined.

4. CONCLUSION

> The problem is not trust . . . the problem is how he will implement what has been agreed upon.
>
> (Arafat 1997)

In the physical realm, a party to an exchange generally is able to use a number of social and personal cues that are present in the environment to assess the likely trustworthiness and reliability of a potential trading partner. In the virtual environment, the deployment of PKIs supported by TSPs and digital certificates is intended to provide a trading party with a measure of trust and confidence in online commercial exchanges. In this context, the role of the TSP is understood as being that of constructing trust in the electronic marketplace as well as encouraging cooperation and mediating direct exchanges between trading parties through the use of digital certificates.

Based on the analysis of the results of a set of in-depth interviews with firms and other organizations that have expressed interest in the development of the trust services marketplace, the evidence suggests that TSPs initially intend to rely upon a variety of existing sources of trust—that is, the transfer of sticky trust—to facilitate market entry. For example, most TSPs are proposing to associate the use of and reliance upon digital certificates with the known brand names of physical products and services that exist in the conventional marketplace. The bundling of these sources of trust into a branded package appears to be important for creating user confidence and as an educational tool for attracting users to participate in electronic exchanges in unknown and uncertain environments.

The evidence also suggests that certification is the core process that will be used by TSPs to manage user expectations about the safety of online exchanges. For TSPs, legitimizing the practice of certification as a common value system was regarded as important for creating the presumption of trust for trading parties. In other words, trust was treated as an interactive process that can be employed to encourage the voluntary cooperation and compliance of trading parties in exchange relations. In contrast to the PKI models that are being endorsed by some governments, or proponents of techniques used to achieve the security of communications networks and computer systems, both sticky trust and the central role of the certification process emphasize establishment of trust as a result of the exchange process itself rather than on the basis of the features that are embedded in technical systems. The analysis also confirms that TSPs tend to perceive digital certificates as offering a new market opportunity and not as a precondition for the security of electronic commerce.

Based on this analysis of the proposed market-making strategies of TSPs, it appears that certification procedures are enabling buyers and sellers who engage in electronic transactions to create reputation or trust profiles in the form of branded certificates. The use of these certificates, in turn, allows users to communicate their intentions and, more importantly, to negotiate an exchange with an electronic handshake. In this sense, then, a digital certificate not only governs the conduct of potential exchanges but also structures and organizes the relationships between the participants. This is a significant finding because it leads to the suggestion that digital certificates and the process of certification are likely to play a fundamental role in shaping the types of relationships and, subsequently, the commercial structures and processes that emerge in the electronic commerce marketplace.

Different perspectives on the best way of implementing the certification process are emerging as the need for TSPs grows alongside the expansion of electronic commerce. Initiatives proposed by some national governments may lock users into a solution that is inconsistent with the ways in which trusting relationships are constructed. Alternatively, competing industry models may result in uncertainty and confusion for users and a proliferation of relatively closed electronic commerce trading environments despite the open characteristics of the networks that support electronic trading. A gap between the certification practices that are being developed by some TSPs and the need for new processes and practices to support the development of new trusting relationships in open environments has been identified. The evidence suggests that the trust required to facilitate the uptake of electronic commerce is multidimensional. It is dependent upon a variety of mediating processes, including the presence of a trusted intermediary. Failure to understand the processes that are mediated by TSPs and that underlie the emergence of electronic relationships has implications for the types of interactions and relationships that will emerge as electronic commerce becomes an increasingly common mode of business conduct. The improved understanding of trust offered in this analysis suggests that investments in PKI systems may be questionable as they are inconsistent with the processes that appear to give rise to trusting relationships in electronic environments. Furthermore, such investment in the early stages of trust service market development may suppress innovation in technical, legal, business, or regulatory environments. The consequences may be significant difficulties for electronic traders who seek to attract and maintain a critical mass of users.

6

The Colleague in the Machine: Electronic Commerce in the London Insurance Market

1. INTRODUCTION

Electronic commerce is an attempt to virtualize the economic activities of firms and markets—that is, to relocate them from the physical world of shops and offices into computer-mediated environments. This contemporary definition from the OECD offers an indication of the scope of electronic commerce and the aspirations of the players.

Electronic commerce refers generally to all forms of commercial transactions involving both organizations and individuals that are based upon the electronic processing and transmission of data, including text, sound, and visual images. It also refers to the effects that the electronic exchange of commercial information may have on the institutions and processes that support and govern commercial activities. These include organizational management, commercial negotiations and contracts, legal and regulatory frameworks, financial settlement arrangements, and taxation, among many others (OECD 1997: 20). Some observers predict profound and wide-ranging social effects from what is variously called the information economy (Castells 1996; Shapiro and Leone 1999), the digital economy (Margherio *et al.* 1998), the weightless economy (Quah 1997), or simply the 'new economy' (Kelly 1998). Frances Cairncross (1998) uses the phrase the 'death of distance' to express the idea of fundamental and pervasive changes in the way that individuals and social groups relate to one another. Manuel Castells (1996: 1) writes of a 'restructuring of capitalism', which is causing corporations to change their organizational model from vertical and bureaucratic forms to horizontal, networked models that focus on customer satisfaction and emphasize strategic alliances with partners. According to the OECD (1997: 24), 'there are strong current indications that massive changes have already begun to occur across the entire business spectrum'. This chapter examines the development of electronic commerce in the insurance industry between 1987 and 1999, and specifically in Lloyd's of London, and efforts to develop the London Insurance Market Network, or LIMNET.

The insurance industry already possesses all the social and economic characteristics of the 'new economy'. First, it is weightless, as the object of its trades (transfer of risk) is entirely intangible and there are no goods to deliver, only documents and money. Secondly, it has been globally organized for centuries, thanks to its origins in ocean-going commerce (Raynes 1964). Thirdly, it relies on social networks: large commercial contracts customarily involve a dozen or more underwriting firms. Finally, it thrives on information and information processing—that is, insurance companies were among the earliest commercial users of computer technology and provided advice and investment to the nascent computer industry (Bashe *et al.* 1986; Yates 1997).

Much recent writing, particularly since the late 1990s, about electronic commerce focuses on the ability of the Internet to transform all aspects of computer use. However, reports from the US Department of Commerce and the OECD acknowledge that larger firms have been using value added networks (VANs) for a number of years (Margherio *et al.* 1998; OECD 1999). These VANs have used proprietary technologies that are relatively expensive to implement, most notably, for example, in the form of electronic data interchange (EDI).[1] Internet-based electronic commerce is expected to build on this trend and accelerate as a result of Internet technologies being cheaper and based on open standards.

Insurance firms have been taking steps towards electronic commerce since the 1980s. LIMNET was a VAN that was created in 1987. It supported EDI for transfer of structured data and email for unstructured exchanges. However, the growth of electronic commerce over LIMNET was more gradual and more difficult than its member firms anticipated. There were notable successes in so-called back office processing such as accounting and settlement and the advice of claims. However, electronic commerce has not radically changed the way that business is conducted in London. At the beginning of the nineteenth century, most contracts were still being negotiated face to face between the representatives of firms. The city streets continued to be filled with brokers walking from meeting to meeting, carrying leather satchels heavy with paper. One of the major investments of the 1990s was the creation of the London Underwriting Centre, a building whose sole purpose is to provide a venue for meetings between brokers and underwriters.

The London Insurance Market's enduring preference for face to face negotiation contradicts the predictions of much writing on the new economy.[2] It also confounds the insurance firms' own expectations about the possible and supposedly inevitable consequences of the introduction of information and communication technologies (ICTs). This chapter examines some of the social and technical factors that have mediated the development of electronic commerce in the London Insurance Market and establishes why brokers and underwriters still choose to meet face to face, rather than to negotiate with their colleague 'in the machine'.

---

[1] EDI is defined as the direct transfer of information from one computer to another without interpretation or transcription by people. It relies on the information being structured according to rules and standards that must be agreed by the participants involved (DTI 1989).

[2] This chapter follows the habit of people who work in and around the London Insurance Market in referring to the market as a single entity—that is, 'the market does this', 'the market doesn't like that'. There is no intention to imply that 400 firms and 50,000 employees ever do act as a homogenous body with a single opinion.

2. ELECTRONIC COMMERCE IN INSURANCE

2.1. *Lloyd's and the London Insurance Market*

The London Insurance Market is the world's leading market for internationally traded insurance and reinsurance risks.[3] As its name suggests, it is based in the City of London, but it is distinct from the UK domestic insurance industry in respect of its membership and the type of business that is conducted. The London Market is a wholesale market devoted to large-scale risks and high-exposure risks. Life insurance and personal insurance policies in general are not transacted in this market unless the persons involved are exceptionally wealthy or have unusual requirements.[4]

In 1997 the total premium income capacity of the London Insurance Market was £14 billion (see Table 6.1). Around 40 per cent of this came from Lloyd's, with the balance coming from limited companies. Business is introduced to the market by around 200 insurance brokers. Lloyd's of London is a self-regulating society whose capital is provided by investors (known as 'Names') organized into syndicates. The syndicates are managed by Managing Agencies, who employ the insurance underwriters and other staff required to operate the business of the syndicates. Until quite recently, all of Lloyd's investors were private individuals who accepted unlimited liability for risks underwritten on their account. Corporate membership (with limited liability) began in 1994. By 1998, there were 435 corporate members providing around 60 per cent of Lloyd's premium income capacity. The remaining 40 per cent was subscribed by 6,818 individual members.[5]

Table 6.1. *Size of London insurance market by income and number of firms*

Type of firm	Number	Premium income[a] (£m)
Lloyd's syndicates	132	5,853
Insurance companies	133	7,387
Marine P&I Clubs[b]	39	772
Total	304	14,012

[a] The amounts shown are the gross of brokers' commissions.
[b] Protection and Indemnity Clubs are mutual associations, usually of shipowners, who indemnify one another.

Source: adapted from Carter and Falush (1998).

---

[3] This definition comes from Carter and Falush (1998).
[4] For example, some astronauts on the MIR space station have held personal accident insurance underwritten at Lloyd's. Among older underwriters, the phrase 'Betty Grable's Legs' is still used as a generic term for unusual personal insurance policies.
[5] Figures from Corporation of Lloyd's (1998) and Carter and Falush (1998).

Lloyd's no longer dominates the world of insurance to the extent that it did in the 1950s or even in 1900, but it continues to be important because of its financial size and the expertise of its underwriters. Lloyd's underwriters have a reputation for being able to insure anything from factories, fleets of aircraft, nuclear reactors, satellites, and thoroughbred racehorses, to the taste buds of the renowned food critic Egon Ronay, to give just a few examples. Collectively, Lloyd's has risk management expertise that no other institution or regional market has been able to appropriate. It is still regarded as the 'market of last resort'. Veteran underwriter Dick Hazell cites the apocryphal wisdom that in Lloyd's 'there is no such thing as a bad risk—only a bad rate' (Hazell 1996: 3).

The complexity and diversity of risks underwritten in the London Insurance Market have made the management of information a major challenge for underwriters and brokers. Many contracts have ten or twenty counterparties and risks shared between 100 underwriting firms are not unknown. This requires a tremendous effort on the part of the market to coordinate information. The information flows involve corporate purchasers of insurance and their brokers in a near-continuous cycle of discussion of requirements and preparation of statistics.

2.2. *LIMNET and the View from 1988*

LIMNET, the London Insurance Market Network, came into existence in May 1987 with the award of a contract to IBM to be the network provider. It was a joint venture involving the four main representative bodies in the market—that is, Lloyd's, the Institute of London Underwriters (ILU) and the Policy Signing and Accounting Centre (PSAC) representing the insurance companies, and the Lloyd's Insurance Brokers Committee (LIBC) representing the brokers.

The first EDI service to be delivered over LIMNET was the PSAC Claims system, which took around eighteen months to develop. The first day of trading was reported in the *Financial Times*:

Monday, November 7, was a day to remember in the London insurance market. For the first time, some insurance brokers were able to do away with the bulging files of paper which normally carry information about insurance claims. Instead, they were able to use computerized electronic data interchange to send claims to 130 insurance and reinsurance companies. It was . . . one of the most tangible signs of the momentum that has now gathered behind 'LIMNET', the London Insurance Market Network, which was launched in May 1987. . . . The eventual aim is to cut costs, speed up services and bolster the London market's competitiveness by placing risks, and processing claims electronically.   (Bunker 1988: 15)

The London Insurance Market's decision to invest in electronic commerce was motivated by a widespread belief that firms could not remain competitive without it. Competitive in this context meant retaining the lucrative and prestigious international business that might otherwise be lost to insurers in North America and mainland Europe.

Insurance rates are cyclical, and, in a 'soft market' when premium rates are low, insurance companies find it difficult to make a profit unless they can reduce their costs in proportion. London in the 1980s was one of the most expensive cities in the world in which to do business. In 1987, the London Insurance Market employed 50,600 people. Of these, 40,600 worked in offices in the 'Square Mile' of the City of London, and probably 30,000 of them in postal district EC3 alone.[6]

Financial deregulation in the securities industry, the 'Big Bang' of 1986, had caused an increase in salaries and rents in London. Deregulation also showed London firms the forces of competition in action as many British stockbrokers were taken over by foreign banks, which reorganized their operations and introduced the latest technologies. Dennis Mahoney, of the international reinsurance brokers Alexander Howden, gave a speech in October 1988 in which he took as his theme 'The Cost of Doing Business': 'In my own company, operating as we do in 73 countries and 230 cities around the world, we see it very clearly. The cost of reasonable office space in London is second only to Tokyo. It has risen dramatically in the past ten years and has yet to show any real signs of slowing down' (Mahoney 1988: 1).

The main focus of Mahoney's speech was not on exogenous costs, but on what he called 'our own self-imposed costs of doing business'. He pointed out that there was little benefit in brokering and underwriting firms attempting to pass costs onto each another when all of them had to make a living from the premium paid by customers. The key to ensuring the long-term profitability of the market was to reform inefficient business practices based on face-to-face trading. He argued that

Most London brokers place a great deal of business in overseas markets. Does anyone here believe that when placing business with markets in Paris, Oslo, New York or Tokyo that the broker goes to see each underwriter with each risk? Is anyone surprised that most brokers utilize telex and fax machines to access more than two or three hundred different underwriters in one day! (Mahoney 1988: 3)

Mahoney, in effect, was saying to his firm's trading partners that London brokers were already using new technologies more efficiently than competitors in other countries.

Most, if not all, of the top ten Lloyd's brokers already have electronic links with their offices around the world. So is it not strange that there are probably more electronic links outside this market than in it? Clearly an underwriter not located in the most expensive location in the world can afford to be more price competitive than one who is. (Mahoney 1988: 3)

Mahoney's comments were echoed by the chairman of an insurance company at another conference. Philip Marcell of the Continental Reinsurance Corporation was speaking on 'What the Network will Mean for the Insurer'.

Practically the only way to cope with these pressures [of rising costs] is to reduce the numbers of people—and reducing these numbers while still maintaining a proper service . . . The obvious way to cope with reducing numbers and maintaining service is the use of more mechanization and, moreover, mechanization not just within a company but in its external relations. (Marcell 1988: 2)

---

[6] The Square Mile is the colloquial name given to that part of London administered by the Corporation of London in which insurance, banking, and commerce are historically located. Although not square, it is almost exactly one square mile in area.

Marcell was quite explicit about the need to replace face-to-face trading.

Networking is concerned with connecting internal systems, so that we can talk to each other electronically without the intermediary of paper, and without face to face contact which is expensive in time and money.... The fact is that the old pattern of claims brokers walking around the market with cases full of advices must be archaic and non-efficient. (Marcell 1988: 4)

It seems that the firms who invested in LIMNET had a clear expectation of the benefits they stood to gain from electronic commerce and clear incentives to make it succeed. The next section examines the difficulties of designing the electronic commerce systems that were intended to replace face-to-face trading.

3. DIFFICULTIES OF SOFTWARE DESIGN

Many accounts of electronic commerce emphasize how much of it takes place—that is, how rapidly it is growing, how many firms are implementing it, and how much money they are spending. It is relatively rare for someone to admit that we still do not know what the statistics about electronic commerce really mean. US Secretary of Commerce William M. Daley did exactly that in May 1999 at a conference on 'Understanding the Digital Economy'. 'More and more people in the private sector are making what amounts to billions of dollars in decisions about e-commerce. And they are doing it without a reliable base of information. We want to better measure what e-commerce means and doesn't mean to the economy—so decisions are not made in a vacuum' (Daley 1999: 1).

One strand of work presented at that conference focused on the implementation of electronic commerce and the effects it has on organizational structure and business models (Kling 1993; Kling and Lamb 1999; Orlikowski 1999).[7] The speakers drew on data from thirty years of empirically grounded research on information technology and organizational change to show that implementing technology and learning to use it is a challenging social process. Just because we know that the technologies are affordable and firms are investing in them, we cannot infer that the predicted benefits are being realized. Organizations are 'imperfect implementers of business strategies . . . especially those that depend in critical ways upon complex information technologies' (Kling 1993: 1). No matter how appealing a strategy may be, and how clearly understood the incentives for implementing it, it takes a lot of work to get the systems 'up and running' (Kling 1993: 1).

The study of firms' experiences when they try to implement new strategies reveals a qualitative story that cannot be seen in the aggregate statistics. It shows us what Orlikowski (1999) calls an 'enacted' view of technological change. In the enacted view, the digital economy is not something 'out there' like a technological meteorite heading towards the world of business.[8] It is 'in here' within organizations and is being enacted,

---

[7] This field of information systems research is known as Organizational Informatics. The term was coined by Rob Kling (1993).

[8] The terms 'enacted' and 'out there' come from Orlikowski (1999).

or worked out, by real people in real time, who are discussing their options and making decisions. Nor is the outcome of technological change inevitable, however much it might look that way from the statistics or the sales literature.

Viewed in the light of the organizational informatics research tradition, the story of LIMNET is a story about the social organization of the London Insurance Market. The development of electronic commerce in London has been motivated by the extent to which people believed that something huge and inevitable was happening in their industry. However, it has also been motivated by people's determination to take responsibility for what was happening and their attempts to shape it to their own ends. In particular, it is a story about the social and political organization of information technology projects and the process of articulation through which software designers establish what is important in a system and what is allowed to count as a solution to a problem.

Software development methodologies are not, on the whole, well equipped to deal with the social and political settings in which they are used. A typical response of software developers is to ignore them. Studies of software design practice have highlighted the inadequate response and indifference of many engineers to the social dimensions of their work. For example:

Engineering has sought to institute a form of reasoning that is objective because it is external; the rationale behind a technical design can be laid out on paper and argued through in a public way, at least within the community of engineers and their expertise. This reasoning is instrumental; starting with a problem to be solved, it does not question the problem but simply seeks the demonstrably most efficient means of solving it. Its claims to social authority lie not in the choice of problems but in the methods for their solution.   (Agre and Schuler 1997: 4)

The methods employed on information technology projects in the insurance industry in the 1980s and 1990s were invariably from the 'structured-techniques' school of methodologies, such as SSADM, JSD, or STRADIS.[9] Structured techniques placed great emphasis on separating the machine-specific elements of a computer system from the description of the functions it was intended to perform. Designers schooled in these methodologies were taught to analyse by abstraction in order to produce designs that could be implemented on many different machines and that would satisfy the requirements of many different users. They were also encouraged to abstract from the details of the social setting in which the work took place. The specification of functional requirements was often called 'logical design'.

Quintas has highlighted the inherent weakness of such an approach. This approach to systems design fails to recognize that the end product of their work is 'not so much an object of representation of "true" user needs, as the result of particular social relations and the exercise of professional power' (Markus and Bjorn-Andersen 1987,

---

[9] The abbreviations respectively stand for Structured Systems Analysis and Design Methodology (SSADM), Jackson Structured Design (JSD), and STRuctured Analysis and Design of Information Systems (STRADIS). The need for a structured approach to design was originally proposed by Stevens, Myers, and Constantine (1974) and developed by Gane and Sarson (1979), Jackson (1975), Myers (1975a,b, 1976), and Yourdon and Constantine (1978), among others.

quoted in Quintas 1996: 95). If the designers feel that it is not their job to comment on or analyse the social relations surrounding the work, how are those judgements made? Are they erased from the design entirely or are they introduced covertly by other means? Quintas (1996: 95) suggests that, 'Those paying the bill may impose their own views on the design process, or impose financial constraints which conflict with the needs of the "real" system users.'

The following extract from a contemporary management handbook shows very clearly how the problem was approached in the 1980s. Senior management, the people who pay the bills, are also expected to represent the organization's culture and value systems as part of the planning process.

The first requirement for a systems planning project is to understand the value system, the culture, and the strategies of the organization . . . how easy this understanding is depends very much on how explicitly top management has stated its strategy. . . . The team should not start the requirements analysis unless there is a clear statement of the scope and objectives of the project . . . Ideally this step will tell the systems developers with some precision how much they can question existing rules or habits so that the team will know to what degree of abstraction the requirements should be carried.   (Flaaten *et al.* 1989: 121, 186–7)

When a methodology is employed that treats cultural analysis as one of the outputs of strategic planning, it will be very difficult to establish an empirically grounded view of the social context of that work. Even acknowledging the rights of management to speak for the whole of the firm, strategic planning is forward looking, abstract, and high level. In the London Insurance Market, it was informed by beliefs about electronic commerce as an irresistible, irreversible force for change.[10]

The need to consider an enacted view of electronic commerce was not formally recognized by the project teams or their sponsors. As a result, there was no mechanism for the community to articulate what was good or bad about the traditional face-to-face method of conducting business in London, other than to say that it was expensive and archaic. Questions about the characteristics a computer system might need to possess in order to replace face-to-face trading were pushed to one side. However, as we shall see, these questions had a way of forcing themselves back into the picture. To illustrate these difficulties, the next section considers the case of a meeting between a senior manager in a Lloyd's managing agency and a representative of a software design team.[11]

4 'WE DON'T WANT A MARINE SYSTEM'

The meeting took place in a large project to develop a general-purpose underwriting system that would be capable of recording the many thousands of insurance contracts

---

[10] Chelmsford (1992) is a detailed account of the creation of LIMNET, written by its first Chief Executive. Barrett (1999) and Barrett and Walsham (1999) discuss the cultural assumptions of different participants in LIMNET.

[11] The representative of the software design team is the author of this chapter. From 1987 to 1989 he was Business Systems Manager with Sturge Holdings plc, which at that time was the largest managing agency in Lloyd's of London, managing £1 billion of premium income a year.

underwritten by all the syndicates under the control of one managing agency.[12] The design team had spent over a year modelling business processes and the use of data, using a methodology that encouraged them to abstract the essential features of any insurance contract no matter what kind of risk it covered or where in the world it originated.

The objective of the meeting was to review the design of a screen for recording and enquiring into information about insurance contracts. The intended value of the 'Contract Details Enquiry' screen to the user community was that it should contain on one screen, or, if printed out, on one side of a sheet of paper, all the information needed for a qualified person to talk knowledgeably about the contract to a colleague elsewhere within the managing agency or in the market. There was a lot of information on the screen and there had been long discussions about which elements of data were regarded as essential or inessential for different groups of users.

The user representative was the Head of Non-Marine Claims (HNMC). The terms 'Non-Marine' and 'Marine' require some explanation, since they have a special meaning in Lloyd's of London and a political history attached to them. The insurance of vehicles and goods in transit is generally known in the insurance industry as Marine, Aviation, and Transport (MAT). In the nineteenth century, before planes, trains, and automobiles, MAT was simply marine insurance. London was the centre of the British Empire and Lloyd's was the world's leading marine insurance market. The last quarter of the nineteenth century saw considerable innovation at Lloyd's, including the first insurance policies covering burglary, hurricanes, and earthquakes, loss of profits following a fire, and employer's liability. The more conservative members of the Lloyd's community looked with distaste upon these 'non-marine' insurances,[13] and in 1885 persuaded the Committee of Lloyd's that the Lloyd's Stamp, a guarantee of solvency, should be affixed only to policies insuring 'vessels, cargoes, and freights'. The battle between the two camps continued for decades until it was settled in favour of the innovators by two Acts of Parliament—the Assurance Companies Act of 1910 and the Lloyd's Act of 1911. Throughout the twentieth century, marine, non-marine, and aviation underwriting at Lloyd's developed on separate trajectories with different professional societies, different business practices, and different traditions.

Claims handling is an important test of an underwriting system. The insured expects customer service and prompt settlements of any amount payable. The underwriter expects the claims department to check that the claim is valid before paying it and to update the loss history for the next year's statistics. The accountant, who has booked the premium, wants to see a net profit on the books after paying the claim and deducting the operating expenses. The system has to handle claims in a way that keeps all these people happy.

A meeting was organized to seek agreement about the design and implementation of the information system. Presenting and defending the design of the 'Contract

[12] Syndicates are the operating units of Lloyd's of London. A syndicate is a collective of many investors and a Lloyd's underwriter accepts a premium on behalf of a syndicate. A managing agency is a limited company that manages the affairs of syndicates. The managing agency employs staff, rents office space, buys computers, and so on.

[13] For readers interested in this story, see Wright and Fayle (1988: ch. 23).

Details Enquiry' screen was the Business Systems Manager (BSM), an information technology professional responsible for liaison between the user community and the software design team.[14] The BSM arrived at the meeting confident that at last the team had produced a design that would meet with approval. The discussion proceeded as follows.

BSM. I think we've managed to fit it all in.
(*HNMC surveys the screen.*)
HNMC (*pointing*). 'What's this?'
BSM. 'That says Hull or Cargo. It's for marine risks. . . . (*He begins to explain.*)
HNMC (*interrupting*). We don't want a marine system!
BSM. Your people won't use it. The mariners will use it. On non-marine contracts that field will be protected based on the syndicate. (*He points to the syndicate number on screen.*)
HNMC. Well, what's it going to do with the claims? We don't want any of our claims going to NN (*he names his opposite number, the Head of Marine Claims*).
(*The BSM, still confident at this point, begins to explain the processing rules the system will use to process claims and route each syndicate's claims advice messages to the appropriate node of the network. He is interrupted.*)
HNMC. And how's it going to calculate the premiums?

At this point the BSM's confidence began to waver. He had assumed that the question about claims was designed to test his understanding of the business rules, but the question about the premium threw him. The rule for calculating a premium was a simple piece of arithmetic that is, take an amount (visible on the screen) and multiply it by a percentage that was also visible on screen. Was the Claims Manager really not convinced of the system's ability to do that? The meeting broke up a short while later.

To the software designers, the generic 'Contract Details Enquiry' screen was an attempt at interpreting, simplifying, and codifying a business practice that succeeded. However, to many in the user community, it was a failure. The design did not meet with the users' approval and the software was implemented only after major reworking and the introduction of separate screens for different classes of business.

The design team had little reason to doubt that a generic insurance underwriting system was the best approach to take. There was a clear imperative to reduce costs and 'rationalize' the business. Their profession had trained them in methodologies for constructing generic models of data and processes by successively refining and abstracting the cases presented to them by the user community. Cases were classified as 'typical' and 'atypical', and requirements were graded into 'must-dos', 'nice-to-haves', and 'rinky-dinks'. When it came to designing a generic 'Contracts Details' screen, they found that the widely diverse practices of the syndicates could all be reconciled with their abstract models as long as each user was prepared to sacrifice some 'nice-to-have' requirements and the 'rinky-dinks' were banished. They designed the software accordingly and the BSM set about negotiating its acceptance.

---

[14] The BSM, the information technology professional referred to here, is the author of this chapter and the account given of the discussion is the author's account of that discussion.

The BSM thought he was acting as a spokesman for the non-marine users and did not expect the screen design to be so firmly rejected. The elements on the screen had been included or excluded based on the strength of feeling among users. 'Must-do' priorities had been respected and he felt the design was a good compromise. He knew that it was not a marine system. It was generic and he had heard the marine and aviation users say that it looked like a non-marine system to them, which had convinced him even more.

Electronic commerce places heavy emphasis on standardization to drive down costs. This was true in the EDI era and remains so in Internet commerce.[15] When they are applied in conjunction with the 'structured-techniques' school of software development, standards lead to generic computer systems and generic business processes. The people who perform those processes are expected to adapt their work to a logical model of the business that has been constructed by software designers based on senior management's statement of its strategic objectives.

The applicability of this approach to electronic placing has been the cause of much anxiety in the London Insurance Market, as quotations from interviews and speeches given by senior market figures reveal. First, Roger Townsend, a director of a major brokering firm: 'Client–broker relationships are not easily standardized: They are not currently well structured and are best dealt with by individual firms on a case-by-case basis. The broker–underwriter placing dialogue is clearly important but is surrounded by tradition and highly sensitive professional activities' (Townsend 1988: 3).

Robert MacKenzie, a director of an underwriting agency and a practising underwriter, was one of the most progressive voices in his advocacy of electronic trading. He recognized, however, the potential of technology to shift power away from the underwriters and place it in the hands of other professions: accountants, actuaries, and technologists.

As soon as you get people in management positions providing information systems, they are going to use their systems to encroach on the way underwriters do things. Lloyd's strength has traditionally come from the fact that the people who are making the important underwriting decisions are the same people who are running the business. They don't answer to anybody. So if something is not within the normal policy guidelines, but is particularly attractive, they can write it.   (Pitt 1988: 37)

One of the market's accountants, Peter Rawlins, who was also chief executive of an underwriting agency, pointed out that 'one of the effects of technology will be to turn underwriting into a younger man's game. The main reason for this is the decline of the apprenticeship system at Lloyd's, a decline hastened by new computer systems designed to remove the mundane tasks which new recruits at Lloyd's have traditionally performed' (Pitt 1988: 37).

The emphasis on the unstructured, case-by-case nature of client–broker relations, and the importance of allowing an underwriter the freedom to make a contract that

---

[15] See, e.g. DTI (1989), Ferné, Hawkins, and Foray (1996), Forrester Research (1996), Lessig (1999), Mansell and Steinmueller (2000: ch. 5), and Timmers (1999).

does not fit within guidelines, may explain why software designers have found it difficult to design electronic trading systems that are regarded by their users as being good enough to replace face-to-face negotiations. There is more, in the insurance professional's view of the world, than can be captured neatly in a 'logical design' or embodied in an EDI message. The quantitative and analytical elements of insurance are supported by a deep layer of subtle, cultural knowledge that enables brokers and underwriters to make decisions that they cannot easily justify to accountants, actuaries, or systems designers.

The remainder of this chapter is devoted to a discussion about what the design team and the business community failed to learn from each other in the case described above, and what this experience can contribute to an 'enacted' view of electronic commerce. In order to take the story further, the next section considers a novel interpretation—that is, that the HNMC met a colleague in the machine. This colleague was not an engineer, nor an accountant, but a marine broker. For well-founded, but hard-to-articulate, reasons, he rejected the 'Contract Details Enquiry' screen in the same way that he or his non-marine colleagues would reject a contract presented by a marine broker.

5. COMPUTERS AS SOCIAL ACTORS

Most systems design, and much of the discourse that interprets it, are predicated on the belief that technological artefacts become part of society either through a process of optimal (or at least second-best) selection, or through a process of negotiation about the meanings and values different social groups attach to them.[16] On the first account, a computer system has no social qualities, while, on the second account, the social qualities are inscribed into the system and the process of inscribing is something to be explained. But what if human computer interaction is fundamentally social and natural? And what if a user's response is underdetermined by any inscriptions or designs? This is the prospect presented by one approach to these questions—that is, social presence theory.

It can be demonstrated that under certain circumstances human beings respond to an electronically mediated experience, such as a telephone conversation or an adventure game on a computer, as if it were in fact a non-mediated, face-to-face encounter. When people respond in this way, they are said to be experiencing 'presence' or 'telepresence' (Lombard and Ditton 1997).[17]

Most research into this phenomenon deals with presence with regard to sensory perceptions—that is, vision, sound, and touch. However, there is a growing body of evidence that suggests that social and cultural phenomena can be simulated in electronic environments. Research conducted within Stanford University's Social Responses to Communications Technology programme by Byron Reeves and Clifford Nass (1996)

[16] See Dosi *et al.* (1995) and Metcalfe and Miles (1994) for a discussion of market selection processes that assume the presence of rational or quasi-rational actors; see Bijker and Law (1992) and MacKenzie and Wajcman (1999) for insights into the way meanings of technological artefacts are negotiated.

[17] Lombard and Ditton (1997) provide a comprehensive overview of research on the topic.

and various others has revealed more compelling evidence of social presence phenomena than the 'interpersonal warmth and richness' that had been detected by earlier studies (e.g. Short, Williams, and Christie 1976). It has been demonstrated, for example, that even experienced computer users respond to computers as social actors.[18] They treat a computer differently if it speaks with a male or female voice, for example, judging that a computer with a female voice knows more about love and relationships than one with a male voice. They show more respect for a television set if they are told it is an expert, and feel their body space is more threatened by large faces on a screen than small ones. People are also polite to computers—for example, subjects were more generous in their praise of a computer that asked them to assess its own performance and gave less generous assessments when the computer was absent. They have also been found to distinguish multiple personalities in a computer running more than one program.

The 'as if' quality of presence—that is, the way that subjects suspend their disbelief—is what makes this concept a useful one for software designers. Participants in experiments have been interviewed after the event to establish that they knew they were using an electronic medium and experimental controls have been applied to eliminate the possibility that subjects' social responses may have been to the designer or programmer of the software.[19] Thus, it makes sense for us to talk about the 'Colleague in the Machine' and perhaps also, the 'Machine as Colleague'. How does this help us understand the conversation between the HNMC and the BSM?

5.1. *What the Designer did not Learn in Time*

Later, too late to keep the project on time and budget, the BSM discussed his meeting over a drink with a colleague from the aviation department and this discussion enabled him to made sense of the HNMC's objections to the system. At that time, the marine syndicates had more capacity than they could use, so they diversified into underwriting non-marine and aviation contracts. Unfortunately, they did not always write them at a profitable rate. Sometimes an underwriter would be invited to take a share of a non-marine contract, only to find that the contract had been led (that is, priced) by a marine underwriter. This was a cue to consider very carefully whether it was a good risk or if the rate was too cheap. In other cases, non-marine claims would receive a call from marine claims because a loss advice had come in on something that had never been within 100 miles of the ocean and the marine department needed non-marine expertise to help it out of its difficulties.

---

[18] Reeves and Nass (1996) is a compilation of findings from several experiments, with a discussion of the implications for media use and design, some of which are highlighted here. The detailed explication of the experiments, methodology, and results may be found in Nass and Steuer (1993) and Nass *et al.* (1994, 1995).

[19] Nass *et al.* (1995) found that subjects' prior experience with computers fails as a predictor of anthropocentric attitudes towards computers. In Nass and Steuer (1993) subjects were all undergraduates who had completed a programming course. As a control, forty-five of eighty-eight subjects were randomly selected and asked whether they thought at all about the programmer or programmers while doing the experiment. Only four subjects indicated that they had any thoughts of this kind. All four believed the computers had been programmed by the same programmer, and all responded to the computers as separate persons.

Once we take the concept of social presence into account, the HNMC's questions begin to make sense. He treated the 'Contract Details' screen as a social entity. It used the language of the mariners, so he treated it like a mariner—that is, with contempt. 'How is it going to calculate the premium?' to him meant, 'Mariners quote bad prices, so I expect this system will do likewise'. 'What are you going to do with the claims?' meant, 'Mariners need a lot of help with claims, and so will this system'. The HNMC saw these cues in the system without them having been inscribed there by designers. They formed part of his accumulated expertise in insurance, and, as evidence, they were stronger and more persuasive for him than any abstract model of the business.

5.2. *LIMNET, the view from 1999*

In 1988, the *Financial Times* was able to report that outline plans had already been drawn up for electronic placing, and 'few now doubt that it will be commonplace in the 1990s' (Bunker 1988: 15). The first version of the Electronic Placing System (EPS) was launched in 1992 but it was not popular. After much further design work, EPS Release 2 was implemented in January 1995 at the end, rather than at the beginning, of the busiest part of the business cycle. The London Insurance Market Report for 1997 recorded that 'a third, even more user friendly, version is due to be released later in 1997', but added with a note of regret that, 'for the vast majority of insurances and re insurance placed on the London Market, EPS has not yet supplanted the traditional system of brokers placing business with underwriters through a process of face-to-face negotiation . . . nor is it likely to do so within the foreseeable future' (Carter and Falush 1998: 44–5).

The same report also recorded that in 1997 the market had employed 39,800 staff. Ten years of electronic commerce had seen a reduction of 10,000 on the 1987 figures, but 80 per cent of that number were still housed in the 'Square Mile'.

6. CONCLUSION: BACK TO THE FUTURE?

The LIMNET era of electronic commerce drew to a close in March 1999, when its sponsors announced that it was to merge with two other insurance industry networks, RINET and WIN.[20] Once again, the *Financial Times* was there to report the event and it is instructive to compare the 1999 announcement with the beliefs that were expressed a decade earlier.

Many of the world's biggest insurance companies and brokers are to create a global e-commerce network for trading risks among themselves in an effort to counter a vicious squeeze on margins.

Premiums charged by insurers and reinsurers have been under pressure in recent years as an increasing number of corporate customers have decided that frequently occurring and relatively predictable risks can be more cheaply retained in-house. Industry analysts also believe that up to

---

[20] The ReInsurance Network (RINET) was an EDI network founded in 1987 by a group of reinsurance companies in continental Europe. It had collaborated extensively with LIMNET throughout the 1990s. The World Insurance Network (WIN) was a private global network established by the largest brokers.

50 per cent of premium payments can be consumed by so-called 'frictional costs'—commission and administration charges.

Previous attempts to link insurers electronically—including RINET, WIN and LIMNET—have suffered from the industry's fragmentation. A negligible proportion of the industry's multi billion-pound business is currently placed electronically.

However, industry leaders believe that the recent trend towards the consolidation and globalization of insurance has given them an opportunity to strip away barriers to the widespread use of e-commerce.

It is difficult to quantify the possible savings that could be achieved by using e-commerce, but a report by the management consultants Coopers & Lybrand, now part of PwC, estimated the London insurance market's annual administration costs of £1bn could be cut by about a third. (*Financial Times* 1999: 1)

It seems that in 1999 the insurance industry was once again feeling optimistic about electronic commerce. The expected benefit was still a reduction in costs and the industry was still looking forward to the day that those benefits would be realized.

What, then, has changed? Two aspects appear to have changed—that is, the technology, and the size and scope of the community. It is conceivable that the new technologies will be different. An OECD report suggests that electronic commerce solutions based around proprietary technologies have only limited capacity to transform firms' business models, sectoral organization, and market structure. However, 'as a standardized system available to all at minimum cost, the Internet takes over at the point where proprietary systems reach the limits of their possibilities' (OECD 1999: 83). This may be the solution the insurance industry has been waiting for.

The new community may be different. Maybe, as the *Financial Times* suggests, globalization and consolidation will overcome the industry's fragmentation. In 1987 the London insurers were trying to implement EDI-based electronic commerce, fearful of competition from continental Europe and North America, and afraid that their quaint English customs would make them obsolete. In 1999, the London, continental European, and North American insurers were all members of the same community. They were all trying to implement Internet-based electronic commerce, and all it seems were equally disappointed by the results.

In 1998, Stefan Sieger, the head of electronic commerce in Swiss Re (Zurich), reviewed the state of industry initiatives across Europe and North America and had this to say:

Despite the increasingly rapid advances made in the field of technology (PC, networking, Internet), companies in the insurance industry are making the most use of the new technical possibilities offered today in their internal operations only, while hardly employing them at all when communicating with their business partners. It consequently comes as no surprise that many of the work processes used in the insurance industry are still inefficient and prone to error, one reason being the necessity for data to be re-entered many times.   (Sieger 1998: 7)

It would be tempting to argue that LIMNET was a local story and one that is unlikely to be repeated elsewhere. It was the story of a community of small British firms and the eccentric institution Lloyd's of London, which had tried to rush into the digital age and found that they had a lot of catching up to do and a lot of political wrangling to endure.

It would be tempting, too, to suggest that this is now behind them for the most part and that their future experience with electronic commerce will be different.

However, research in the tradition of organizational informatics suggests that we must acknowledge that all stories are 'local' stories. Electronic commerce is not something 'out there' that is external to businesses and that is unavoidably coming closer. It is inside the world of business and businesses are enacting it on a day-to-day basis in their actions and decisions and in their negotiations and conversations. Knowledge exchanges need contexts in which the evidence has agreed significance and in which meanings become personalized and rich. The conversation between the HNMC and the BSM was one such conversation. It was not a conversation that would have taken place in Chicago, or Frankfurt, perhaps not even anywhere outside Lloyd's. It was a very local affair. However, it is an example of an encounter between different communities that had different working practices and knew different things. This conversation was a search for meanings and involved a discovery of unintended meanings. Dozens of similar conversations are taking place daily in which business people who are skilled and experienced and who have strong motivations and incentives are exploring new high-performance computing applications and trying to work out how they can use them effectively in their work practices.

One belief that continues to motivate the London Insurance Market in its collective efforts towards electronic commerce is the conviction that insurance will be a single global market. That belief has motivated firms to form a standards community that is reaching out to more and more partners in Europe, North America, and around the world. But, as the community becomes larger and involves more expertise and a greater variety of traditions, the need to find and articulate shared meanings will not diminish; rather, it will intensify.

The insurance industry's predominant vision of electronic commerce is influenced by the industry's search for efficiency gains. However, the work of implementing electronic commerce—that is, of putting the vision into practice—is a search for shared meaning. The defenders of face-to-face trading in London talk about flexibility and sensitivity, about client relationships, and the need to handle business on a case by-case basis. Too often in the 1980s and 1990s, it appeared that the searches for meaning and for efficiency were getting in each other's way.

7

Repersonalizing Data in the Banking Industry

ANDREAS CREDÉ

1. INTRODUCTION

Commercial banks—the banking companies that receive deposits and provide lending and other financial services with the purpose of making a profit—display a number of contrasts and seeming contradictions. The sector exhibits continuous technological change coupled with massive-scale, long-standing traditions, and a surprising degree of permanence. The banking sector is at the forefront in using new technologies, but it is also one of the oldest industries, with many market practices dating back to the merchant bankers of Florence, Venice, and Genoa who financed intra-European trade in the eleventh century. Banks have made enormous investments in computer hardware and software, yet they remain very labour-intensive service organizations that are highly dependent on a large number of costly, skilled staff.

How can these apparent contradictions be accounted for? To what extent do recent trends provide insights into new patterns of intermediation in the relationships between savers and lenders? Will the use of information and communication technologies (ICTs) provide a basis for the virtual provision of most, if not all, services offered by banks? In this chapter, the practices of a group of major commercial banks are considered to assess how ICTs have been used to manage information and how they may have contributed to distinctive knowledge generation activities in the banking sector.[1]

Intermediation is the term used in the banking industry to describe the way in which commercial banks and other savings institutions act as intermediaries between savers and borrowers. Savings intermediation occurs because individual savers find it easier to entrust their money to a bank than to manage all their investments directly. These relationships involve complex exchanges of information. The institutional forms of intermediation that have been devised through the centuries define which organization serves as an intermediary and for whom (Kracaw 1980). The application of ICTs provides the means to automate aspects of the information exchange process and this is why advances in these technologies are cited as leading to 'dis-intermediation'. For

---

[1] This chapter is based on an examination of twelve major international commercial banks based in London (see Credé 1997).

example, Bryan (1988, 1993) suggested that new technology applications would lead to the elimination of the need for financial institutions to perform the role of intermediary. The evidence in this chapter suggests instead that there will continue to be selective adoption of ICTs in the banking industry and the main emphasis will be on enabling human communication in its various forms, rather than on information processing.

This view is developed to augment the 'information-processing' view of knowledge management (Brown and Duguid 1998; Nonaka 1994), which tends to focus mainly on the automation of information-related processes (Barras 1990; Eliasson 1990; Lamberton 1971). This chapter illustrates how the financial services intermediary plays an important role in enabling the 'repersonalization' of data that underpin the intermediation process in an important segment of the banking industry. The analysis highlights the complexity of the knowledge production process and suggests that, through their selective use of ICTs, banks are likely to retain leadership in certain core business areas, such as credit assessment, despite major advances in the use of ICTs to automate information processing.

2. THE COMMERCIAL BANKING SECTOR

In terms of their core processes of savings intermediation and payment transfer services, commercial banking organizations have no overriding economies of scope or scale to protect them from challenges to their pre-eminence in world markets. Nevertheless, the last thirty years of the twentieth century witnessed the emergence of some of the world's largest commercial organizations in the banking and financial services sectors. Major commercial banks like Citicorp expanded to become truly global corporations with a direct presence in more than 100 countries. In the manufacturing sector, the emphasis in the 1990s was on divestment and on downsizing corporations to their core activities, but commercial banks continued to expand and to become larger, both through organic growth and corporate acquisition. Throughout the 1990s and in the first years of the twenty-first century there was a series of mega-mergers involving major commercial banks and related financial institutions. All indications are that this will continue.[2] Unlike other strongly technology-based sectors, like computer software and semi-conductors, the global leaders in banking seem to have fought off the threat of new entrants successfully. The top fifty players are remarkably unchanged from those of several decades ago (although their relative ranking has in most cases changed). For the majority of major banks, organizational routines have been accumulated over a corporate history that extends back a century or more.

Banking is the world's largest industry when measured in terms of the market capitalization of its firms. The value of the world's listed bank shares exceeded $1.5 trillion in 1996.[3] This is more than the total for the telecommunications and energy sectors combined. Commercial banks are principally service organizations. Operations are highly labour intensive with staffing costs accounting for over half of all direct costs.

---

[2] For further details, see Bryan (1993).
[3] Calculated from figures published in the *Financial Times* (1997).

Yet, in their more recent past, commercial banks have been among the earliest adopters of advanced ICTs. The sector accounts for a major proportion of the total worldwide sales of computer hardware, software, and related services. Commercial banks also represent one of the most important groups of corporate customers for telecommunications, networking, and communications companies.

Many commercial banks have led the way in the application of new technologies. Banks in Europe and North America were among the first commercial organizations to introduce computerized accounting systems. Similarly, banks were at the forefront in introducing electronic networks for payments and messaging, including the introduction of automated teller machines (ATMs) (Pennings and Harianto 1992). More recent innovations include the complex technologies incorporated into call centres, the use of smart cards, and the application of new multimedia systems for selling and distributing bank services (B. Johnson 1995). These applications of ICTs have contributed to substantial changes in the way financial services are provided to customers in both the retail and business banking contexts.

Despite the growing variety and sophistication of financial services, all the major commercial banks continue to derive more than half their operating revenues from their roles as intermediators between savers/investors and borrowers.[4] In this respect, their core activities have remained unchanged for over a century. Furthermore, while technological change is normally accompanied by changes in market share and the arrival of new entrants, the commercial banking sector is characterized by major institutions with very long histories. These institutions have maintained a more-or-less dominant position in their respective markets. As financial markets have become more global, each of the major banks has built on its existing level of international exposure and maintained a strong position in its home market.

If the core activities of banks are increasingly being mediated by the use of ICTs, then some radical changes in the structure and organization of banks might be expected. Banks might also be expected to experience a decline in their market presence as the traditional intermediary role is replaced by automated systems and services offered by other kinds of institutions. The economic view of the financial system is that it is a mechanism for bringing together savers and investors to use resources more productively. Financial intermediation creates an environment in which both savers and investors are potentially better off than if all investment were to be self-financed. In this context, commercial banks operate by taking in deposits—that is, allowing savers to defer consumption to a later date—and by lending these funds to borrowers at a margin. Furthermore, since savers have a preference for liquidity—that is, they seek to deposit funds short term—investors will generally wish to borrow for as long a term as possible. Banks balance these two conflicting requirements by the transformation of short-term liabilities to long-term assets.

Information asymmetries have been identified as being responsible for the economies of scale that encourage savers to entrust their savings to a bank rather than

---

[4] Operating revenues represent the combined total of net interest margins plus trading profits and fees received for banking business that is not directly loan related (Credé 1997).

attempt to assess the potential of each investment (Diamond 1984). Banks are also expected to be more efficient at monitoring borrowers' repayments and creditworthiness after a loan has been granted. Commercial banks, therefore, are said to exist because they are perceived as the most efficient information processors compared to any alternative. This is the way that banks traditionally have contributed to economic activity. The role ascribed to ICTs, according to this view, is principally one of automation. Information processing can be made progressively more efficient and less costly by substituting human labour with machines. Information that previously had to be handled in print format is becoming progressively digitized, so that it can be more readily stored, communicated, and processed. As information-handling costs rapidly decline, Barras (1990) has suggested that most kinds of information will become tradable commodities. If this argument is applied to the banking sector, then banks could be expected to take a leading role in the creation of new types of automated—information-processing—services for finance and banking to support intensive and global knowledge management activities. In this chapter, the characteristics of one of the commercial banks' core businesses—credit assessment—are examined to detect whether there are limitations to the automation of information processing in this important area.

3. BANKING AND THE SELECTION OF ICTs

In this section, we draw on the results of a study undertaken in 1996 of how a group of major commercial banks with operations in the United Kingdom have used ICTs to manage information, specifically with respect to one of the core bank processes—namely, credit decision making (Credé 1997). The twelve banks included in the sample for this study represent world-class institutions. They were selected by virtue of their size, global presence, and ability to provide representation of all the major world economic regions.[5] The objective of the study was to identify key common features in the production and exchange of information that might transcend culturally and historically specific differences. The research design enabled a comparative examination of the sample of banks to determine how ICTs were being used and to detect how the applications were related to changes in processes of information production and exchange for credit assessment that are central to the banks' core activities of intermediation.

All the commercial banks in the sample at the time the study was undertaken in 1996 were making substantial investments in extensive communications networks with the aim of enhancing and maximizing various types of information processing and communications *within* their respective organizations. These investments often constituted a major element of total expenditure on ICTs. But, notwithstanding this investment, there was also an equally important and ongoing requirement for investing in highly skilled staff who would be capable of making professional judgements. There was an expanding, and large, very highly paid workforce in commercial

---

[5] The twelve banks were Bank of America, Chase Manhattan, Citibank, Chemical Bank, Mitsubishi Bank, Sumitomo Bank, Dai-Ichi Kangyo Bank Ltd (Japan) (DKB), HSBC, Deutsche Bank, UBS, NatWest, and Société Générale.

banking. Over half the operating expenses in each of the banks in the sample was taken up in salaries and staff benefits. This investment in staff appeared to be consistent with the highly personal nature of information production and exchange within certain areas of the banking sector. The existence of a large staff and the communication-intensive nature of the work in core bank areas suggest an ongoing and extensive requirement for effective communication so that activities can be coordinated at the organizational level. In fact, sophisticated telecommunications services were essential to the workings of the banks in the sample and each was employing up to 100,000 staff and sometimes more.

Commercial banks maintain large numbers of individual offices, both domestic and overseas, as part of a growing international network. Major banks continue to expand their respective offices abroad rather than choosing to service them from a central location. The banks included in this study had a large number of newly opened offices and were continuing to invest in new buildings. This expanding network of local offices, coupled with a large workforce comprising highly paid professional staff, provided a further indication of the priority that investment in advanced communications networks was being given. It also suggested the existence of a requirement to maintain close personal contact with customers as well as within the teams that service them.

If the effects of ICTs are truly as transformative for existing organizations as the information-processing model suggests, we might expect to find progressive automation of the work of skilled banking staff in core areas such as credit assessment. This might be expected to result in lower overall employment levels, reductions in staffing costs, and increases in incremental expenditures on ICTs. However, commercial banks are continuing to spend a major part of their resources on staff.[6] While there has been rationalization in domestic branch networks, the major banks are expanding employment in their foreign operations, as well as for their specialist functions. Overall, staff numbers either rose or remained constant over the decade of the 1990s.[7] This is in sharp contrast to other major corporations that have dramatically reduced employment levels over the past ten years (Cappelli 1999).

The analysis of the banks included in the study confirmed that they are devoting substantial resources to establishing sophisticated parallel systems of communication using email, groupware, and videoconferencing. Each of the banks was following a similar trajectory, both in terms of its broader approach to communication services and, specifically, in terms of the expansion of email services. Basic investments in ICTs tended to have started with sophisticated private automated branch exchange (PABX) voice systems, facsimile, and telex and then to have progressed to email, groupware, and videoconferencing. Similarly, the implementation of email had started with locally based email, followed by expansion to intra-branch, then inter-branch, and ultimately, fully operational global email networks. While the rate of penetration of personal computers (PCs) in the sample of banks had been slower than for many other major commercial companies, once introduced, networking had greatly accelerated the overall rate of installation.

---

[6] Staff costs represented, on average, more than 70% of total costs for the major commercial banks included in the sample.

[7] In aggregate, staff numbers increased by 26% for the twelve banks between 1991 and 1995.

In making new investments, the emphasis appeared to be on improving the overall environment for communication and, in particular, on reducing the barriers of time and distance for each of the banking organizations. There were signs of a growing emphasis on teams working at the global level. Such transactions involve close coordination between different departments in their respective branch locations, often thousands of kilometres apart. This was reflected in the increasing importance that was being placed by the banks on 'global reach' and was consistent with the strategies that have been announced publicly by major financial institutions over the past several years. For example, the theme in the Citicorp (1994: 3) *Annual Report* was 'around the world, around the clock . . . the Citi(bank) never sleeps'. The Union Bank of Switzerland (UBS 1995: 12) stated in its *Annual Report*, 'All our activities are geared to the needs of the customer . . . We manage our businesses globally in the interests of the customer and coordinate our expertise across the Group.' Similarly, HSBC (1995: 7) in its *Annual Report* said that 'As one of the largest banking and financial services organizations, the HSBC Group prides itself on its management's ability to think globally but to act locally in the best interests of its customers'.

In each of the banks, new communications services were being introduced and maintained in parallel with existing networks and services. For example, those banks with email networks continued to be heavy users of facsimile. Although the interviewees reported that email had often found ready acceptance precisely because it complements other services that mediate communication, there was also a marked preference for voice rather than data communications. Several banks in the study were using regular telephone conference calls for coordinating teams that were spread across several locations. Despite the inconvenience of the need to allocate a fixed time slot, communication in this form was found to be more effective than available text-based alternatives. Similarly, the introduction of email appeared to be having only a minimal impact on some of the formal credit procedures. Even as new electronic services were being introduced, the arrangements for circulating formal documents within the banks often had remained unchanged, because of the need for decision-makers to see annotated comments and original approval signatures.

Each of the banks was devoting significant expenditure to external telecommunications services as well as to a variety of internal ICT-based systems and services. In the case of some of the banks, detailed figures were available from financial statements. In 1995, for example, Bank of America had spent $359 million on external telecommunications services, equivalent to more than its total gross expenditures on ICT equipment—that is, including both net additions and replacements, and representing just under 10 per cent of its total staff costs. Figures available for Chase/Chemical (accounts combined to reflect the merger) and for UBS showed that they spent $333 million and $264 million, respectively, in the same period (UBS 1995).

In summary, the empirical evidence from the banks included in this study suggests that investments in ICTs were made selectively. There appears to have been a strong emphasis on applications that would support various forms of human communication using voice or audio-visual services. Particularly striking were the large investments in voice, facsimile, email, and, increasingly, videoconferencing. Expenditures on telecommunications services, on average, represented more than half of all gross ICT equipment expenditures by these banks.

In the next section, we examine the credit review process within these banks to determine whether ICTs were being employed to support this type of banking activity. If the information-processing view is adequate to explain the role of new ICT applications, we would expect to see signs of the increasing application of ICTs within the banks to support the automation of this information-intensive activity.

4. CREDIT PROCEDURES IN COMMERCIAL BANKS

In-depth interviews were conducted with employees from each of the banks in the sample to establish the details of the credit review process, including the documentation requirements that must be satisfied.[8] The focus was on processes involving relatively more complex credits, including project finance facilities and complex corporate loans, as compared to standard loan products for consumers and small businesses. The results of the interviews indicated that broadly similar procedures were being applied within each of the banks.

Credit decisions generally were the responsibility of a designated credit committee, which took a number of different forms. Committees often included the most senior managers available at branch, regional-office, or head-office level, but they also included representatives of the credit department. Decisions normally required a consensus, but there were some differences between the banks in this area.

Credit decision making is, by its nature, a key process in all banks. Nearly every transaction is linked to a credit approval process in some way, whether it represents settlement limits for making payments, contingent liability under a derivative trade, or a loan commitment. While there were some differences, the results of the study reveal a striking consistency in approach, despite the fact that the home countries of the banks in the sample were different and they might be expected to have devised somewhat different organizational routines.

The credit approval process for major corporate credits was typically a bottom-up process that was initiated and controlled by a middle-ranking lending officer at either the industry, sector, or regional level. An analyst working with the account manager would complete most of the necessary creditworthiness analysis and, in most cases, would also write the report. Credit applications were prepared under the supervision of a designated account manager and the analytical input was provided at a less experienced level of the organization.

Once an application had been completed, the data were presented to the individuals who had been delegated the relevant approval authority. The report document itself played a central role in the overall approval process and followed a predefined format that was more or less formalized. The credit application would be circulated in paper form to those individuals with the authority to give approval and comments would be

---

[8] A total of seventy-one in-depth interviews were undertaken for the study, of which half were with employees of the twelve banks.

sought. Changes and amendments would be agreed and the credit approval document would then be circulated at the next level. Each individual reviewing the document would confirm approval by a signature. Thus, the final contents of the credit application reflected the various informal negotiations and discussions that had taken place during the approval process.

Within each of the banks in the sample, the credit approval process was an iterative one. The process resulted in a negotiated document rather than a distinctly 'authored product'. Comments were passed down to the lending officer who was responsible for making any changes. In all the banks, financial data were an important element of the overall credit assessment process. However, the banks placed as much emphasis on the process of presenting the financial data using a common format as they put on the analysis of the data. Each bank used slightly different formats and these conformed to a particular 'house style'. The need to fit information into specific categories often created difficulties, particularly when these categories were included in a pre-designed form based on the accounting standards of the home country. The emphasis on using a common template made it feasible for those components indicating financial strength that were valued particularly by each bank to be more readily recognized. One interviewee commented that the analysis based on the forms was essentially a 'statement of facts' and did not itself contain any insight into the financial state of the company seeking a loan. Nevertheless, the use of common formats for the presentation of financial data did permit senior managers to form an instant 'shared' picture of a company's financial position. This picture was then incorporated into the overall credit assessment of the financing proposal. In general, financial analysis of a company seeking a loan took the form of a static analysis and considerable attention was given to balance-sheet items and relatively less attention was given to dynamic cash-flow analysis and cash-flow projections.

In each of the banks, credit decisions were the result of a negotiation process involving a hierarchical chain of managers at progressively higher levels of seniority. They were the products of a bureaucratic process that, for the more complex credits, could involve up to fourteen individual levels of decision making before the necessary corporate authority could be given authority to proceed with a particular financing proposal. Once all the internal authorities had been documented, procedures for monitoring and accounting for the transaction would be followed, often supported by ICT systems.

Analysis of the fine detail of the banks' procedures indicates that credit decisions involve a complex intermediation process. Interaction between the approval levels and the originating departments was almost exclusively in the form of face-to-face personal contact or by telephone. Instead of using ICTs to automate the processing of credit assessment information, ICTs were being used primarily to support the human communication aspects of decision making and they were permitting these to take place without the spatial and temporal constraints that would otherwise apply. The credit decision-making process seems to reflect distinctive kinds of intermediation that take place in banks. These depend heavily on the authentication and validation of customer information rather than principally on the automation of information processing.

Economies of scale and specialist knowledge traditionally have been seen as giving banks the necessary comparative advantage over other organizations to perform credit intermediation (Diamond 1984). However, this examination of the credit review process suggests that commercial banks are engaged in producing credit information from confidential customer data rather than mainly in processing information that is 'out there' waiting to be processed. The results of the research showed that the credit review process is not a simple case of collating different elements of information in support of an argument for or against making a loan so that a collective decision can be taken. Instead, it is built around a complex process of giving meaning, significance, and relevance to customer data. In this study, the results showed that very little emphasis was given to automating the process of weighting existing information that may be favourable or unfavourable to a positive decision to approve a proposed transaction. Instead, credit applications were prepared, and then reviewed and analysed by individuals within the approval hierarchy. Key elements of the credit presentation were filtered through the credit review process and, through this negotiated process, these elements were assigned different values. The credibility of the proposal was established through the series of personal contacts and exchanges within this process and all of these personal exchanges involved varying degrees of personal trust. Thus, the credit assessment process within the banks required personal authentication and validation and was an integral element of the credit review process.[9]

5. COMMERCIAL BANKS' HANDLING OF CODIFIED DATA

The study of credit assessment procedures within the banks in the sample provides a basis for examining the sources of data that were used in the credit review process. Much greater emphasis was placed on data sources that each of the banks considered to be under its own control. For those with the authority to participate in the credit assessment process, the most important data were those obtained from the customer. This was particularly so for proprietary and privileged data that could be obtained as a result of the bank employee–client relationship. Although other external data sources were available to the credit assessors, data requirements generally had to be satisfied by the client. Relatively little investment was being made in other forms of access to company information. Furthermore, where external sources were used as part of the decision-making process, they served primarily to validate the information that had already been gleaned as a result of previous customer contacts.

Most of the resources, in terms of staff time, were taken up in the preparation of the credit review document and this was devoted mainly to the presentation and subsequent amendment of internal data. The emphasis was much less on research to obtain other external sources of corroborating information for the analysis of certain key

---

[9] Further evidence of the importance of specific institutional factors and the complex processes involved in acquiring the necessary skills for credit assessment is suggested by variations in the development paths experienced by major commercial banks. David Rogers (1993) points, for example, to differences in the processes between Citibank and Chase Manhattan.

issues that might arise as the negotiation proceeded than on informal sources of information and 'tacit' knowledge.[10] The procedures that were followed to obtain a credit approval allowed very limited time for detailed analysis of 'codified' data or for any form of 'research' on a particular credit issue.[11] Instead, considerable time was spent on reviewing proposals that were almost entirely based on the package of data that had been provided by the customer. Where there was inclusion of other externally based data sources, these served largely to validate the customer data rather than to extend or deepen the analytical basis for the decision.

In effect, the available effort was being devoted to applying the bank's existing lending expertise to assessing the validity of the representations that were made by the prospective borrower. The emphasis, therefore, was on validation rather than on analysis of information generated by automated information processing. The principal purpose of the credit review process was to *authenticate* data won as a result of the privileged relationship with the customer rather than as a result of a more broadly based analysis incorporating a variety of different data sources. Once a credit facility had been agreed, the signed legal contract in the form of a credit agreement placed requirements on the borrower to continue to make privileged data available as well as to confirm that these earlier representations had not changed materially in any way.

Given the emphasis on the customer as the key source of data in the credit review process, in each of the banks there was a corresponding requirement to maintain confidentiality at all levels in the organization. This requirement, in turn, generated a 'secrecy culture' where access to data was strictly controlled. Any effective sharing of privileged data was restricted or blocked to avoid the risk of misuse. This was perceived by the majority of interviewees in this study as being necessary, not only to avoid conflicts of interest, but also because the customer had supplied confidential data on the understanding that it would only be utilized in the context of that particular account relationship. Specific experiences with individual borrowers and particular details of their commercial operations, therefore, were usually dealt with informally rather than in a more formal or codified way.

As data were compartmentalized at every level within the banks, they remained specific to each department. Requests for data had to be processed through a human filter. Various participants in this study stated explicitly that the resulting inefficiencies in data processing had to be tolerated because any procedures that would make data more easily available for credit assessment would mean that they could no longer be controlled. The overriding concern of the interviewees was that multiple exchanges of data could create potential conflicts of interest and result in privileged data being released to unauthorized third parties. The suggestion that the more widespread availability of customer-related data throughout the banks might facilitate learning as a result of data sharing was not supported by the interviewees. They regarded the potential for learning between staff as a result of the ICT-supported automated circulation of data as of

[10] Tacit knowledge can be defined as the 'disposition to codify and abstract' (Boisot 1995); see also Senker (1995).
[11] See Cowan and Foray (1997), David and Foray (1995), and Foray and Lundvall (1996).

much less importance than ensuring the confidentiality of the data used in the credit review process.

In many of the banks, data about clients were stored in paper form. Each department was responsible for keeping its own client files. Files were entrusted to 'guardians', who, in the case of the principal accounts, were generally the relationship managers. File notes would often be generated electronically but would then be printed out and circulated in hard-copy form. Despite the potential contribution that this information could make to the learning process, credit applications were maintained in a customer file and the data were not captured in electronic form with the exception of a few details that were transferred to an automated loan booking system. Even where copies of individual credit applications were held electronically on a file server, only members of the same department normally had access. Individuals from other departments had to make specific requests for copies to be provided should the need arise.

In each of the banks, responsibility for customers was organized on a departmental basis. Account managers were given responsibility for specific customer relationships that were also the responsibility of a particular department. Client data were controlled within each departmental area and required 'negotiated access' before the details could be made available. If another department required data on a particular customer or project, the standard procedure would be to call the client officer to make the request. While a request for data was refused only infrequently, it always required an element of negotiation and human exchange. The starting point in all such exchanges was that data were only made available on a 'need-to-know basis'. It was incumbent on the person making the request to demonstrate an overriding need to have access to the data. The importance of respecting these institutional processes was readily accepted by the interviewees in this study and these arrangements were acknowledged as legitimate barriers to learning within the banks. The bank employees appeared to accept that the dangers of misuse from more open exchanges of data were greater than the possible benefits that could be obtained from wider information sharing and that this was simply a characteristic of the banking business.

Barriers to the movement of credit assessment data were also reflected in the design of systems to deliver electronic information to the desktop. The 'secrecy culture' was pervasive in the banks and it applied to a variety of data sources that could have been delivered to potential users directly and more effectively using various ICT systems. For example, information about internally set portfolio limits—that is, the maximum amounts a bank is prepared to lend to different sectors in the economy and to different countries—was controlled by the credit department. Although these data were of the utmost importance in marketing new services to customers, individual relationship managers were required to seek negotiated access to these data.

The relatively low value placed on external data sources by all the banks in the study was reflected in the limited resources devoted to this function and the *ad hoc* nature of research and analysis that was undertaken when specific customer-provided sources of data were lacking. In contrast to research-based organizations, such as universities or consultancy companies that have a requirement for externally sourced data, banks devote a much smaller proportion of their resources to facilitating the acquisition of the kinds of external information that might facilitate learning about credit decision making within their organizations. The interviewees indicated that very few external data sources were available and that, in most cases, when they were available, they too required negotiated access. Most of the banks had a corporate library that controlled access to periodicals, special reports, and external electronic sources purchased by the bank. However, the role of the library was generally restricted to a document delivery function—that is, to supplying copies of clippings, requested articles, and limited online searches. Expenditures on external data were strictly controlled and, generally, were among the first items to be cut in times of budgetary stringency.

In summary, although large quantities of 'codified' customer-relevant data were prepared and collated within the banks, access to these data was strictly controlled and, frequently, available only as a result of negotiated access. Non-specific external data relevant to credit decisions played an important role in validating customer-related information, but these forms of data were used less frequently as a basis for making credit decisions. In the next section, we consider these banking practices and procedures in the light of alternative expectations about the implications of the use of ICTs to create changes in the intermediation functions that, historically, have been provided by banks.

6. 'REPERSONALIZATION' AND BANKING INTERMEDIATION

When knowledge management is regarded as the result of a socially constructed process, the information-processing analytical framework needs to be supplemented. The role of ICTs in mediating the production and exchange of information cannot be understood adequately when the main focus is on the automation of the information exchange. The potential for the application of ICTs and the implications for organizational procedures, such as those established within the banking industry for credit assessment, are better understood when the characteristics of specific information production and exchange processes are taken into account. These characteristics are neglected when the emphasis is on the capacity of ICT systems progressively to reduce the costs of information processing as input into decision-making processes.

The evidence from our research on banking procedures for credit assessment suggests that, while the use of ICT application to support more cost-effective processing of data is an important consideration, it is not the only factor governing the selection of ICT applications within banks. Another major factor appears to be the extent to which applications of ICTs will enable the 'repersonalization' of data in a way that is consistent with certain socially defined organizational routines and procedures. 'Repersonalization' refers to the reduction in the amount of codification required for data to be communicated in a way that retains some aspects of the tacit context that is required for its interpretation. As indicated above, the cost structure of commercial banks is heavily weighted in favour of staff costs that, in turn, reflect very high levels of remuneration in relation to other sectors of the economy. Similarly, the process of making credit decisions depends, ultimately, upon authenticating and validating customer-originated data. Access to codified information or the capacity for electronic processing of these data may play an increasingly important role as ICT systems become more attuned to the socially constructed processes they are intended to support.

The 'repersonalization' of data is also an important aspect of how banks handle the already codified data that are amenable to distribution and exchange using ICT systems. In the light of the priority attached to the need to respect commercial sensitivities, most customer data are not shared on a bank-wide basis. Data sharing within the banks in this study appeared to require a complex process of negotiated access before any information was made available. As a result, the learning process within the banks appeared to be highly individualized and involved a personalized process based upon communication between designated, trusted, bank employees. These staff were playing a crucial role as intermediaries in credit decisions. At least in the mid-1990s, information-processing ICT-based systems were perceived as being able to do little more than 'automate' the holding of certain parts of the necessary credit information. ICTs that could be used to support enriched communication were favoured by the bank interviewees to support the credit assessment process.

The main role of bank employees in this context can be understood by reference to a principle of 'economizing' (Boisot 1995). Boisot has argued that human beings need to economize on the processing of raw data. He has suggested that the human brain is extremely powerful and able to detect complex patterns and relationships. However, it is not very well suited to processing raw bits of unconnected data. Thus, intermediation in the context of economizing involves choosing to economize on time and effort by entrusting intermediaries who are able to *reduce* the amount of basic information processing that is required. Intermediation can be regarded as a very effective strategy to minimize the costs of processing data. However, intermediation depends crucially, as the study of credit assessment suggests, upon a willingness to trust the ability of individuals to authenticate and validate information. Intermediaries can provide a means of authentication and validation for various parties by reducing the need for the processing of undifferentiated data. The intermediaries make selections and choices and rely largely upon their experiences and skills to do this. By enabling economizing to take place, intermediaries provide a ready means of adding validity and significance to data.

In the financial services sector, and specifically in commercial banking, intermediation is especially important. As we have seen, banks depend heavily on data that are provided to them by their customers. The particular features of these private, privileged customer data, which provide the essential input into bank decision making about creditworthiness, require several forms of intermediation. The data are often not externally tested, nor do the participants in the decision-making process appear to perceive a need for corroboration from external sources. All data that become part of the decision-making process appear to undergo a process of authentication and validation that depends on extensive interchanges between bank professionals and the preferred mode of communication is that which is deemed to maximize trust. The professional judgements that are the product of bank-created knowledge appear to provide banks with a significant comparative advantage over potential contenders for a share of their market. This is reflected in the long-standing ability of commercial banks to retain market dominance in many areas of banking despite attempts by other organizations to usurp some of their traditional territory. The information-processing framework with its emphasis on automation suggests that the application of ICTs might erode this advantage, making it easier for other organizations to play the intermediary role historically provided by banks.

It is important to distinguish between those conditions where ICT applications are regarded as contributing to the improved conduct of bank processes and those conditions where they are not. In some cases, the banks have used ICTs to create a very 'rich' environment for internal as well as external information exchange—for example, the expansion of voice telephony and audio-visual conferencing. These are applications that tend to augment the capacity for the 'repersonalization'—that is, the authentication and validation—of data. By emphasizing 'repersonalization', the focus of analysis shifts to the ways that ICT applications can help to enrich human communication processes that would otherwise be conducted through face-to-face meetings to achieve negotiated understandings between individuals drawing largely upon uncodified sources of information.

New, digitally-based communications technologies often facilitate human communication by reducing the costs and extending the reach of communications and information services. Some applications, such as audio-visual and advanced multimedia services, permit the 'richness' of human communication to be preserved. The banks in this study all demonstrated a very high degree of selectivity with respect to the implementation and use of ICTs both within the organizations and to support communication and business processes between their branches. Thus, it appears that variations in the way ICT applications are selected, applied, and used can be explained by analysis of the specific organizational factors and routines that are employed. Analysis of the potential cost savings in information processing that may be achieved by introducing automated data-processing systems do not take full account of the key features of the social processes that influence important aspects of the bank's core business of credit assessment. The information-processing model suggests that initial resistance to the smooth adoption of cost-reducing technologies and their applications will be overcome, especially when strong competitive forces emerge (Barras 1990). However, this model does not seem fully to explain developments in the case of the commercial banking sector. The distinctive patterns of ICT use that were found in this study are persistent in spite of continuous innovation in ICT applications to support information processing. The information-processing model, on its own, does not explain the persistence of the selective use of ICTs for particular applications within banks.

Tacit knowledge appears to play a key role by providing the context within which codified information can be rendered meaningful and, therefore, can influence the selection of ICTs. Human beings interact with their social and physical environments by codifying and abstracting from raw data. As Boisot (1998: 38–9) puts it:

complexity absorption leads to a steady accumulation of tacit, experiential knowledge inside an organization. Although not necessarily ineffable, such knowledge can only be articulated and communicated with difficulty. It therefore tends to remain locked up in the heads of its possessors. . . . for this reason, productive organizations find it a better strategy to invest in the articulation of knowledge and complexity reduction than in the accumulation of tacit knowledge and hence in complexity absorption. Articulate knowledge can be shared and can thus be used to facilitate the coordination of productive activity.

However, since the articulation of knowledge happens in a social context, the way abstraction and codification of data take place is influenced strongly, and is itself influenced, by the institutional and organizational routines that are created.

Studies that emphasize the particularities of how ICTs mediate the exchange of information in organizational settings such as banks yield insights into the potential of innovative applications to establish rich, 'human' environments for the communication of data. The focus in this chapter on how ICTs enhance the production of information from raw data brings an immediate benefit. It helps to indicate the potential for the use of alternative ICT applications within organizational contexts on the basis of the extent to which the applications are likely to support the repersonalization of data. ICT applications are not simply a set of process technologies that contribute to substantial reductions in the costs of handling information, although this aspect must be taken into consideration. They are selected and used depending upon the specific conditions that are required for producing information from data within and between organizations. In most of the literature on knowledge management (Boisot 1998), these highly specific processes of information production from data are hidden within simplified accounts of data, information, or knowledge 'use'.

ICT applications do have the potential to automate data-handling processes, but much of the literature on the diffusion of ICT applications does not focus on how information is produced within organizational contexts. ICT applications are treated as if they were homogeneous in the extent to which they facilitate 'repersonalization' of data, and their appropriate use within a given organizational context is often misunderstood. Anomalies and problems experienced in implementing certain ICT applications come to be regarded as targets that may be eliminated through the introduction of some form of hypothetical unitary 'best practice' that will yield reduced information costs. However, the evidence from this study of the banking sector points to another solution. ICTs, like other technological innovations, are likely to be selected by users to reflect the specific characteristics of information and exchange processes that occur at the institutional level (Nelson and Winter 1982). The factors contributing to ICT selection processes in commercial banks are rendered more transparent when the importance of the institutional environment within which information and, ultimately, knowledge are created is taken into account. The analysis of these selection processes is enhanced considerably by focusing on the technical features of ICTs that facilitate communication in ways that mimic direct human interaction—that is, where the need for codification of data is minimized and where data can be 'contextualized'. In contrast to the predominant emphasis in the 'knowledge-management' literature on the role of ICTs in automating information processing and on the progressive codification of knowledge, the 'repersonalization' framework emphasizes how ICT applications facilitate the richness of multi-channel, 'repersonalized' communication. Some ICT applications appear to give these tacit and experiential aspects of information generation greater, rather than reduced, significance.

The intermediation roles that may be augmented by specific ICT applications are best considered in three ways. First, ICTs permit a huge increase in the volume of data that can be processed in its raw state, thus providing a basis for 'economizing' and for the automation of data processing. However, this is only one feature of the ICT selection criteria. ICT applications also embody the capability to 'informate'—that is, to create new data as a result of their application. Zuboff (1988) argues that informating demands a new set of skills and competencies in organizations. Taking pulp-mill case studies as an example, she describes the impact of process control automation on the workforce and how previously learned tacit or 'action-centred skills' were replaced with a requirement to acquire a set of 'intellective' skills. Whereas an operator previously physically squeezed the pulp to determine its consistency, new skills of abstraction, explicit inference, and procedural reasoning were required to operate the same plant from a central control room. The key elements of the production process could no longer be felt, heard, smelled, or touched. They had to be 'visualized' by using the information available from a set of visual display units. It seems reasonable to suggest that the acquisition of 'intellective skills' will be important in other contexts as well. For example, intellective skills require training in the use of ICT systems as well as an understanding of organizational processes and routines that apply in a given context. This has implications for the control of information within and between organizations and represents a second feature that plays a role in the ICT selection process.

Where ICTs are used to augment the capacity to 'repersonalize' data that are exchanged in the process of communication between decision-makers, they can be applied to support processes such as those involved in the credit assessment process. This aspect indicates that a third important feature of ICT selection is the extent to which a given application supports the 'repersonalization' of data. If an ICT application cannot be used in this way, its application is unlikely to be valued, even where it is acknowledged that its use may reduce costs.

Advances in ICT applications are enabling communication processes to retain more of the tacit components of information exchange, thus reducing the need for a prior understanding of the context in which information has been generated. It is worth noting that two of the most successful ICT innovations in terms of the revenues that they generate have been the telephone and the facsimile machine. Both permit human communication to take place in a relatively rich way as compared to other existing text-based media. In the case of the facsimile machine, the technology released written communication from the straitjacket of the telex. The facsimile permitted typed documents to be annotated, to be supplemented by drawings, and, if required, to be transmitted in hand written form. In the commercial banking sector, time-sensitive communication that was once almost exclusively telex-based has switched nearly entirely to facsimile.

While tacit knowledge can be held in its own right in the minds of bank staff, information representing abstracted and codified data requires a tacit context if it is to be applied. The tacit dimension of communication is an indispensable element of information production and exchange. Following the definition of knowledge as a disposition to act, tacit knowledge can be understood as a 'disposition to codify and abstract' based upon prior experiential learning (Boisot 1995). Information refers here to data that have been rendered meaningful through the codification and abstraction process. In this sense, only the components of knowledge can be tacit, since, by definition, all information involves a certain level of codification and of abstraction. Tacit knowledge will always define the context within which information can be understood.

Consideration of the importance of tacit knowledge complements the explanatory power of the information-processing model. As Polanyi (1966: 21) points out:

a mathematical theory can be constructed only by relying on prior tacit knowing and can function as a theory only within an act of tacit knowing, which consists in our attending from it to the previously established experience on which it bears. Thus the ideal of a comprehensive mathematical theory of experience which would eliminate all tacit knowing is proved to be self-contradictory and logically unsound.

A component of tacit knowledge is what Polanyi terms 'indwelling'—that is, the ability to maintain awareness of the broader context while considering the particular. Polanyi (1966: 23) describes this as tacit knowledge that 'dwells in our awareness of particulars while bearing on an entity which the particulars jointly constitute'. In order to accumulate this tacit knowledge, it is necessary to learn the component parts, confident in the knowledge that what was meaningless to start with has, in fact, a meaning that subsequently can be uncovered through a process of negotiated meaning that is performed by intermediaries. This idea is particularly helpful in understanding what the 'repersonalization' process entails. By extending the concept of 'indwelling', it can be seen that the 'tacit authentication' of data involves trust relationships and that this offers a way of understanding the dynamics of the repersonalization process.

Trust plays an important role in defining the nature of information exchange at the level of the institution. Greater levels of trust are required to convert data into usable information when the data have few reference points and they are very diffuse. Tacit authentication is a process that enables specific kinds of information exchange processes to be supported or mediated by ICT applications. The specific applications must involve tacit authentication procedures if they are to be integrated with the organizational routines, such as those involved in credit assessment, in the commercial banking industry.

A further feature of the tacit authentication process is related to organizational learning in institutions that depend heavily on authenticated data. Herbert Simon suggested that individuals are constrained by their ability to interpret complex reality—that is, they are boundedly rational (March and Simon 1958). This means that learning is socially constructed inasmuch as what is learned is profoundly associated with the conditions in which it is learned. The learning that takes place by an individual within an organization is highly dependent upon what is believed or already known by other members of the organization. Learning is constrained by a complex web of tacitly shared meanings. Tacit authentication is a key component of the process of evaluating the significance of new data and giving them meaning. The evidence from this analysis of commercial banks suggests that they rely heavily on a special kind of data. These data are often highly privileged, private and confidential, customer data. The particular characteristics of these data require a special process for transforming raw data into usable information and, in this process, tacit authentication plays a key role.

7. CONCLUSION

The analysis presented in this chapter examines intermediation within the banking industry and suggests that this process involves the repersonalization of data. Banks perform their role as savings intermediators, principally, by 'adding trust to data' or by engaging in processes that lead to the tacit authentication of data. This has important consequences for key bank decision-making processes, for the selection of ICT applications, and for the operation of communications networks.

Commercial banks depend on the confidential and proprietary data that are made available to them by their clients. These data are provided on the basis of trust, but they also require internally generated tacit authentication before they can become usable information for the purposes of credit assessment. Specific ICTs play a role in this intermediation process to the extent that they are used to preserve and enable a rich, highly tacit context for human interaction. This requirement influences the selection of ICTs by biasing the adoption of new applications in favour of those that enable the 'repersonalization' of data required for credit assessment. Banks have successfully adopted digital communications technologies and information-processing systems in areas of their business where greater emphasis is placed on the capacity to store and process data. But in core areas such as credit assessment the issues of trust and security are paramount. As a result, banks tend to favour directly controlled, proprietary communications networks, even when these networks may result in higher data handling costs.

The process of concentration that has characterized the emergence of key global financial centres in London, New York, Tokyo, and, to a lesser extent, Hong Kong and Singapore can be understood in a new light based upon the analysis in this chapter. Despite advances in ICTs and the extension of global networks that make it possible, in principle, to deal and trade in financial securities from any remote geographical location, London and the other global financial centres are maintaining and extending their hold on the market. In London, close human contacts and traditional forms of communication are assured by the concentration of financial services in and around the area referred to as the 'City Square Mile'. Despite its attendant high costs, this area contains the highest concentration of international financial institutions in the world. London has become the world centre for foreign exchange trading, Euro-bonds, and international equities.[12] It also has become the centre for international lending, with the number of foreign banks located in London steadily increasing over the past twenty years. Clearly these conditions are favourable for the repersonalization of data through tacit authentication and for the organizational routines that are believed by many of those within the banking community to be necessary for credit assessment. Whether the banks' pre-eminent position as intermediators will be sustainable in the future is likely to depend upon the extent to which future innovations in ICTs preserve the richness of communication that is required to perform core intermediation functions.

The analysis of intermediation through the lens provided by the 'repersonalization' of data has received limited attention in the research literature on organizational change, market restructuring, and the services provided by banks and other financial service companies. It presents a more complex picture of the use of ICTs than that offered by the information-processing model, which offers a partial view of the potential of ICT applications. The 'repersonalization' perspective draws on findings emerging from several

---

[12] For further details on London's role as a global financial centre, see Her Majesty's Treasury (1996).

social sciences including economics, sociology, and cognitive psychology, as well as the expanding literature that deals with 'knowledge management' (Davenport and Prusak 1998; Winslow and Bramer 1994). By promoting greater recognition of the complexities inherent in the production and exchange of data and information, this perspective may help us move towards more insightful analyses of transformations in the way knowledge is generated and applied. Extended to other aspects of banking and financial services, further research along these lines would also help to shed light on whether the major commercial banks are likely to retain their predominant role in the face of further technological innovation.

PART TWO

BUILDING CAPABILITIES FOR KNOWLEDGE EXCHANGE

8

Co-Design in Action: Knowledge Sharing, Mediation, and Learning

JANE E. MILLAR

1. INTRODUCTION

This chapter is concerned with the role of mediation in the co-design of business processes and the information systems that support them. Various forms of mediation, including intermediation and remediation, are shown to influence knowledge sharing and learning through collaborative participation in these contiguous design processes. The analysis focuses on the interrelationships between business process design, information system design, and the design of practices to stimulate learning and knowledge sharing in organizations. Building on theories of learning drawn from the cognitive-science discipline, concepts of 'strong' and 'weak' learning are developed and a notion of the 'zone of proximal development' is invoked to highlight the role of mediation in explanations of learning effects. These concepts are used to analyse and explain the results of co-design in two firms. The analysis identifies several factors that appear to have influenced the different outcomes of co-design activity in each of these firms.

Product and service markets are increasingly volatile and unpredictable. The outputs of production must meet increasingly short time scales and high-quality standards. These outputs must also be produced at a relatively low cost so that they can be sold on the market at competitive prices. The requirements for firms to be agile, adaptive, and responsive to market demand have led many of them to reflect on their strategic operations and, where necessary, radically to change or to re-engineer their business processes.

Business process re-engineering has received a mixed press. For some, it has become a 'euphemism for mindless downsizing' (Hammer and Stanton 1999: 108); for others, it is believed to have had a more positive impact on customer responsiveness, costs, revenues, and stock values. Through business process re-engineering, some business leaders have come to see their organizations as 'flexible groupings of intertwined work and information flows that cut horizontally across the business, ending at points of contact with customers' (Hammer and Stanton 1999: 108). Much of the literature on process re-engineering recommends that flexibility can be achieved through the creation of cross-functional, task-oriented teams of professionals that can be configured and deployed to work on important projects. It is assumed that knowledge sharing and learning among such teams will enable them to combine their expertise and apply it as necessary. Process re-design alters the way that knowledge and information are deployed in firms.[1] As a result, process re-design is often accompanied by information systems re-design and, in co-design, these two activities are closely coupled. In effect, co-design involves the collaborative participation of technical and business specialists in the contiguous design of business processes and the information systems that support them.

This chapter is based on the results of a comparative study of the ways two firms have implemented co-design in order to create process change (Millar 1996). This has had different outcomes for the firms. In the two firms, infrastructure and management support for co-design were aimed at stimulating the knowledge sharing and learning that are required for the creation of new information and business systems and a transition to fundamentally different trading practices. The analysis focuses on the interactions between attempts to stimulate knowledge sharing and learning, and the new business processes and information systems that were created as a consequence.

In the following section, the dominant drivers of co-design implementation in firms are examined. The knowledge management challenges that are associated with this practice, and particularly those that relate to knowledge sharing and learning, are examined in Section 3, where the notion of the 'zone of proximal development' is developed to highlight the role of mediation (intermediation and remediation) in the learning process. Rapid Application Development (RAD), an example of infrastructure support for co-design in firms, is also introduced in this section. In Section 4 the results of a comparative study of the use of RAD to stimulate knowledge sharing and learning during co-design are presented. The results are used to contest a commonly held belief that identical 'best practice' techniques to support co-design will have similar (beneficial) outcomes for different firms (Davenport 1993; Wood and Silver 1995). In effect, these authors argue that the results of the implementation of RAD in firms where the activities supporting production have been subject to mechanistic control will be no different from the results in firms where organic control has been practised.[2]

In the final section, the factors that appear to influence varying outcomes from the implementation of identical techniques to support co-design in different firms are examined. These show that the strong learning effects that are often assumed to result from co-design activities are neither automatic nor costless. Instead, they require

---

[1] A distinction is drawn here between information and knowledge. Information refers to raw data that have been organized in some way in order to give them meaning and relevance in a particular context. Knowledge derives from information (about experiences, norms, and so on, as well as information about events and facts) that has been evaluated and is then used as a basis for decision making and action. The evaluation process that transforms information into knowledge involves making comparisons, and contrasts, drawing analogies and making associations, and other similar mechanisms.

[2] According to Burns and Stalker (1966), mechanistic management systems employ a hierarchical structure of control and specialized differentiation of tasks. They are characterized by a tendency towards vertical integration and greater importance is given to knowledge and capabilities that are internal and local as opposed to those that are non-local and cosmopolitan. Organic forms of management are appropriate to changing conditions that present management with new challenges for action. They are characterized by a network control structure, lateral communication consisting of information and advice (as opposed to instructions and decisions), and collaborative work on production problems. These forms have been discussed by Gibbons *et al.* (1994).

extensive planning, preparation, and wide support often beyond the boundary of a single firm. These activities involve setting up elaborate systems of mediation to support co-design that include resources specially configured for learning and knowledge management.

2. CO-DESIGN PRACTICE

Analyses of the dynamics that are shaping product markets and that are responsible for reconfiguring production processes often emphasize the significance of knowledge and information in commercial activity. Firms are relying more and more on information systems to propel and support the flow of information that is necessary for production. These information systems include formal and informal social networks as well as technical systems based on information and communication technologies (ICTs). ICTs, and in particular, software, can support and potentially extend the range of process design options that are available to firms. For example, software can assist firms in their efforts to meet changing market requirements by enabling them selectively to reconfigure and automate support for the production process. The re-design of business processes to bolster production and the creation of information systems to reinforce revised business processes are increasingly linked or closely coupled together.

Knowledge production within firms has traditionally depended upon technically 'rational' (Schön 1991), functionally specific, departmentally bounded activities that are associated with the production of homogeneous knowledge, or what Gibbons *et al.* (1994) have called Mode 1 knowledge. For example, knowledge inputs to the in-house design and development of information systems to support production have typically been produced by staff in computing departments who have had little or no interaction with the users of those systems. But Gibbons and his co-authors argue that such discipline-bound knowledge production activities are being supplemented by practices that support the creation of heterogeneous, trans-operational, or Mode 2, knowledge. The production of this kind of knowledge is grounded in, and reflects, ongoing experiences and practices of production. Mode 2 knowledge production practices aim to stimulate social interaction and learning among loosely coupled communities of 'reflective practitioners' (Schön 1991; Weick 1976). These are temporary groupings of staff that are brought together for specific purposes and/or projects that cross several functional departments and may extend beyond the boundary of a single firm. The expected benefits of Mode 2 knowledge production activities have led many firms to redesign their business processes quite radically in order to support interdependencies and interactions between the traditional functional groupings that are characteristic of Mode 1 professional practice.[3]

For firms that have retained their linear, discrete Mode 1 knowledge production practices, the need to establish a close coupling between information systems and business

---

[3] For example, in terms of overcoming the difficulties of design tasks that involve ill-structured problems (Simon 1973), or capitalizing on opportunities for innovation and strategic flexibility (Sanchez and Mahoney 1996).

process design has, in practice, often resulted in information systems design initiatives that lead to business process change. Technology-led process redesign has not, however, been an unqualified success (Benjamin and Levinson 1993; Davenport and Prusak 1998). The information systems software that has been produced is often delivered late, is over budget, and is prone to many faults. Pessimists often maintain that to generate value from an IT project it is necessary to underestimate the benefits and overestimate the costs of obtaining those benefits.

Such problems have been blamed on the immaturity of software development practices, which are often regarded as being craft based and undercapitalized. Technical solutions have been sought, for example, through the use of new and improved programming languages and software development tools. Techniques and methods have been introduced in order to simplify, standardize, and impose some control over software professional work and to make the quality, cost, and delivery of its outputs more predictable. Yet analysis of common software problems has shown that the majority of them are not due to a lack of advanced development techniques. Instead, problems appear to stem from a lack of extensive communication between technical information systems specialists and the intended users of their systems (Easterbrook 1991). In effect, the failures of functionally bound, Mode 1 professional software development practices are driving efforts to re-engineer the software development process so that it can better support the production and use of trans-operational or Mode 2 knowledge.

These failures have reinforced contemporary trends towards co-design—that is, the collaborative involvement of technical and business specialists in the contiguous design of business processes and the information systems that support them. Co-design processes require firms to mobilize and combine their knowledge assets in order to find new ways to reveal and exploit latent sources of agility and responsiveness. These processes depend upon the production of trans-operational knowledge and this requires collaboration, knowledge sharing, and learning within hybrid, and often distributed, communities. During co-design, these communities participate in the reconceptualization of the design of information flows within production processes. The learning processes and knowledge production activities that their participation depends upon are examined in greater detail in the next section.

3. LEARNING AND THE PRODUCTION OF NEW DESIGN KNOWLEDGE

Co-design involves the collaborative production of business process and information system design knowledge. It requires the active and managed social participation of all stakeholders that are potentially affected by the business processes that are designed in parallel with the information systems to support them. Management practices during co-design are responsible for the knowledge sharing that occurs among the stakeholders, for the production of new knowledge, and for its appropriation by the wider business community. The techniques for achieving co-design involve stimulation of knowledge production, knowledge sharing, and learning. Learning in this case involves the creation of new knowledge out of existing knowledge that has been intermediated—that is, made accessible and relevant—through joint social activity (Blackler 1993; Lundvall 1992*a*; Vygotsky 1978).

Intermediation during learning is an interactive process that bridges human activity (including mental activity) and the cultural, historical, institutional situations in which it is situated. In effect, it gives social meaning to that activity. It achieves this through a process of social apprenticeship that offers insights into, and the experience of, new ways of working and acting with knowledge. Intermediation, therefore, refers to a particular way of organizing and managing knowledge production that carries with it the potential to shape and transform social activity.

In co-design, intermediation relies on apprenticeship in mature forms of co-design practice for effective learning outcomes. Mature co-design practices are not learned automatically or costlessly. Rather, they evolve through managed participation and progressive engagement in mature practice. Managed participation provides guidance in the language, cultural tools, norms, and values of professional co-design practices. Learning is, therefore, highly dependent on the specific management techniques that are used to support intermediation during joint social activity.

During apprenticeship, joint social activity enables a transformation in socially embedded knowledge into usable counterparts that enable their individual bearers to adapt their existing knowledge in order to exercise greater control over their behaviour. Knowledge management techniques that support intermediation aim to achieve the alignment that is required between new social/collective knowledge and existing personal knowledge in order for new meaning to be established and for learning to take place. These techniques attempt to accomplish alignment by loosening the association between the new knowledge that is to be shared and the idiosyncratic features of the context in which it was created.

As Wertsch (1985) notes, Vygotsky considered these management techniques to involve the 'principle of decontextualization of mediational means', a principle that Vygotsky (1978) argued was primarily responsible for developmental change interpreted here as 'strong learning'. Mediational means include the language, cultural tools, norms, and values of mature and professional practice. The application of the principle of decontextualization is believed to strip mediational means of their contextual specificities and to enable cognition to focus on conceptual generalities in any given context. It encourages the expansion, abstraction, reduction, and generalization of these new concepts, and their comparison and contrast with existing knowledge that has been generated elsewhere. It is expected to facilitate the creation of an 'associative organization' or network structure of knowledge whereby newly acquired knowledge and previously existing knowledge are brought into alignment and connected.

Knowledge that originates in any given social context is progressively internalized during the learning process. The internalization process implies a shift of control or responsibility over knowledge from the inter-psychological—that is, jointly controlled by professional and apprentice co-design practitioners through managed intermediation—to the intra-psychological—that is, independently controlled by the apprentice. As Cole (1985) suggests, during the learning process, the knowledge to be shared traverses a space, or a zone of proximal (or next) development (ZPD), 'where culture and

Jane E. Millar

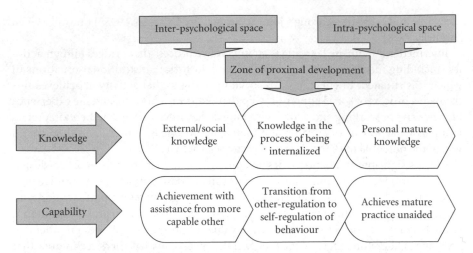

Figure 8.1. *Knowledge and capability in the zone of proximal development*

cognition create each other', between the domains of the inter- and intra-psychological. Fig. 8.1 shows the relationships between the zone of proximal development and inter- and intra-psychological spaces, and illustrates their implications for knowledge and capability development.

For Cole, the ZPD is the site where newly produced social knowledge is prepared for integration with existing knowledge. As a result, attempts to intervene and to provide effective guidance for knowledge sharing and learning need to be directed to the knowledge that is traversing the ZPD. Thus, according to this view, knowledge-management initiatives need to be pitched beyond existing knowledge and capabilities and to support the development of concepts and behaviour that are in the process of being internalized. Lave and Wenger (1991: 53) suggest that the outcome of managed intermediation is that shared knowledge will be transformed into new personal 'situated' knowledge that is enmeshed in the learning context.[4] This new knowledge may alter the way that the world is perceived by its bearers and challenge the way that they habitually apply it to the problems that they encounter.

In co-design practice, learning is often managed by an expert intermediary, who is responsible for guiding knowledge sharing and directing the production of new design knowledge.[5] This requires the expert to create a system of mediation (intermediation and remediation), including language, cultural artefacts, and embodied social rules, that is appropriate to the collaborative co-design task. In this system, remediation is a

---

[4] Lave and Wenger (1991) use the roughly equivalent term 'legitimate peripheral participation' to describe the apprenticeship process through which novice participants progressively engage with mature practice during learning.

[5] The requirement for expert guidance in knowledge sharing is partly due to the fact that learning is a conflict-ridden process and, according to Davenport and Prusak (1998), conflict is a threat to knowledge-sharing efforts.

necessary component because the existing knowledge that participants bring to the co-design initiative has been learned through intermediation and needs to be reactivated and reinterpreted to allow it to be used in a different context.

According to Lave and Wenger (1991), understanding the collaborative production of new knowledge requires particular attention to three principal, and inseparable, features of the learning situation. These features are: (1) guided joint social interaction— that is, management techniques to direct knowledge sharing and learning; (2) structural resources to support the learning process; and (3) the specific characteristics of a given socio-institutional context.

The first two of these features are part of the new system of mediation that has to be created to support co-design. The third forms part of the general-purpose system of mediation that is the active cultural medium for work practice. Figure 8.2 shows how these features influence the production of knowledge during co-design.

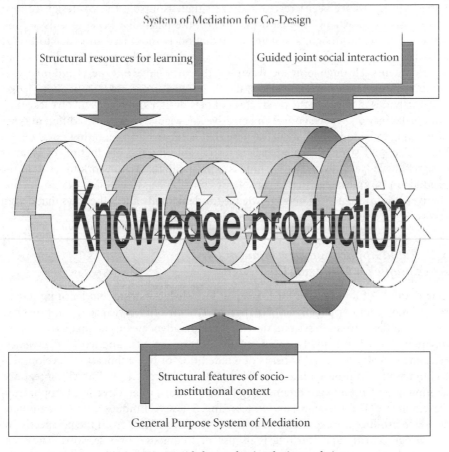

Figure 8.2. *Knowledge production during co-design*

3.1. *Weak Learning, Strong Learning, Creative and Novel Design Knowledge*

New knowledge includes knowledge that is creative as well as novel. The production of creative knowledge requires that one particular understanding of a problem domain is fundamentally transformed into another. Because of its dependence on conceptual transformation, creative knowledge cannot be produced before a particular learning event (Boden 1990). In contrast, novel knowledge is knowledge that has not been, but, in principle, could have been, produced before a particular learning event—for instance, as a result of other learning events. The production of novel knowledge does not depend upon transformed understanding, but relies instead upon encouraging new combinations of existing knowledge.

Participants involved in co-design are experts and novices in particular and different domains of knowledge. They bring their existing knowledge, albeit individually and separately, to the co-design initiative. These 'preinventive structures' (Sternberg and Lubart 1999: 7) have been created and are situated outside the co-design context and in other contexts of use that depend upon the particular backgrounds of their bearers. During co-design, a system of mediation is mobilized in order to bridge between the different types of existing knowledge that have been produced in various work domains. Through joint social activity, the knowledge of experts and novices is combined to generate novel knowledge and then transformed into creative knowledge.

All activities that lead to the production of new design knowledge involve learning, but the distinction between novel and creative knowledge enables two different types of learning to be identified: weak learning that is capable of generating novel design knowledge, and strong learning that is responsible for the production of creative design knowledge. Only strong learning is expected in this framework to result in the production of knowledge that is capable of contributing to genuine developmental change. This is because the new knowledge fundamentally alters the ways that design tasks are perceived and that design knowledge is applied.

3.2. *Rapid Application Development (RAD): A Technique for Managing Co-Design*

One example of a technique for managing knowledge sharing and learning during co-design is Rapid Application Development. RAD is a life-cycle process approach to business and information systems development. Its advocates argue that its use is synonymous with productivity gains, quality improvements, and cost effectiveness (Dudman 1994*a*: 30). RAD consists of a federation of five application development techniques: rapid prototyping, Specialists With Advanced Tools (SWAT), integrated Computer-Aided Software Engineering (iCASE) tools, interactive Joint Application Development (JAD) sessions, and timeboxing. These techniques are used simultaneously to produce the expected results. A key feature of RAD from the perspective of knowledge sharing and learning is its use of interactive JAD sessions. These are communication-intensive group workshops during which business members of a heterogeneous community of practitioners lead the production of business design information. The sessions typically are mediated by a facilitator who is often expert in both business and data modelling. The facilitator guides and directs participants' expressions of their existing knowledge of business process characteristics into an evolving model of a business task domain by applying the suite of RAD resources to the design task. This model not only forms the basis for the design of new business practices but also defines the essential features of a software system that would be required to support them. SWAT teams of software information systems professionals analyse the business knowledge that is produced during these sessions. They capture the elements that will be supported by the software and work with those elements to produce a rough 'first copy' of the software system.

The RAD approach to business process and information systems design is one that appears to provide infrastructure support for Mode 2 types of knowledge production practices. The approach builds upon guided joint social activity among heterogeneous communities of practitioners who are involved temporarily in the design and development of competitive software solutions that are responsive to situated business problems. The application of RAD aims to stimulate the rapid production of new, creative and novel, trans-operational knowledge by providing infrastructure support for knowledge sharing and the learning process. The facilitator role of managing the co-design process by stimulating and guiding knowledge sharing and learning is likely to be essential to its potential success.

The media hype that has surrounded RAD suggests that there is one best solution to the problems of software development.[6] It also creates the expectation that organizational process change is most likely to be effected by establishing cross-functional teams of software and business professionals—that is, by migrating to a new, Mode 2, pattern of knowledge production.[7] The results of the research that are presented next appear to challenge these views.

4. CO-DESIGN IN COMMUNITIES OF PRACTICE

A comparative analysis of the experiences of two firms (Firms A and B) provides a basis for assessing the extent to which different organizational contexts influence the outcomes of co-design processes that employ identical RAD-based techniques.[8] Firm A

---

[6] The RAD process was first formalized in the mid 1980s by DuPont and called Rapid Iterative Production Prototyping (RIPP).

[7] See e.g. Dudman (1994b), James (1994), Sharpe (1994), and Wood and Silver (1995).

[8] The empirical research was undertaken between 1990 and 1993 by the author, who was a participant observer of co-design activities on three projects, two in Firm A and one in Firm B. These activities spanned requirements analysis and high-level information system design. Data-collection methods involved the analysis of discourse, activities, and firm- and project-specific documentation, and forty-five semi-structured interviews (see Millar 1996). The data were organized, analysed, and interpreted in line with the conventions of Grounded Theory (see Glaser and Strauss 1967, Millar 1996, and Strauss and Corbin 1990). The study analysed the relationship between learning and technology transfer, and issues relating to training and learning to design and use software-based tools, techniques, and methods in the context of software and business process change.

was involved primarily in food manufacture and Firm B was involved in distribution. The firms were selected on the basis that they employed the same configuration of RAD techniques, including software-based tools (a particular iCASE tool), communications techniques (JAD), and notation methods (information engineering) in order to support the co-design practice. Also both had invested in the services of the same company to supply these tools, techniques, and methods, and to provide expert facilitation during the co-design process. In this chapter the analysis focuses specifically on the relationship between identical efforts to stimulate knowledge sharing and learning within the co-design communities in the two firms and the design of business processes and information systems that resulted from them.

4.1. *Knowledge Sharing and Mediated Learning during Co-Design*

Investment in identical RAD tools, techniques, and methods and their use to provide infrastructure support for co-design produced different results in each of the firms in this study. In Firm A, RAD-based techniques were superimposed on an organizational context that can best be characterized as mechanistic and as being geared to the production of Mode 1 knowledge. Here, the attempt to introduce co-design was not successful and there was little or no transition to the style of knowledge production that has been characterized as Mode 2.

In Firm A, knowledge sharing did result in learning and the production of new knowledge but this new knowledge was defective; it was damaged intentionally by co-design participants who colluded to combine their existing knowledge in inappropriate ways. This led to the design of a business process model that was unsuited to the business practice. According to the information system professionals and the external service providers who were involved in co-design, if this had gone undetected it would have generated a redundant information system. The new design knowledge that was produced appears to have resulted from a process of weak learning—that is, the knowledge that contributed to the business process design could have been produced, in principle, before the co-design initiative. It had not been produced, mainly because it was not suited to support the professional work practices of the community. This experience indicates that learning, which is often seen as a positive and progressive process, can produce the 'negative' or erroneous results that are often associated with trial and error.

In contrast, in Firm B, where organic management and Mode 2 style knowledge production activities were the norm, the implementation of co-design resulted in the production of new and creative knowledge that could not have been produced without the co-design initiative. This 'strong learning' effect seems to have been based on an understanding of business practice that was uniquely created through guided joint social interaction during the co-design initiative. It led to the production of a genuinely new representation of business practice that, in the views of the participants in the process, could be given effective information system support.

Examples of co-design practice are used here to illustrate the factors that appear to have interacted to facilitate knowledge sharing and learning in the two firms in ways that influenced the outcomes of the implementation of RAD.

4.2. *Co-Design in Action: Weak Learning in Firm A*

Firm A was rooted in traditional industrial manufacturing and it had implemented practices that are normally associated with mechanistic management and that are characteristic of Mode 1 knowledge production. These practices were generally considered by the managers of the firm to have secured the company its long-standing market dominance. Tradition, and the security offered by it, were often contrasted by the managers with the insecurity that accompanies the introduction of new technologies and the changes that have come to be associated with their use in firms. New technologies, therefore, were regarded as being detrimental to the maintenance of this firm's market position and corporate survival. The perceived irrelevance, and even damaging impact, of ICTs on this firm's core business practices and their reluctant acceptance by managerial level staff as tools, for example, to 'oil the wheels' of established business practice were reflected in a lack of integration between technology strategy and business strategy in the firm. The poor quality of information systems that had been produced in-house simply confirmed their irrelevance for management and testified to the in-house information systems department's professional ineptitude. The need to establish communication links with members of other business departments within the firm in order to achieve effective information system design was seen as providing further evidence of the department's technical incompetence.

The contraction of employment in Firm A further reinforced the desire to constrain the application of ICTs within the business. The threat of redundancy, together with the business and information systems department members' general lack of knowledge of, or involvement in, high-level management decision making, generated additional insecurity. Theories of conspiracy concerning the deployment of new technologies and their impacts on employment were rife at the time the research was undertaken. For example, business members viewed the use of RAD as a crude effort to expose them and rob them of their functional knowledge and as an attempt to automate their expertise. None of the participants in the co-design initiative had an interest in changing established working practices, partly because such change was perceived as a precursor to redundancy.

Employees in Firm A had generally been isolated from involvement in corporate decision making as a result of the hierarchy within the firm and the lack of transparency in decision-making processes. These features of company structure had given employees protection and defined the legitimacy of management in the past. However, these features also shielded the staff from opportunities for behavioural self-regulation and prevented them from accepting responsibility for their own problem solving. The self-regulation of behaviour and acceptance of a transfer of responsibility for problem solving are both salient features of the establishment of the trans-operational learning that is required for mature co-design performance. The co-design participants' lack of experience with these activities appeared to prevent them from taking an active role in their own learning processes. This produced a major 'scaffolding'[9] effect that precluded the establishment of productive relationships with the new techniques and work practices that were associated with the co-design initiative.

The specialized, functional fragmentation characteristic of traditional manufacturing practice was imposed on co-design projects by the firm's senior management. One project had initially been selected by an external service provider as a resource that could be employed to engage community members progressively in mature co-design practice. Fragmentation reduced that project's scope, and it was replaced by several smaller projects that mapped directly onto existing business functions. This denied the information systems and business department participants the opportunity to develop a trans-functional perspective on the process design problem, which involved many of the firm's activities and mitigated against the establishment of a trans-operational learning environment.

The limitations that were placed on the project's scope were intended to reduce the complexity of the task. However, the trivial scope of the revised projects simply aggravated the growing conflictual relationships between members of the information systems and business departments. Communication-intensive RAD techniques are geared towards encouraging learning by rapidly and progressively evolving appropriate and detailed transformations in the participants' conceptual representations of the business operations. The limited scope of the new projects meant that intricate, and often tedious, base-level details had to be dealt with before participants were fully engaged in mature co-design practice. These details often concerned areas where conflict was likely to be experienced. Conflict between members of the information systems and business communities was aggravated further by insensitivity regarding the interaction between the scope of the projects and the selection of business members who were taking part in the co-design initiative. The revised projects encompassed the operational features of an existing information system that had been designed and built by those business members.

Business participants in the co-design initiative became resistant to attempts to stimulate knowledge sharing. The traditional rationale for their business practice and their allegiance to it was reinforced by employment insecurity. This, together with the insensitive scope of the projects, generated hostility. Business members sought to protect the value of their personal expertise through deviant behaviour. The external facilitator, together with the information systems professionals in the co-design community, reacted defensively to business members' expressions of hostility. This reinforced departmental allegiances and re-established divisions between the business and technical functions and personnel within the firm. The participants in the co-design process eventually ceased to work as a single trans-operational team. The following provides excerpts from discussions during the co-design process to illustrate this point.

---

[9] The scaffolding metaphor is used here to refer to the structures that are used to provide supported situations in which people's current skills and knowledge are developed, and can be extended.

BUSINESS SUBJECT 1. What are these two doing (points to Information systems subject 3) and (Information systems subject 4)? I mean, there's four of you and you're mostly quiet and there's six of us and we're defining what a 'brand' is! We know what a brand is. We haven't the time. I'm not sold on the way we're doing this. Is it because we're training you (indicates to Facilitator 1)? What's your input? You've been writing down all the pearls of wisdom.

FACILITATOR 1. We can't, unfortunately, write things down in the uncoordinated way they've been coming out.

BUSINESS SUBJECT 1. Perhaps that's because we don't get where you're coming from. I think we've given a lot of input here for no return.

Confrontations like this characterized subsequent actions and interactions within this co-design community. Following one confrontation, a business representative assumed leadership of the business participants and actively resisted their participation in the joint production of trans-operational knowledge by expressing distaste for the act of joint social activity with the external facilitator.

BUSINESS SUBJECT 1. I don't like your words.
FACILITATOR 1. Well I'm trying to speak English.
BUSINESS SUBJECT 1. Well I speak English.
FACILITATOR 1. Then we'll get along then!
BUSINESS SUBJECT 1. (*Exaggerated hysterical laughter.*)

The protection offered by the conventional rationale for the firm's manufacturing practices encouraged business members of the community to unite and prevented early detection of their attempts to sabotage the establishment of a trans-operational learning environment through the production of false, discipline-bound knowledge. This was possible because the specialized nature of the discipline-bound knowledge was not equally shared between the heterogeneous participants who were involved in the co-design process. Undetected by members of the information systems department and the external service provider, distorted knowledge—that is, false information about business processes and their interrelationships within the firm—would probably have discredited the rationale for the co-design process. Discovery of the sabotage further fragmented community identity along disciplinary lines between business and information systems department members and gave additional strength to the predominant Mode 1 rationale.

Shared disciplinary identity between the information systems professionals and the external facilitator appeared to reinforce the facilitator's technical orientation and his tendency to align his views with those of the information systems representatives. For example:

FACILITATOR 1 (*to Information systems members during coffee*). Can someone drug [Business subject 1] while I poke her eyes out?
INFORMATION SYSTEMS SUBJECT 2. I think the big problem with this is, you're a professional and we know it. Even if the group thinks it's crap. (*Information systems subject 2 refers here to business members, whom she considers to be a distinct group within the community of practitioners.*)

The facilitator further encouraged the separation of identity between software and business department participants and dissuaded them from accepting the business community subject's authority over the co-design initiative.

FACILITATOR 1 (*to Information systems subject 2*). [Business subject 1] said some things I just
    didn't hear. I asked her to repeat it twice. Once she did and once she didn't. And sometimes she
    said some things that aren't English—that have some kind of hidden meaning.

For the remainder of co-design activity in this firm, business and information systems professionals and the external facilitator pursued their separate, short-term,
functionally specific interests. Business participants' interests were secured through
their continued resistance to communication with the other participants and their
covert production of anomalous knowledge. Their hostility to trans-functional collaboration meant that the facilitator and participants from the information systems
department were forced to resort to more traditional work practices for information
system design, which involved creating their own representations of business processes
without the co-participation of the business members.

4.3. *Co-Design in Action: Strong Learning in Firm B*

Firm B was aiming to secure and extend its market share by providing services that
both anticipated, and were responsive to, customers' needs. This was being achieved by
enhancing the range of services provided—for example, by offering new services or
adding functionality to existing services. Changing the firm's deployment of ICTs was
seen by the firm's management, as well as by members of the information systems and
business departments, as one effective means of altering the way that services were
being provided.

Firm B was committed to the strategic exploitation of interactions between software
systems and business processes. Its dependence on business process flexibility and creative change in order to protect and extend market share appeared to provide a supportive environment for the introduction of RAD-based techniques for co-design
activities. Employees across the firm and those working in its supply chain organizations were united in their support of the pursuit of common competitive goals through
the application of ICTs. As a result, participants in this co-design initiative were drawn
from its retail outlets and supplier networks.

Transparency in decision making within Firm B had helped to minimize the more
negative effects of hierarchy on interdepartmental information sharing in the firm and
had combined to achieve a balance of authority and legitimacy with considerable
employee autonomy and empowerment. This had contributed to the creation of a culture in which participants were accustomed to taking a leading role in managing their
own learning experiences, regulating their problem-solving behaviour, and accepting
responsibility for performing tasks that were transferred to them. This culture seemed
to have prepared the participants well for their active and positive involvement in the
co-design activity.

Business and information systems department participants had made extensive
preparations for implementing the co-design practice. For example, they had
appointed a process champion to raise corporate awareness of the co-design initiative
and to establish momentum for collaborative participation.

INFORMATION SYSTEMS SUBJECT 1. [The external service providers] are good. Especially [Facilitator 2] because he's so experienced across a range of development projects. Because he's outside [Firm B], he can command attendance and he doesn't impose his own perspective on a system. In order to get this going I had to sell RAD at many levels of the organization. He's [referring to Facilitator 2] the best, but you can imagine, he's not cheap. I had to start at the top and work down, carrying them all the way.

Early involvement had prepared participants for the adoption of new relationships to practice and the techniques that would accompany successful co-design implementation. For example, members of the information systems department traditionally had developed systems by imposing their own solutions on business problems. They had established ownership of what the business participants considered to be low-quality software applications that had failed to provide support for their strategic operational requirements. This had reduced the collaboration and communication between the information systems department members and the business community.

INFORMATION SYSTEMS SUBJECT 1. This method of working is long overdue. So far, we've been fire-fighting with inappropriate systems being developed because there have been no RADs—no view from business members. IT staff have had ownership of projects, not the business people. . . . It's about the management of change. It's about culture, ways of working in the organization depending on your background.

Low morale among the information systems professionals who had been deeply frustrated by their wasted software development efforts, together with the information systems manager's desire to enhance the professionalism of software practice, had created a context within which members of the information systems community were eager to embrace change. It was anticipated that creative and appropriate strategic business opportunities would be revealed and stimulated by the use of RAD-based techniques. These positive expectations framed the selection of the task project to be developed through co-design and encouraged participants to accept the risk of co-design failure.

INFORMATION SYSTEMS SUBJECT 1. We've been looking for the right pathfinder project for a while now. This project isn't perfect; it's too high profile, so it's a very high risk if it fails. It'll look very good if it works though!

The role of the external facilitator in managing joint social activity during the co-design process appeared to have been critical to the successful establishment of a trans-operational learning environment in Firm B. The facilitator was very experienced in implementing systems to mediate change. He quickly assumed his role as a charismatic leader of the group. The participants accepted their roles as well as his authority and their relationship evolved into an instance of mature co-design practice. This outcome further legitimated his facilitator role in managing and directing their learning processes.

The facilitator set clear boundaries and goals for the co-design task. He also 'contextualized' and 'depersonalized' the participants' involvement in the co-design process. This encouraged their enrolment in the community and strengthened the establishment of a group identity.

FACILITATOR 2. The first problem concerns the customer and about lines displayed but not stocked. Quickly dealing with this will allow us to add value to the customer. We are looking for a solution for the business—not for you, but for the business. We don't want you to provide us with a cul-de-sac. It will be a neat trick if you can do it! Good luck.

The facilitator was clear to state his alignment with the information systems community, but he used this to amplify the information systems department members' lack of business understanding, to establish empathy with the business members, to enrol them as participants in the co-design initiative, and to prepare them for a transfer of responsibility for knowledge production. 'The IT department knows nothing about (target task domain). We aim to get that information from you, get it on the board and analysed in real time. I'm not an expert in this area, you are.' The business participants soon became aware of the extent of the facilitator's business knowledge and this contributed to his ability to win their respect.

The facilitator monitored his use of language during the co-design process. For example, 'We will endeavour as far as possible to avoid using IT words, so if a strange word comes out—shout at me!' He deployed his use of language, in conjunction with the structure of the co-design session, to shepherd the members towards increasing engagement with the discourse of mature co-design practice. This was effective in encouraging their collaborative participation. The nature of the knowledge that they produced during co-design did appear to be progressively transformed. Knowledge sharing, learning, and changes in the conceptual understanding of the project task enabled a variety of representational schemes to be applied to business functions and tested for their appropriateness. It also encouraged consideration of alternative competitive opportunities for the firm. The evolution of participation in the co-design initiative was important, not only for generating a trans-operational learning process, but also for stimulating collaborative creativity and developing alternative views of the firm's competitiveness. For example, the facilitator tended to structure and pace the evolution of complex debates by introducing an important, and potentially conflict-laden topic before a break. Alternatively, as the example below shows, he sometimes began to negotiate a complex topic and then would interrupt the debate by presenting less conceptually demanding information.

BUSINESS SUBJECT 3. Are we looking for solutions where a product may be stocked usually but the store may not have it right now?
FACILITATOR 2. Yes. (*He explains why the company needs to support this.*)
BUSINESS SUBJECT 3. What about where you've got, because of store size, fewer products in smaller stores so some items will not be in stock in these stores which will be stocked in larger stores—but these are not special order items.
FACILITATOR 2. Yes, it all needs supporting. This will potentially support all the products in the catalogue.
BUSINESS SUBJECT 7. What about products which don't have bar codes?
FACILITATOR 2. That could be a problem. How many lines aren't coded?
BUSINESS SUBJECT 7. About 28, 000!
FACILITATOR 2. So, that's something that needs to be catered for. Let me take one step back . . . I want to explain the RAD process.

This incremental approach allowed one layer of the scaffolding supporting the learning process to be constructed and secured at a time. Each layer of scaffolding provided a foundation of 'already-known' knowledge for the new concepts that were being developed. This approach enabled bridges to be constructed to link more familiar, already known, concepts with novel or creative ideas. In this context, established knowledge, and its associated scaffolding, played the role of providing familiar bridging concepts that were available to be built upon on return to the debate. The earlier debate was used to provide a context for the subsequent negotiation and the topic under negotiation was progressively decontextualized. In addition, by carefully managing the interaction in the 'zone of proximal development', the participants did not engage in premature debates on complex subjects that could have discouraged them from further participation. This process appeared to stimulate their enthusiasm for continued participation in the co-design initiative.

The facilitator used rich communicative resources to structure and pace the members' evolving relationships within co-design practice. Perhaps the most significant resource was the use of humour to introduce temporary relief from the pressures of negotiation. For example, in a debate concerning customer-ordering procedures, the business representatives couched the discussion in examples. The facilitator guided their knowledge production activities through a process of transformation that was required in order to construct a business process design that would encompass all their different examples. Humour served to depersonalize and decontextualize the evolving, collective, creative representation of the business process. The participants' use of examples drawn from their own experiences suggests that they were playing an active role in structuring their own resources for learning. This mutual scaffold-building process also suggests that they had established a communal identity and had accepted, and even begun to engage in, the transfer of responsibility for co-design practice.

FACILITATOR 2. So, we need to check if a deposit is required.
BUSINESS SUBJECT 7. If it's in the catalogue, I'm not worried about a deposit.
FACILITATOR 2. We need to check special circumstances then.
BUSINESS SUBJECT 5. Isn't time the variable? So, if delivery is in 7 days, then they can get payment in full then.
BUSINESS SUBJECT 1. The important thing that's come out is that it's at the store level, that decision.
FACILITATOR 2. (*plays with his stack of felt pens*). This indicates how bored I'm getting! OK lets have an 'assumptions' list. (*Writes on flip chart: 'Ability to define deposit criteria needs to be set at the store level'*) Right, in terms of the big picture, is that what customer ordering is all about?
BUSINESS SUBJECT 5. The levels are not equal, like, 'PLACE ORDER WITH SUPPLIER' is a very complex process. There may be many suppliers—such as 'I'll have this wallpaper and that wallpaper.'
BUSINESS SUBJECT 3. What about different times of delivery as well, such as 'I don't want the conservatory furniture before the conservatory'.
FACILITATOR 2. (*adds 'SEQUENCING OF ORDER' to the evolving business model on the white board*). Hm (*laughs*) . . . Lovely idea that! Wallpapering a greenhouse!

Other examples of the facilitator's use of contextualization and decontextualization to stimulate knowledge sharing occurred during the negotiation of the back-ordering process. Here, the facilitator oversimplified the back-ordering process in order to provoke members into adding details and to encourage a further transformation of their representational model of the business process:

FACILITATOR 2. Let me just check my understanding against your understanding. The problem is that existing back orders aren't already accounted for. So, you've got a hole in the shelf and someone says 'Oh! I'll have some of that 'cos it's all gone.' And he goes and orders it and you say, 'You can have the next one in then' and someone else comes in and says 'I'll have some of that hole', etc.

The selection of the members of the co-design team, and, in particular, the involvement of representatives from organizations in the company's supply chain network, also seem to have been important for the success of the co-design activity. This enabled a realistic consideration of, and debate about, the potential benefits for the competitiveness of the firm from the application of the business process design that was under development. Further, it allowed the re-engineering processes that were associated with the co-design initiative to be begun, despite the fact that this would involve major structural changes in the firm's business practices. For example, changes in the minimum ordering process to enhance competitiveness were made during co-design and these had implications for the supplier firms' profits, product delivery schedules, including arrangements for loading delivery vehicles; payments and established accounting practices; and product packaging. The new practices required alterations in established work practices in retail outlets, for instance, changing the way that products were identified, and they involved internal redesign to increase storage space.

The successful resolution of these problems during the co-design initiative was considered by the participants to be a creative achievement with major implications for the firm's competitiveness. The evolution of a mature co-design practice was sensitively negotiated by the facilitator in order to enable community members increasingly to participate in the process of creating and transforming collective social knowledge about particular domains of practice. For example, one debate about the ordering process was resumed four times. Each re-negotiation transformed the representation of participants' understanding of the process. These transformations were accompanied by simultaneous changes in the way that the participants' knowledge was applied to representations of that business process.

The facilitator modelled the evolution of community members' knowledge in information engineering (IE) notation on a white board in real time. Iterative negotiations and incremental modelling of the information system went hand in hand. The facilitator interspersed exposure to the software design process (Example 1) with other cognitive tasks such as process naming (Example 2).

Example 1

FACILITATOR 2. Let's progress the model a bit more. I won't go through too much 'cos it gets boring. The reason for doing this is because it lets the IT guys [*sic*] get a really good structure going. I know it's mind numbing but it pays off later.

Example 2

FACILITATOR 2. OK, 'CONFIRM QUOTE?' 'CONVERT QUOTE?' 'CONVERT QUOTE TO ORDER?' which?

BUSINESS SUBJECT 7. 'CONFIRM QUOTE AS ORDER'.
FACILITATOR 2. Oh! Nice one, I like that! You've played this before. OK Next one. 'STORE DELIVERY'—only I don't like that name 'cos it's confusing everyone. Four words (*starts to mime*). It's 'CONFIRM something' isn't it?
ALL. Yes!
FACILITATOR 2. So, 'CONFIRM' is good. Confirm what? Second word (*reverts to mime*).
BUSINESS SUBJECT 7. 'CONFIRM PRODUCT AND DELIVERY DATES?'

The allocation of roles and responsibilities for the co-design activity was established prior to the initiative and this seems to have helped to deflect any potential friction. For example, conflicts that were not related to the task at hand were channelled through participants who had accepted responsibility for resolving these issues outside the JAD session. Conflicts relating to the co-design process were directed to an internal process champion and resolved within the session. In addition, participants agreed to sponsor the further discussion of issues concerning the competitiveness of the firm that were raised during co-design. This extended the boundary of the initiative beyond its immediate context and secured the links between decisions taken during the co-design process and their eventual business implementation.

5. CONCLUSION: OUTCOMES OF CO-DESIGN ACTIVITY

Analysis of the co-design processes that were experienced by the two firms helps to illustrate the indivisibility of the co-design practice and the learning process. It highlights the extent to which knowledge sharing, learning, and knowledge production within firms is situated in particular contexts. In the case of Firm B, learning was mutually constituted during co-design activity by the systems of mediation that were employed to organize and manage the process. These included management practices to support knowledge sharing and learning, and resources to support joint social activity and their implementation, together with a sensitive awareness of the context of organizational change and associated cultural features.

The outcomes of co-design exhibited 'weak-learning' effects in the case of Firm A and 'strong-learning' effects in the case of Firm B. Therefore, it seems clear that the use of techniques such as RAD to support co-design does in fact produce variable outcomes when they are applied in different contexts. These outcomes should be expected to generate creative or novel design knowledge depending on how the systems of mediation combine to deliver learning effects.

A comparison of the experiences of the two firms enables the identification of factors that seem to have interacted to yield the results of co-design practice in each case (see Box 8.1). The factors shown in Box 8.1 mutually influenced the outcomes experienced by each of the firms reported in the previous section. In Firm A, these resulted in both weak learning and the production of novel design knowledge. In Firm B, they resulted in strong learning and the production of creative design knowledge.

The results of the study reported here suggest that infrastructure support for co-design activity is a necessary, but not sufficient, condition for generating strong learning effects and an anticipated shift to the knowledge production practices that are

Box 8.1. Factors responsible for outcomes of co-design in firms

Categories	Dimensions and factors
The general purpose system of mediation	
Socio-institutional context	Market stability/volatility
	Closeness/distance of supply chain relationships
	Socio-political transparency/opacity
	Strategic integration/disintegration for technology and business
The new system of mediation for co-design	
Learning resources to support co-design	Preparing the host culture for change
	Identifying roles and responsibilities
	Establishment of a system of conflict resolution
	Defining the team structure
	Scope of the task environment
	Choosing the right facilitator
Management techniques to direct knowledge sharing and learning	Promoting knowledge retrieval
	Scaffolding for new understanding (creating mental associations)
	Constructing mental bridges (drawing mental analogies)
	Effective use, timing and structure of dialogue to stimulate learning in the zone of proximal development
	Transferring responsibility for new practices of knowledge production (sponsors for change)
	Managing and directing conflict

Source: based on Millar (1996).

characteristic of Mode 2. The implementation of identical techniques to support knowledge sharing and learning during co-design activity appears to lead to different outcomes depending on the operational contexts of the firms. For example, in a firm where mechanistic control and Mode 1 knowledge production practices are the norm, a different outcome is likely from that for a firm where an organic form of management and Mode 2 knowledge production practices are customary. The empirical study of the actual experiences of firms also suggests that the systems of mediation that support Mode 1 knowledge production may actually inhibit the spread of both those systems of mediation that support the co-design process and encourage a transition to Mode 2 knowledge production. This is because the traditional Mode 1 practice appears to impede the generation of the trans-operational knowledge that is required to achieve a reasonable return on investment in a co-design process that yields new business practices and information systems that are likely to contribute to strengthening the firm's competitiveness.

The implications of this analysis are, first, that 'best-practice' models of co-design involving the stimulation of joint social activity among temporary teams of heterogeneous professionals may not be suited to all organizations. Secondly, techniques that support Mode 1 and Mode 2 knowledge production may be mutually incompatible, although this observation should be subject to further systematic analysis of a larger sample of firms. Thirdly, the strong learning effects that are assumed to result from co-design and the implementation of Mode 2 knowledge production practices do not emerge automatically and are not costless. Instead, they require extensive planning, preparation, and wide support both within the firm and, in some cases, beyond the boundary of the firm. Successful co-design requires the establishment of elaborate systems of mediation including specially configured resources for learning and knowledge management. Sensitive implementation of systems of mediation for co-design and their progressive integration with the general purpose system of mediation that supports the organizational culture of the firm are important features of the activities that are likely to sustain momentum for process change within firms. A lack of investment in infrastructure to support co-design is likely to create inefficiencies and drain resources by prolonging outmoded work practices and by reducing the commercial exploitation of new tools, techniques, and methods for information systems and business process design.

9

The Distribution of Spatial Data: Data Sharing and Mediated Cooperation

UTA WEHN DE MONTALVO

1. INTRODUCTION

The urgent need to encourage the sharing of the substantial stocks of data that are accumulating as a result of the use of geographic information systems (GIS) in industrialized and developing countries is on the agendas of many public- and private-sector organizations. It is acknowledged that increased sharing of the spatial data required by these systems would help to reduce duplication of resources and to defray the costs of collecting and using such data. Nevertheless, the data-sharing practices of those involved in the GIS community have received very little attention in the technical literature or in studies of the social determinants of GIS use (Campbell 1991; Masser and Onsrud 1993; Onsrud and Pinto 1991; E. M. Rogers 1993).[1]

These systems represent a growing application area for information and communication technologies (ICTs).[2] They are regarded by GIS software producers and by many existing and potential users as being a very important component of the information environment. This environment may be improved by the use of data held by GIS so that decision-makers in developing countries have an improved basis upon which to take decisions about how best to allocate resources to address a wide range of problems (Schwabe, O'Leary, and Sukai 1998). The sharing of computerized information resources—in this case, the spatial data required as inputs to GIS—is difficult to promote because of the variety of stakeholders involved. Their interests in achieving greater data sharing vary enormously. Encouraging and facilitating increased spatial data-sharing involves national, regional, and global initiatives that aim to secure cooperation between stakeholders so that the potential benefits of innovative GIS applications can be made available to a wider number of users.

The facilitation of coordination and interaction between the array of actors with overlapping roles that are involved in collecting and using spatial data is treated in this chapter as a form of intermediation. Intermediation may occur as a result of actions

---

[1] This chapter is concerned with data-sharing issues relating to *digital* spatial data but, hereafter, reference is made only to spatial data.
[2] See Dataquest (1996) for estimates of market growth.

taken by institutions that seek to alter the perceptions of other actors about the potential benefits of enhanced spatial data sharing.

There are numerous studies on the diffusion and use of new technologies such as GIS. Many of these have been conducted within the 'diffusion of innovations' framework originally proposed by Everett Rogers (1968) and subsequently updated (E. M. Rogers 1995), and these and other studies are grounded in a wide variety of disciplines. The main focus of this work has been on how the process of diffusion can best be described. This chapter focuses on a particular aspect of the preconditions that provide a basis for new technology-users to behave in ways that are likely to increase the potential benefits of the application of a specific technological innovation—that is, GIS.[3] In this case, the analysis focuses on the determinants of a willingness of spatial data-users to share their data across organizational boundaries.

Research on interorganizational information flows and the economic and social determinants of whether information system-users are likely to seek to retain information within organizational boundaries or to enable it to flow beyond those boundaries is equally vast (see Goldhaber and Barnett 1988; Macdonald 1992; E. M. Rogers 1982). Much of the discussion in the literature is located within the field of the economics of information (Antonelli 1992; Arrow 1984) or within the management of interorganizational alliances, partnerships, and network arrangements fields (Alter and Hage 1993; Galaskiewicz and Zaheer 1999; Harrigan and Newman 1990; Porter 1991). In both instances, the primary concern is with the potential benefits and costs of treating information as a private or public good. Research in these traditions has not been concerned with the underlying motivations that precipitate action that results in the release of information beyond the boundaries of an organization. Such motivations are associated with attitude and belief systems, which, in turn, are influenced by a large number of cultural, social, political, and economic conditions. In this chapter, a framework for analysing these attitudes and beliefs in the context of GIS use is developed and applied to offer insight into the preconditions for sharing the complex datasets collected for use in GIS applications.

Initiatives are being introduced to create an improved infrastructure to enable data sharing, but the promoters have yet to acknowledge the behavioural determinants that need to be considered if their efforts are to be successful. These behavioural issues are just as important as ensuring that the necessary technical standards, procedures, and policies for spatial data interchanges are in place. The analysis in this chapter is based on an empirical study on the incentives and disincentives that seem to give rise to, or to curtail, data sharing by stakeholders in the public and private sectors. The results provide a basis for suggestions about how the relationships between the actors contributing to the use of GIS might be more effectively organized.

The next section introduces basic characteristics of GIS, examines the extent of the spread of GIS in developing countries, and highlights features of the spatial data that are required for GIS use. In Section 3 the concept of institutional intermediation is introduced and developed to provide a framework for examining the developments in

---

[3] For reviews of the diffusion literature, see Coombs, Saviotti, and Walsh (1987), Lissoni and Metcalfe (1994), J. S. Metcalfe (1988), and J. Sarkar (1998).

the GIS community that are emerging at the national, regional, and global levels. In Section 4 a case study is presented that focuses on the National Spatial Information Framework (NSIF) in South Africa. Specifically, the effectiveness of the NSIF is examined to identify some of the determinants of the willingness of individuals to engage in spatial data-sharing. Spatial data-exchange initiatives are shown to play an important role as institutional intermediaries in the use of GIS. However, if that role is to be effective in encouraging greater data sharing, it will be necessary for policy-makers to understand how a number of complex motivations influence decisions about whether to engage in this kind of sharing activity.

2. WHY DO SPATIAL DATA MATTER?

Geographic information systems (GIS) are ICT applications that depend upon digital spatial data.[4] GIS were developed initially within the discipline of geography during the 1970s as the result of projects to digitize maps of geographical terrain. GIS technology involves software and hardware computer applications that evolved rapidly during the 1990s and that are in widespread use outside the discipline. GIS applications combine spatial and various kinds of socio-economic data from numerous data-sets. They are used to support a growing number of activities such as planning, policy making, and monitoring with regard to natural and other types of resources and infrastructures such as transport networks, telecommunications networks, and waterways. The combination of spatial and socio-economic data in these systems allows the resulting composites of information to be used, for example, to add a geographic dimension to development planning. GIS may be used to make the information visually accessible for many users and they may be designed so as to locate the data in a context that lends itself to meaningful interpretation from the perspective of the user (Swarts 1998). The application of GIS is widely believed to have a potentially very significant role to play in facilitating 'the task of planning and determining priorities in assigning specific resources to specific areas' (Schwabe, O'Leary, and Sukai 1998: 4). The use of GIS offers a means of structured storage, retrieval, and manipulation of information and, therefore, the opportunity to use changes in this information in decision making in various kinds of planning activities.

2.1. *The Use and Spread of GIS*

GIS are employed by the public and the private sectors in the industrialized countries to support a wide range of activities including town planning, marketing, and environmental monitoring. In these countries, GIS applications are available at relatively low cost to the user compared to the levels of skill and equipment that are available in many developing countries. The trend towards Internet- and intranet-embedded GIS applications and the integration of map information into management information systems are factors that are helping to reduce the price of GIS applications. In

---

[4] Spatial data specify the location and characteristics of physical phenomena and may be held in analogue or digital formats.

developing countries, GIS are increasingly being used by the public and private sectors and by development agencies. In the 1990s GIS requirements for computer-processing power declined substantially and personal-computer-based applications became commonplace.[5] Box 9.1 illustrates the variety of GIS applications being utilized for development planning and by development projects.

The spread of GIS applications has been relatively rapid. Van Helden (1999) refers to 461,300 licensed users worldwide with North America accounting for 71%, Europe 14.6%, Asia 6.7%, South America 4.1%, Australia 2.2%, and Africa 1.3%.[6] In terms of economic value, Dataquest (1996) estimated the GIS market worldwide to be worth $2.6 billion in 1995, a 16.5 per cent increase over 1994. More recently, a strong growth rate has been confirmed for the GIS market worldwide (Dataquest 1999).

Although the share of developing-country implementation of GIS is much smaller than that in industrialized countries, there is evidence that it is growing (Christiansen, Christ, and Hansmann 1997; Nuttall and Tunstall 1996). This may be due to their incorporation within development planning initiatives and their use by development

Box 9.1. Areas of GIS application in development projects

Regional rural development/resource management
Forestry/nature protection
Environment and resource protection
Projects falling into more than one category
Plant production/plant protection/agrarian research
Animal husbandry/veterinary/fishery
Water supply and waste management
Education science
Urban planning/urban development
Food security project/emergency refugee help
Energy and transport
Agrarian policy
Health
Irrigation

Source: based on Christiansen, Christ, and Hansmann (1997).

---

[5] Three categories of GIS can be distinguished (Wilson 1999): scientific and research applications, enterprise applications, and desktop applications. In the sophisticated applications category, the leading GIS in scientific and environmental research applications is Arc/Info by Environmental Systems Research Institute (ESRI). Arc/Info can run on a variety of systems (e.g. Unix, AIX, or Windows NT 4.0). Requirements for Arc/Info 8.0.1 to run on Windows NT 4.0, for example, are Pentium (or higher), at least 128 Mb RAM, 200 Mb Paging File, 2.5 Gb free disk space, plus a database management system such as Oracle 8 Server. In the basic, desktop applications category, MapInfo is one of the leading companies. System requirements for MapInfo Professional are: 486 PC or Pentium, at least 8 Mb RAM, VGA monitor, 60 Mb disk space, and Microsoft Windows 9x/NT.

[6] Compared to an estimated 93,000 GIS sites worldwide in 1995 (Burrough and McDonnell 1998).

agencies in projects concerned with geographical issues of location and distance. The generic nature of GIS applications and the evidence of the spread of GIS in developing countries in recent years strongly suggests that these applications are perceived by stakeholders in these countries as being potentially useful in their various circumstances and with respect to a wide variety of development goals and aspirations.

The availability of spatial data is a prerequisite for the effective use of GIS. The spread of these applications within industrialized and developing countries is producing a surge in demand for digital spatial data from many organizations that are not able to obtain these data on their own for both economic and technical reasons. The characteristics of spatial data are examined next, together with the way that these characteristics affect their availability.

2.2. *Spatial Data Characteristics*

Spatial data specify the location and characteristics of physical phenomena.[7] Although GIS-users generate data that are specific to their areas of application, they nevertheless depend upon the availability of certain kinds of data-sets that contain a wide range of complementary spatial data, as shown in Box 9.2.

Spatial data have several characteristics that appear to set them apart from other types of data that may be collected. For instance, the collection process for spatial data is complex and expensive, involving a variety of technologies and procedures ranging from ground-based traditional survey methods and photography, to remote sensing, aerial photography, and global positioning systems. Furthermore, once established, spatial data-sets need to be maintained and updated frequently or replaced to reflect changes in the physical environment due perhaps to transport infrastructure and urban expansion or natural disasters.

Across the spectrum of use of GIS from mapping to development planning, GIS applications have fundamentally different requirements in terms of the levels of data accuracy and complexity that must be achieved (Abbott 1996). Fig. 9.1 shows the four main types of data and their characteristics with respect to the relative amounts of detail required in the accuracy of spatial data referencing and with respect to the complexity of the data-sets. Problems often arise with respect to verifying the accuracy of data, and incompatibility of collection procedures can create uncertainty about locational attributes of the data owing to differences in the standards for geographical referencing systems; variations in scale and map-scale-dependent accuracy and resolution factors; variations in the parameters for generalization owing to sampling errors; boundary and location data errors; and differences in the timing of observations. All these factors affect the reliability of spatial data (Coppock and Rhind 1991; Flowerdew and Green 1991; Openshaw, Charlton, and Carver 1991). The scale and error characteristics of a spatial data-set must be well documented to ensure that the

---

[7] Although the terms 'spatial (or geographic) data' and 'information' are frequently used interchangeably, this chapter will refer to spatial data. Spatial data-sets, once used in particular contexts, constitute spatial (or geographic) information.

Box 9.2. Types of spatial data

Geodetic
Land surface elevation/topographic
Bedrock elevation
Digital imagery
Government boundaries/administrative boundaries
Cadastral/landownership
Transportation/roads
Hydrography/rivers and lakes planimetric
Ocean coastlines
Bathymetry
Physical features/buildings
Place names
Land use/land cover/vegetation
Geology
Real estate price register/land valuation
Land title register
Postal address
Wetlands
Soils
Register of private companies
Gravity network
Zoning and restrictions
Source: based on Onsrud (1999).

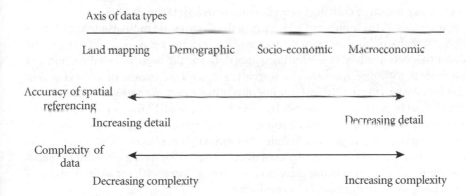

Figure 9.1. *Relationships between data types, accuracy, and data complexity*
Source: based on Abbott (1996).

data can be interpreted in a way that is consistent with the reliability established by the originator. Such data about spatial data or 'metadata' are especially important because the GIS software can conceal information about the level of uncertainty associated with a data-set from the user who is relatively unfamiliar with its operation.

Like other types of large-scale data-sets, as well as being affordable spatial data need to be up to date and reliable if they are to be a relevant source of information for their users. Many of the skills required to collect, acquire, integrate, and update the data-sets are highly application specific. This means that, although the prices of GIS applications software and hardware are falling, the costs of skill and training development across the spectrum of GIS-related data-collection activities and of using the systems are substantial.

In parallel with the increasing number of public- and private-sector users of GIS, most national mapping agencies that were traditionally involved in surveying and mapping national territory for military and government planning purposes have digitized their operations. These agencies are major suppliers of national spatial data-sets, although they operate on very different scales. In some developing countries, the national mapping agencies—for example, the National Statistics, Geography and Informatics Institute (INEGI) in Mexico and the Chief Directorate for Surveys and Mapping in South Africa—respectively, have invested heavily in the digitization of their map inventories and are using advanced geographic information technologies (D. G. Clarke 1997; Jarque 1997).[8] Nevertheless, because of the large number of potential application areas, the demand for different types of data-sets cannot be met by these organizations alone. Many users of GIS, such as oil companies and utility companies, which have the financial resources and spatial data-capturing skills, engage in the collection as well as the use of spatial data. They may make some or all of their data-sets available, at a price, to users external to their organizations in order to recover some portion of their investment. In this case, they act both as users and suppliers of geographic information data-sets. Other types of firms integrate and then sell existing data-sets as a means of creating value (Burrough and McDonnell 1998).

In summary, as the number of GIS-users in developing and industrialized countries increases, many public- and private-sector organizations are having to confront the substantial costs involved in generating spatial data and the need for collaboration and cooperation with one another. The search for innovative means of achieving cost reductions is creating incentives for numerous initiatives at the national, regional, and global levels. In Section 4, the particular experience of South Africa in such initiatives is examined. In order to understand some of the factors that mitigate against the kinds of cooperation that may give rise to scale economies in the collection and use of spatial data, it is necessary to look at the types of organizations that have an interest in spatial data sharing. The next section, therefore, outlines a framework for analysing a particular instance of 'institutional intermediation'.

[8] The Mexican National Geographic Information Systems (MNGIS), for example, encompasses capture, production, organization, integration, analysis, and presentation processes. The modernization programme has been carried out in each of the ten regional offices of INEGI in Mexico. The system includes advanced software and equipment, and seventy-three specialists were trained over a period of nine months (Jarque 1997).

3. INTERMEDIATION

The idea of 'intermediation' has been employed rather extensively within several academic disciplines. For example, in the library and information science discipline, information specialists (whether human or technical—for instance, expert systems) are classed as 'intermediaries' between people and the information they make available (Drenth, Morris, and Tseng 1991; White 1995). In trade and commerce, institutions such as banks and insurance companies are considered to 'intermediate' between buyers and sellers (Allen and Santomero 1997). In some cases, the traditional roles of these companies are being challenged by the arrival of suppliers of electronic commerce applications that seek to establish themselves as the new intermediaries of commercial transactions (Hawkins, Mansell, and Steinmueller 1999). In the context of development initiatives, non-governmental development organizations and donor organizations are often described as intermediaries, which seek to support and mobilize the internal resources of communities within developing countries (Lee 1998). The environment in which intermediation processes occur is a significant consideration for the analysis of such processes and it is important to differentiate between those processes that occur through interpersonal communication and those that occur at the organizational level (Schmittbeck 1994).

3.1. *Intermediation in the Context of GIS*

Common to all these conceptualizations is the notion that intermediation takes place when a particular type of actor (or intermediator) plays a role in (or intermediates) the interaction between two or more individual or organizational actors. The social and economic incentives to engage in an intermediated relationship are an important factor when examining the determinants of newly emerging, existing, or disbanded interactions involving third parties. In the case of abandoned interactions or 'disintermediation', individual or organizational intermediaries may be eliminated or they may be replaced by new intermediaries that are better placed to nurture and sustain certain types of relationships (Hawkins, Mansell, and Steinmueller 1999). For example, it has been suggested in the context of developing countries that the increasing availability of information sources via electronic networks may threaten to reduce or eliminate the significant intermediary roles of existing information specialists (Fourie 1999). However, the key issue in the present context of the analysis of the determinants of more extensive spatial data-sharing is not whether ICT applications create a basis for new forms of intermediation or disintermediation, but whether the way the intermediation process is organized is giving rise to intermediating agents (individual or organizational) that can play an effective role in achieving the desired outcomes for data sharing. An analysis of this issue provides a basis for assessing their roles and the extent to which changes in policy intervention may alter existing data-sharing practices. In this chapter, the focus is on emerging processes of intermediation between spatial data-users, GIS producers, and spatial data-suppliers. The principal issue that is examined is the extent to which the national spatial data initiatives are playing an effective intermediary role in enabling the wider distribution and availability of relevant, reliable, accurate, and affordable spatial data.

3.2. *National Spatial Data Infrastructures*

The activities of national spatial data infrastructures embrace both the context of collaboration between different stakeholders and the distribution of spatial data that individual GIS implementations require in order to function. The extent to which these initiatives are able to facilitate spatial data sharing can be more effectively assessed by focusing on the intermediation process rather than on the organizational structures and functions provided by these infrastructures. The term 'spatial data sharing' is commonly used in the GIS community, but it is defined here to refer precisely to making the digital spatial data used in GIS accessible to or from other parties. Data sharing or exchange relationships may involve barter arrangements, financial payments, or payments in kind.[9]

The Open GIS Consortium of public- and private-sector organizations is primarily concerned with the technical standards for spatial data transfer between different GIS software vendors (Open GIS Consortium 1999). However, the spatial data infrastructure initiatives that are emerging at the national level in industrialized as well as developing countries (and at both regional and global levels) are more comprehensive in their focus.[10] For example, the national spatial data initiative in the United States is described as involving 'the technology, policies, standards, and human resources necessary to acquire, process, store, distribute and improve the utilization of geospatial data' (Federal Register 1994: sect. 1a). Based on a worldwide survey of such initiatives, Onsrud (1999) suggests that they typically embrace the following components: core data-sets, metadata, data standards, and clearing-house arrangements.

The role of national spatial data initiatives can be considered at several different levels of intermediation. At the technical level, intermediation entails the negotiation of compatibility standards such as those for spatial data collection and exchange, and for metadata. Standardized documentation about spatial data is collected in the form of metadata describing the characteristics and indicating the owner of the spatial dataset. Standards must also be agreed for the collection and exchange of spatial data-sets, and their applicability for users, data-producers, and other organizations in the GIS 'value chain' must be established. Clearing houses are a central part of the infrastructure. These are envisaged, and in many cases already implemented, as being focal

---

[9] This definition is derived from a review of the literature on spatial data sharing. See Maguire (1991), Onsrud and Rushton (1992), and Onsrud and Rushton (1995).

[10] A survey by Onsrud (1999) received responses about national spatial data infrastructures from Antarctica, Australia, Canada, Columbia, Finland, France, Germany, Greece, Hungary, India, Indonesia, Japan, Kiribati, Macau, Malaysia, The Netherlands, New Zealand, Northern Ireland, Pakistan, Russian Federation, South Africa, Sweden, the United Kingdom, the United States. Regional initiatives include those in Australia and New Zealand, the Asia Pacific region, and Europe. At the global level, regular conferences concerned with a global spatial data infrastructure have evolved into an umbrella organization for national and regional initiatives called Global Spatial Data Infrastructure (GSDI).

points for identifying and accessing spatial data-sets that meet the needs of users. The clearing-house organization generally provides a database of metadata that may be accessed via an Internet site and, in some cases, on CD-ROMs.

Within the framework of national spatial data infrastructure initiatives, a variety of economic and political issues typically needs to be resolved. For example, agreements must be in place with regard to data pricing structures and usage policies. Core data-sets may or may not be made available free of charge to the user organization; copyright agreements must be in place; and the statutory requirements for the provision of metadata must be acceptable to the many different actors involved. The core data-set generated by a national initiative is important in so far as GIS users in all application areas require access to certain common spatial data-sets and, fundamentally, to a common coordinate system to which other data can be referenced. Although core spatial data-sets are generally recognized as an important national resource in support of GIS implementations, the terms and conditions for their provision continue to be debated. These vary considerably between countries and range from provisions for the recovery of the direct costs of data distribution, to recovery of the full costs of data collection, to profit-making models.[11]

The drive for these cooperative data infrastructure initiatives stems from the growing importance of spatial data and the need for government intervention to coordinate spatial data acquisition and availability (Masser 1999). Public-sector agencies are major collectors of data, and providers or users of information that has spatial components (Burrough and McDonnell 1998). In developing countries public-sector illustrations of the use of spatial data-sets include the development of social and public-service indicators of crime (for example, the density of criminal events), health (for example, urban and rural fertility rates), and education needs (for example, sanitation, learner/teacher ratio, electricity supply). The role of such a national spatial data infrastructure as a potential intermediary between the interests of the various stakeholders in the collection and use of such data in a developing country is illustrated in the next section based upon an analysis of the evolution of the National Spatial Information Framework (NSIF) in South Africa. This example suggests that, even when the various levels of intermediation are explicitly addressed within the framework of such an initiative, as a result of the intermediation processes that are put in place, this may not be sufficient to yield the desired outcome.

4. SPATIAL DATA DISTRIBUTION IN SOUTH AFRICA

In South Africa the public sector has invested substantially in GIS applications. These applications are used widely in the government sector at the national, provincial, and municipal levels (van Helden 1999), by parastatal organizations such as Eskom (electricity), Telkom (telecommunication), and CSIR (scientific and industrial research),

---

[11] Full cost recovery refers to making spatial data available at a price that allows the originator to recover all the costs incurred in producing the data-set. Prices set at the marginal cost of reproduction entail charging only for time and the medium—e.g. CD-ROM required for providing a copy of the spatial data (sub)set.

and by academic institutions. The private sector is also quickly catching up with the public sector in this area.[12] A particularly prominent example of a GIS application is its use by the Independent Electoral Commission (IEC) in collaboration with the Chief Directorate for Surveys and Mapping, the Surveyor General, and Statistics South Africa. A combination of demographic information (census data), topographical data (the geographical position of rivers, roads, and contours), and cadastral data (land parcel delimitations and other boundaries created by human intervention) was used to demarcate electoral wards and voting districts for the 1999 elections (Lester 1999; Martin 1999). Howell (1999) argues that considerable financial savings were achieved by knowing *where* people would vote in the general election.[13] There were also time savings as a result of the use of the GIS information in the organization of voting such that the time voters had to queue at polling stations was reduced. However, despite the apparent success of this GIS application, it is not clear which users might access this data-set or when and at what cost the data will be made available for future use and value-adding activities.

Spatial information management initially received attention in South Africa in 1988 with the establishment of the National Land Information System (NLIS) facilitated by the Chief Directorate for Surveys and Land Information, the national mapping agency in South Africa. As the use of GIS in the public sector became increasingly pervasive, the Reconstruction and Development Programme (RDP) Ministry in the Office of the President commissioned a report to consider the alignment of the use of GIS for development planning by the public sector at national level (Abbott 1996). The report was based on interviews with government and provincial departments and para-statal organizations and it reviewed national and provincial GIS initiatives that were linked to development planning and that were under way in 1996. The report argued that the distribution of spatial data was crucial to ensure that the alignment of GIS initiatives would enable a variety of users to benefit from cooperative measures. Subsequently, in 1997, the Department of Land Affairs created a directorate for the NSIF, which, along with a Committee for Spatial Information, was endorsed by the Cabinet in 1999.

4.1. *The National Spatial Information Framework (NSIF)*

The NSIF was regarded by the government in South Africa as an infrastructure that needed to be provided and maintained in order to pursue social and economic goals (Gavin 1998).[14] The specific aims of the NSIF were

[12] Personal conversation with the two market-leading GIS software suppliers in South Africa, Computer Foundation and GIMS Ltd, March 1999.

[13] e.g. ZAR 34 million in staff costs, ZAR 9 million for ballot papers (Howell 1999).

[14] Efforts to promote spatial data exchange were under way in the region before the establishment of the NSIF. For example, the joint public/private sector initiative Southern African Metadata (SAM) consortium was implementing a clearing house for metadata (R. Smith 1998). A second initiative was run by the South African National Defence Force (SANDF), which maintains a directorate for the coordination of spatial data exchanges (Fourie 1998). Neither of these two initiatives has the scope of the NSIF, which has built on the technical achievements of the other initiatives by purchasing their products—for example, metadata database software.

to expedite the distribution and facilitate the integration of disparate geographic data sets in order to enable planners to make sound development decisions based on accurate, complete and current geographic information. Further, the NSIF will also increase efficiency and savings in government by eliminating duplication in the capturing, storing and maintenance of geographic data sets. (Gavin 1998: 18)

The NSIF encompasses the typical elements of national spatial data infrastructures identified above—that is, a catalogue of metadata, an Internet-enabled clearing-house facility, standards for metadata and spatial data collection, and a number of core datasets. When the initiative was launched, it was intended to give explicit attention to GIS capacity building through the NSIF to enable identification of people with the necessary expertise and skills in using GIS applications so that they could use their experience to assist others. In 1998, efforts were also underway to develop policies on copyright, the pricing of data-sets, and the liability and responsibility of data providers, and on metadata and data-exchange standards as well as the classification and coordination of a system standards that would serve to promote data sharing NSIF 1998*b*). A comprehensive list of interim policy guidelines on access principles, custodianship, privacy principles, and pricing were made available by NSIF (1998*a*).

The NSIF directorate recognized that it would require the involvement of non-governmental actors such as private-sector firms, academics and parastatal stakeholders if it was to succeed in creating an environment conducive to data sharing as a result of the establishment of the new infrastructure. A number of workshops with public- and private-sector, parastatal, and academic institutions were held throughout 1998 to deal with specific components of the NSIF and representatives from all spheres were encouraged (in the NSIF workshop announcements) to participate in the policy and standards-making task teams.

Despite the NSIF's avowed interest in a broadly based multi-stakeholder infrastructure initiative, most of its initiatives have focused primarily on the potential public-sector benefits of GIS applications. For example, NSIF described its primary goal as 'achieving a coordinating system and procedure in the capture and management of spatial information through the NSIF [which] is ensuring that information is available and can be utilized for *development planning by government*' (NSIF 1998*b*; emphasis added). In line with this view of its remit, only public-sector representatives were included on the National Committee that was given responsibility for the implementation of the NSIF. This restriction on stakeholder participation was put in place because the NSIF's focus on spatial data is part of a larger initiative, the Government Information Project (GIP), to facilitate effective information management in government (Gavin 1998). As a result, the development of the NSIF has been dominated by the interests of public-sector actors, despite the fact that there are expectations that the NSIF arrangements will be endorsed by the private sector as well as by various public organizations (NSIF 1998*b*). Because the scope of the NSIF remit and objectives has not been clearly defined, the new organization may find it difficult to play an intermediating role that will succeed in motivating the private sector to integrate its GIS-related activities with the national initiative.

The location of the NSIF within the Department of Land Affairs could create difficulties for the policy-making capacity of the NSIF. The Department of Land Affairs
also contains two of the most important spatial data suppliers in South Africa—the
Chief Directorate of Surveys and Mapping, which supplies the topographical data-
sets, and the Chief Surveyor General, which maintains the cadastral database. The
policy-making capacity of the NSIF with regard to crucial issues such as the provision
of core data-sets and pricing policies for accessing these data-sets may be hampered,
therefore, by having to contend with the potentially conflicting interests of the other
two directorates.

This illustration of the evolution of the NSIF in South Africa suggests that a wide
variety of technical, economic, political, and social issues is involved when the distribution of spatial data is considered. If the NSIF is to succeed in performing an effective
intermediary role between those organizations with an interest in spatial data-sets, it
will need to encourage a shared vision and new forms of behaviour within a disparate
group of actors with overlapping roles—that is, the GIS community. (The relationship
between the NSIF and these actors is depicted in Fig. 9.2.)

An environment will need to be created in which individuals in public- and private-
sector organizations that deal with geographic information will wish to participate in
data sharing across organizational boundaries. At the time of writing in 1999, the willingness of members of the GIS community to engage in spatial data sharing was being
more or less taken for granted by the NSIF. The main efforts were directed towards the

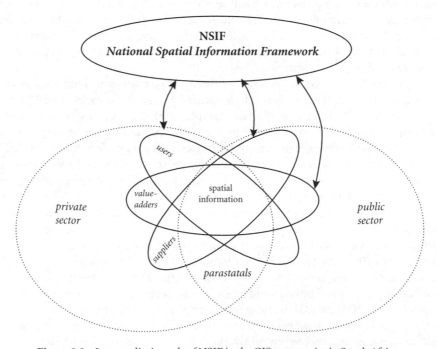

Figure 9.2. *Intermediating role of NSIF in the GIS community in South Africa*

implementation of clearing houses, standards, pricing, and copyright policies. This situation is not peculiar to the South African spatial data infrastructure initiative; it also applies to initiatives at the national, regional, and global levels. The main objectives are the promotion of economic development, the stimulation of better government, and the fostering of environmental sustainability (Masser 1999). However, these other initiatives are similarly failing to address the issue of the propensity to engage in data sharing within a given set of institutional arrangements.

The South African NSIF initiative has recognized the need for a 'spatial data-sharing culture' (Clarke *et al.* 1998). Although this is an important step, those responsible for the NSIF have not elucidated what this might mean in practice, nor have they examined whether the motivations of individual actors within organizations favour participating in, or abstaining from, spatial data sharing.[15] The social network of data-users, producers, and other commercially oriented organizations that comprise the GIS community has been involved in workshops, facilitated by the NSIF, that have provided opportunities for their representatives to voice concerns about spatial data availability and the potential problems involved in data sharing. In the course of these workshops, the identification of the spatial data 'needs' of the users and the 'supply' of spatial data-sets by the producers to respond to these 'needs' was emphasized. The underlying willingness and capability of personnel from the organizations comprising this network to participate in spatial data sharing was simply outside the remit of these workshops. Nevertheless, making use of GIS data in a way that is responsive to the desired outcomes of cost-effective provision is dependent upon the willingness of the GIS community of organizations to make their data accessible across organizational boundaries. An organization that promotes behaviour that will encourage spatial data sharing by creating incentives for data sharing and that tackles existing disincentives may be expected to perform a more effective intermediating role between the GIS community members.

4.2. Determinants of Spatial Data Sharing in South Africa

Providing a starting point to the issues that an intermediary organization would need to address to foster a 'spatial data sharing culture' requires empirical investigation of both the incentives and disincentives to engage in behaviour that would enable spatial data sharing as they are perceived by members of the GIS community in South Africa. Using the organizing framework developed by Ajzen (1991) termed the 'theory of planned behaviour',[16] the author of this chapter undertook a study to determine the

---

[15] Personal conversations with NSIF staff, Mar. 1999.

[16] The theory of planned behaviour is taken from the social-psychology discipline and applied to the issue of spatial data sharing. Its main theoretical constructs are the elements of attitude, social pressure, and perceived behavioural control that together determine the willingness of individuals (or organizations) to engage in a particular behaviour. As the theory of planned behaviour aims to explain, rather than merely to predict, a behaviour, the antecedents of attitude, social pressure, and perceived behavioural control are traced to the underlying foundation of beliefs. Beliefs about the behaviour are deemed to provide the basis for perception.

main factors that are believed to influence the propensity of individuals to engage in data sharing. Interviews were conducted with twenty stakeholders in the GIS community in South Africa in March 1999.[17] In essence, the theory of planned behaviour provides a framework for defining a set of attitudes and beliefs held by individuals at any given point in time, which, taken together, provides a relatively robust indication of their intentions to act in a certain way. In the present context, the aim of the empirical research was to provide an initial indication of the key variables or factors that might be expected to influence such intentions. In circumstances characterized by the reduction of formal institutional and legal barriers to spatial data sharing, people whose attitudes and beliefs on balance are favourable to data sharing may be expected to act out their intentions by sharing their data if they are provided with an incentive to do so. However, if individuals' intentions are characterized by a resistance to data sharing for any number of reasons, the removal or reduction of formal restraints through the provision of standards, pricing regimes, and copyright protections is unlikely to change their attitude, and therefore can be expected to make little difference to their actual behaviour in the form of a proactive willingness to engage in data sharing.

The empirical research yielded a large number of determinants likely to influence data-sharing intentions. Analysis of the interview data was undertaken with the objective of identifying the main components of a model that could subsequently be applied to assess the willingness of individuals within GIS community organizations to exchange spatial data.[18] These determinants are presented according to the three distinct components of the model (see Fig. 9.3).

The *first* component that emerged from analysis of the qualitative interview data, attitude, consists of the specific consequences and possible outcomes that organizations perceive as resulting from spatial data sharing. The *second* component, social pressure, captures insights about influential individuals and pressures within organizations and how their expectations with regard to engaging in spatial data-sharing are perceived by other individuals within the organization. The *third* component, perceived behavioural control, includes the skills, capabilities, and other factors that individuals within organizations perceive may be necessary to engage in spatial data sharing and about the extent to which individuals believe they are in control of, or immediately responsible for, data sharing.

Analysis of the views of those included in the interview sample confirmed the working hypothesis that the nature of the intermediation processes between the various

[17] As identified by Abbott (1996) and confirmed during the interviews phase, there are distinct groups within the GIS community in South Africa that stem from the wide range of possible applications of GIS and the actual spread of GIS in South Africa. The distribution of the interviews across these different sectors was as follows: national government (5), provincial government (2), local authorities (0), parastatal organizations (3), academia (3), private sector (0), GIS industry (3), non-governmental organizations (1), other (3).

[18] Specifically, the interview data were analysed to detect specific beliefs about spatial data sharing. The results were organized to produce a model of willingness to engage in spatial data sharing. The first component included beliefs about the positive or negative outcomes that might result from spatial data sharing across organizational boundaries. The second component included beliefs about the expectations of important individuals. The third component included beliefs about skills, resources, and opportunities. The full details of the methodology are available (see Wehn de Montalvo 2001).

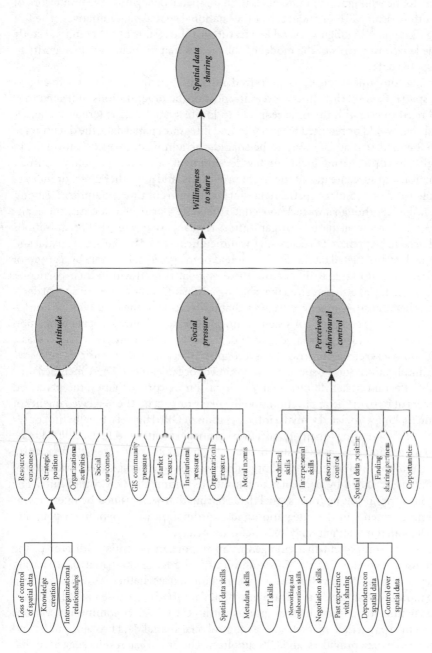

Figure 9.3. *Model of willingness to engage in spatial data sharing in South Africa*

actors in the GIS community requires that much more careful consideration needs to be given to the willingness of individuals to share spatial data within the framework of the South African NSIF, or, indeed, by other national spatial data initiatives. The following discussion highlights several key features of the results of the empirical study and the key components of the model of the factors that influence the propensity to share spatial data.

The *first* component consists of perceived costs and benefits associated with engaging in spatial data sharing. Interviewees in most of the organizations indicated they would need to establish the time required to locate spatial data externally, whether trained staff would be required to exchange and integrate spatial data, the data storage requirements if spatial data were to be shared, the administrative costs involved in arranging for data sharing including the development of a strategic policy, contract negotiation and agreement, and copyright provisions, and any other direct or indirect costs incurred as a result of spatial data sharing in terms of time, equipment, labour, and data costs. Although it was acknowledged that data sharing may increase the quantity of spatial data available to an organization, it was also expected that organizations would need to be prepared to deal with the difficulties created by the use of spatial and attribute data and metadata standards that had been agreed collectively by a group of organizations but that may differ from those standards used within their particular organization. It was also perceived that sharing might affect the quality of spatial data-sets in either a positive or a negative way, that is, in terms of the quality of positional accuracy, attribute accuracy, consistency, completeness, and lineage, depending upon whether errors and gaps in shared spatial data-sets were introduced or reduced as a result of the data exchange process. It was recognized that spatial data sharing would have certain implications for the specific functioning of organizations. These included how it might affect an organization's ability to focus on its core activities, the perceived usefulness of its own GIS application, and the likely impact on the perceived quality of decision making associated with the interpretation of GIS data-sets. Finally, the consequences of data sharing for an organization's strategic position were considered to be important as interviewees suggested that sharing might affect the spread and control of ideas, and the interdependence of organizations, and encourage a redistribution of the influence wielded by particular organizations within the GIS community. The social outcomes resulting from spatial data sharing that were valued by interviewees were improvements in the integration and coordination of development planning initiatives and the resultant range of benefits to society.

The *second* component that emerged as an important potential indicator of the intentions of members of the GIS community to engage in spatial data sharing included the specific individuals who appear to influence whether individuals within an organization will decide to engage in spatial data sharing. Those named by interviewees tended to be individuals or groups within the GIS-user community, various actors in the spatial data market, such as commercial spatial data brokers, public and private spatial data providers, and GIS suppliers. The empirical results suggested that organizational pressures are likely to play a role in the propensity to engage in data sharing depending upon how pressures from management, other departments within an organization, and certain 'champions' of spatial data sharing are perceived by individuals within the organization. Specific institutional pressures that were perceived to favour spatial data sharing in the South African context were mentioned by interviewees, including the interim guidelines published by the NSIF, the establishment of spatial data agreement(s) applicable to their organizations, and the political initiatives of government.

The *third* component to emerge from the study consisted of the skills and resources that were perceived as being important determinants of whether or not individuals within organizations were likely to engage in spatial data sharing. For example, a range of GIS-specific skills were highlighted by many of the interviewees such as assessing the quality of spatial data, handling different data formats, mastering standards for spatial and attribute data, selecting the required data from databases, for example, as single observations, subsets, or as a thematic layer, and integrating data from diverse sources. The integration of spatial data involves converting and manipulating data, verifying its geometric and attribute quality, maintaining its consistency with related data, and updating data layers independently of each other. Another set of skills that were perceived as relevant to spatial data-sharing behaviour was the ability to interpret metadata using metadata interfaces and catalogues, capturing these data and applying standards, and maintaining and updating the data. Finally, the general levels of computing skills were also raised as a factor that might be expected to influence individual data-sharing behaviour. In this area database administration skills, facility in using the Internet to locate data sources or to distribute data, and capabilities for transferring data to and from different media, for example, CD-ROM, were considered to be important likely predictors of willingness to engage in data sharing.

A second set of capabilities was deemed to be an important likely determinant of sharing behaviour. These were related to interpersonal skills. Aspects of networking were deemed to be important such as the capabilities to establish and foster networks of contacts, identification, and attendance at meetings to enhance networks, being able to keep abreast of the activities of members of a network, and letting others know about one's own work. The extent of collaborative working and experience of interdisciplinary teamworking were also expected to be important predictors of data-sharing behaviour. Another set of interpersonal skills that was ranked fairly highly by many of the interviewees was related to negotiation skills, and, in particular, the capacity to negotiate a 'win–win' data-sharing situation whereby all sharing arrangements would be equally beneficial to all the parties involved. Effective negotiation skills with respect to the pricing of spatial data, and data-ownership agreements including copyright and liability agreements mandating an acceptable level of performance in the delivery of services or the quality of products were also considered to be important indicators of a willingness to engage in data sharing. Finally, previous experience of sharing spatial data was expected to establish a precedent for individuals' future engagement in this behaviour and for that of the organization as a whole.

Human resources as well as time and financial resources were also deemed to be potentially important to spatial data sharing. For example, having staff with the time available to engage in data-sharing activities, the availability of funding, and the details of organizational guidelines such as the existence of a data-sharing policy were perceived by the interviewees as important likely determinants. Another consideration that emerged and one that is unlikely to be entirely under the control of either an individual or an organization was the identification of the 'right sharing partners' in terms of their willingness, responsiveness, and reliability and also the compatibility of their application of data. In general terms, the 'organizational fit'—for example, the type of organization that the interviewee would feel comfortable with—was considered to be very important. The extent to which an organization may depend on other organizations for access to spatial data and to which it is able to control data use and interpretation once they have been shared were also perceived as important factors in determining whether data-sharing behaviour is likely to occur.

The spread of GIS applications is being accompanied by demand from a variety of GIS community members for collaboration with regard to the exchange of spatial data-sets that would otherwise be duplicated or unavailable for cost reasons, thus creating bottlenecks in the further development of innovative applications. A significant barrier to collaboration lies in the variation in the incentives of actors whose public and private organizational boundaries are perceived to influence decisions about whether to engage in spatial data sharing. If initiatives such as the NSIF in South Africa are to play effective roles as institutional intermediaries, the empirical evidence indicates that they will need to take into account the likely behaviour of individuals within organizations that comprise the GIS community. In the event that the complex determinants of such behaviour are not examined it is likely that the willingness of organizations to make spatial data accessible to or from other parties will be misjudged. There appears to be a clear role for intermediary organizations to negotiate a national consensus on standards, procedures, and policies for greater spatial data sharing in the interests of all the stakeholders in the GIS community, as well as to undertake assessments of the factors that are likely to influence their willingness to engage in spatial data-sharing behaviour.

5. CONCLUSION

By focusing on intermediation processes that can be supported by organizations that are charged with implementing national spatial data information infrastructures, it has been possible to identify some of the underlying factors that appear to influence the coordination of actors who are involved in the development of GIS applications. The analysis in this chapter has highlighted many of the potential shortcomings of measures that are being taken to implement these infrastructures. These shortcomings are related to a failure on the part of key stakeholders and decision-makers to address the incentives and disincentives that influence the individual's willingness to share GIS-related data across organizational boundaries.

The analysis suggests that, to encourage data sharing, innovative forms of intermediation are needed that take into account the factors that influence sharing behaviour. The stakeholders in the GIS community have very different interests in whether spatial data are retained within their organizations or whether, and on what terms and conditions, these data are made available for exchange across the organization's boundaries. National spatial data initiatives that take these factors into account will be essential if the use of GIS applications is to expand in developing countries and to provide an improved information environment that helps to overcome human and financial resource limitations in meeting social and economic development goals. However, organizations within the national spatial data infrastructure framework need to be sensitive to variations in the perceptions of incentives and disincentives for data sharing and their likely influence on behaviour. This is so, even when technical, procedural, and formal policy arrangements are in place. Effective intermediation between the stakeholders in the GIS community appears to be crucial because of the need to locate data-sets that exist outside the national mapping agencies that played a key role in the collection and use of spatial data before the advent of digital GIS applications.

The evidence from this study supports several conclusions about the strengths and weaknesses of the NSIF in South Africa. If it is to serve as an effective intermediary institution, it will be necessary to define the scope of this initiative more clearly and to embrace private-sector stakeholders, not simply as observers, but as participants in negotiations over standards, pricing, intellectual property rights, and other issues. In addition, if the NSIF is to serve as an effective intermediary institution for the GIS community, it will need to be independent from spatial data suppliers in the same government department. This is important to ensure that the policies it establishes are not perceived as being overly influenced by those departments' interests.

Spatial data-sharing initiatives entail more than coordination and the development of metadata, clearing houses, and standards. An important aspect, and one that must be expected to vary for each national data-sharing initiative, is the data-sharing culture and the extent to which it discourages or fosters views that are favourable towards participation in spatial data-sharing arrangements. The behaviour model developed in this chapter can be applied in other contexts to examine the extent to which data sharing is likely to occur. The results offer guidance about changes in the spatial data information infrastructure that may provide favourable incentives for increasing data sharing. In order to investigate whether spatial data sharing is likely to follow from initiatives that promote agreements on various aspects of data exchange, there is a need to identify the way actors who are expected to engage in such behaviour regard the potential benefits and threats.

Key attitudes and beliefs emerged from this study. These provide new insights into what participants in the GIS community believe they will gain or lose as a result of sharing their organizations' data. They suggest what participants believe that opinion leaders expect of them. They also provide insights into the extent of the control that individuals in the GIS community believe they have over whether actively to engage in spatial data sharing. Further research will be needed to assess whether the features of the motivations uncovered in this study are common to different national contexts. It will also be important to examine whether the policies recommended here encourage greater enthusiasm for sharing GIS data that may contribute to an improved information environment for decision-makers in developing countries.

10

Master of my Domain: The Politics of Internet Governance

DANIEL PARÉ

1. INTRODUCTION

As we move into the first decade of a new century, the evolving capabilities associated with Internet working are becoming increasingly central to the way business and everyday life are organized. The rapid spread of the components of the Internet infrastructure and the enormous growth in the range of its applications mean that the manner in which it is governed and the outcomes of alternative types of governance regimes will have major consequences for a growing number of users. Controversies have erupted internationally over the appropriate forms of governance for the Internet and these have their counterparts within regional and national contexts. This chapter offers an in-depth examination of one such controversy and its resolution within a national context.

During the mid-1990s the Internet-addressing regime for the United Kingdom underwent a period of institutional reconfiguration. This process culminated in the establishment of Nominet UK, the domain name registry responsible for administering and managing the *.uk* top-level domain (ITU 1999*b*). Domain names are unique identifiers and their creation and registration are among the 'few centralized points of authority in the supposedly open, decentralized world of the Internet' (Mueller 1997). Domain name registries may be regarded as intermediary organizations. They coordinate and administer the linking of identities in the physical realm with a specific type of virtual identification—a domain name. The conflicts associated with the restructuring of the management and administration of the *.uk* name space were not restricted to the technical features of the UK registry system.[1] Instead, they encompassed a transformation and reorganization of the institutional framework of Internet governance in the United Kingdom.

---

[1] This chapter is based on a review of primary archive documents, including the UK Naming Committee discussion-list archives; personal interviews undertaken between April and October 1998, and email exchanges with members of the UK Internet industry, government policy-makers, and other individuals who are known specialists on the Internet between November 1997 and April 1999; and the results of a questionnaire distributed to members of the UK Internet industry in November 1998.

The ongoing efforts to restructure the global Internet-addressing regime have given rise to numerous controversies, some of which are also visible in the conflicts associated with the UK case. At both the international and domestic levels, conflict is often generated by the actors' perceptions of the goals of the Domain Name System (DNS) and how these goals might best be achieved.[2] The fundamental issue in these disputes is the management of value allocations and choice alternatives. The disputes that have coincided with attempts to reconfigure the manner in which the DNS is coordinated appear to be directly and/or indirectly related to the economic significance of the technology involved. Within the UK context the manner in which a diverse set of actors interacted to resolve these controversies has affected the perceived flexibility and efficacy of Nominet UK as an intermediary between those who manage and administer one dimension of the Internet architecture, and those who seek to use the Internetworking infrastructure.

The issues that have given rise to controversy over the restructuring of the DNS have provided a focal point for discussions about appropriate forms of governance for the Internet. However, the processes through which changes in governance are being negotiated and implemented have received relatively little attention. In fact, the bulk of Internet-related research on governance tends to treat Internetworking as a conceptual black box. Consequently, relatively little attention has been given to the role of power and politics in influencing the success or failure of organizational and administrative innovations with respect to Internet-addressing regimes. The analysis in this chapter is developed to illustrate how the collective and individual actions of industry players, Internet and non-Internet organizations, government authorities, and specific individuals have influenced the scope and direction of change for a key dimension of the architecture supporting internetworking.

The discussion presented in this chapter focuses on the events leading to the formation of Nominet UK. This is a story about the mediation of the contending interests of various organizations and individual actors. The features of the mediation process are examined by focusing on the organizational, commercial, and technical factors that appear to underpin the social dynamics of the governance process. A detailed analysis of these features reveals how the dynamic relationships between social actors led to the establishment of an intermediary organization that is now responsible for coordinating and administering the *.uk* name space. In the case of Nominet, informal processes of interest mediation appear to have coalesced in a manner that legitimated the outcome of a protracted period of institutional reconfiguration. Analysis of this process of institutional transformation offers considerable insight into the dynamics of organizational change in the context of 'cyberspace' or the 'virtual society'.[3]

---

[2] The Domain Name System is a distributed database within which each unit of data is indexed by a unique name. These names are paths that classify computers on the basis of an inverted 'tree' scheme. The term 'name space' refers to this treelike structure. See Albitz and Liu (1997: ch. 2) and Rony and Rony (1998: ch. 3).

[3] Gould (1996*b*: 199) defines these terms as referring to 'human and computer interactions across open networks and without reference to geographical location (and therefore legal jurisdiction) or real-world social understanding'.

2. THE RECONFIGURATION OF THE UK DOMAIN NAME REGISTRY SYSTEM

2.1. *Formation of the UK Naming Committee*

This story begins at a time when UK domain registrations were handled by an organization composed primarily of a relatively small group of academics working in computer networking research, the Joint Network Team. From the late 1980s up to 1993, this team was responsible for overseeing the development of the Joint Academic Network (JANet), a major internetworking backbone in the UK.[4] With the growth of commercial internetworking in the UK, increasing numbers of organizations sought to register names in the *co.uk* sub-domain. EUnet GB Ltd, a private commercial Internet Service Provider (ISP) founded by a group of academics in 1993, assumed responsibility for maintaining the *.uk* root name server on a voluntary basis for the Joint Network Team.[5]

In early 1993 UnipalmPIPEX, at the time the only UK Internet Access Provider with its own international transmission capacity, approached the Joint Network Team to express its concern with this arrangement. PIPEX claimed that it was unfair that one of its competitors, EUnet GB, should control the domestic domain name registry.[6] Shortly thereafter, Demon Internet,[7] the first commercial Internet Access Provider to offer low-cost Internet dial-up access in the UK, also approached the Joint Network Team with similar concerns about the way the *co.uk* sub-domain was being managed.

---

[4] In hierarchical networks, the term 'backbone' refers to the top-level transmission paths that other transit networks feed into; see www.whatis.com/backbone.html (accessed 1 Apr. 1999). The Joint Network Team coordinated network addressing in accordance with its Name Registration Scheme (NRS). This was a centralized naming system for UK universities that operated as an equivalent to DNS. The primary difference between NRS and DNS was that, under the former, all entries into the registration database were made in UK domain order—e.g. *uk.ac.sussex*. At that time, a gateway between the NRS and the DNS was maintained at University College London, which translated names between the two addressing systems once in every twenty-four-hour period. This service allowed American network-users to see normal DNS names for UK hosts, while UK network-users were able to read DNS names in NRS format.

[5] Name servers are programs that store information about the domain name space and are employed to perform name-to-address mapping (Albitz and Liu 1997: 21–4). Before April 1993, EUnet GB, owned by the University of Kent at Canterbury, had been trading under the name UKnet and providing email, news, and full Internet access to more than 800 UK sites. It was also a founding member of the independently run EUnet Europe, serving as the UK backbone portion of that organization's network. In July 1995, EUnet GB was acquired by Performance Systems International Inc. (PSI).

[6] PIPEX was founded in January 1992 with fifty-six employees by Unipalm, a UK-based company that produced computer networking products based on the Internet Protocol (IP) suite. In March 1994 it was floated on the London Stock Exchange and, in late 1995, it merged with UUNET, which subsequently merged with Microwave Communications Inc. (MCI) Worldcom in December 1996. See www.uk.uu.net, (accessed 1 Apr. 1999).

[7] Demon Internet Ltd was founded in June 1992 by Demon Systems Ltd, a UK-based firm that specialized in software production. It sought to expand the market for Telnet, email, Gopher and File Transfer Protocol (ftp) services to the public, as these services were largely restricted to academic and research environments. Demon Internet grew from a subscriber base of 100 to in excess of 180,000 dial-up subscribers in May 1998. It was subsequently purchased by Scottish Telecom, the telecommunications division of Scottish Power, for £66 million. See www.demon.net (accessed 1 Apr. 1999).

A series of informal discussions between representatives of these four organizations produced an agreement that domain name registrations 'should be handled in a more democratic way' (email interview, 2 Sept. 1998). Under the new arrangement, authority for coordinating and administering the *co.uk* sub-domain was delegated to a Naming Committee made up of three commercial and two non-commercial Internet Access Providers. However, the Joint Network Team, which subsequently evolved into the UK Education and Research Networking Association (UKERNA), maintained a right of veto over decisions reached by this committee.[8]

The five founding Naming Committee members were: The Joint Network Team (Network Registration Scheme) Administrator;[9] JANet; EUnet GB Ltd; Demon Internet Ltd; and UnipalmPIPEX Ltd. In late 1993, BT Internet Services became the sixth organization to join the Naming Committee.[10] Through the voluntary activities of representatives from these organizations, the Naming Committee operated as a self-governing body with no formal structure or chairperson. In essence, the 'committee' consisted of a group of closed electronic discussion lists, where domain name registration requests were processed and naming-related issues were discussed. Commenting on these events, one founding member noted that these operations were very 'amateurish' and that 'everything was happening so quickly, and no one had really thought it through. Basically the system was embryonic' (interview, 3 Sept. 1998).

In October 1994 the London Internet Exchange (LINX) was established to provide a point of physical interconnection for ISPs to exchange Internet traffic through cooperative peering agreements.[11] In order to ensure that member organizations were serious backbone providers, membership of this association was made contingent upon meeting two criteria.[12] First, applicants were required to own a permanent, independent, international connection to the Internet. Secondly, members had to sell Internet services, including at least one public service allowing customers to connect to the Internet.

---

[8] In April 1994 the Joint Network Team became the JNT Association, which now trades as UKERNA. This organization's mandate focuses on the management of the UK's Higher Education and Research Community Network Program. It also manages the *.ac.uk* name space, and, with the UK Central Computer and Technology Agency, the *.gov.uk* name space. See www.ukerna.ac.uk (accessed 1 Apr. 1999).

[9] The role of the NRS representative was to provide continuity between the segment of DNS under commercial control and those segments that remained under the direct control of the Joint Network Team.

[10] BT Internet Services included all British Telecom Internet products before BTnet became a trademark name in January 1995. Prior to the launch of BTnet, BT Internet Services consisted of a team of six individuals responsible for most aspects of the provision of corporate Internet services. BT Internet Services joined the Naming Committee through an informal process. The technical services manager, who was well known throughout the UK Internet community, sent an email message to a representative of the Committee requesting information about how to register a name in the *co.uk* sub-domain for a client. The individual who was contacted responded by informing him that he would add BT Internet Services to the Committee's discussion lists because it was a known Internet Access Provider (interview 5 June 1998 and 9 Sept. 1998).

[11] LINX was incorporated as a non-profit association of ISPs in December 1995. A single network connection to LINX is sufficient to carry traffic generated by any LINX provider, thereby eliminating the need to have an extensive network of links to each provider. See www.linx.org (accessed 1 Apr. 1999).

[12] For a detailed list of LINX membership requirements, see the LINX Memorandum of Understanding, www.linx.org/mou.html (accessed 1 Apr. 1999), and LINX Articles of Association, www.linx.org/manda.html (accessed 1 Apr. 1999).

The five founding LINX members were the same five organizations that had founded the Naming Committee in October 1994.

Although the Naming Committee and LINX were completely separate organizational entities, representatives of the former decided that LINX membership requirements should also be used as the criterion for establishing full membership of the Naming Committee.[13] Consequently, each company that joined LINX simultaneously became a full member of the Naming Committee. This decision appears to have been taken primarily on the grounds that LINX membership was seen as the easiest and most efficient way of establishing the eligibility of prospective members of the Naming Committee. One former representative of the Committee pointed out, however, that the decision to establish joining criteria was made 'because it was to our [member organizations'] advantage, but from a long-term business point of view this was a definite weakness because it was an arbitrary action' (interview, 3 Sept. 1998). The consequences of this arbitrary action for the restructuring of the naming system are examined in the next section in order to highlight how informal processes of interest mediation between social actors gave rise to an administrative innovation that altered the way in which the *.uk* domain was managed.

2.2. *The Dynamics of the UK Naming Committee*

Two aspects of the Naming Committee's operations need to be singled out. First, the activities of Naming Committee members' representatives were subject to very little scrutiny by senior management in their respective organizations.[14] Secondly, in its early days the UK domain name registration system was regarded by members of the committee as being relatively flexible and reliant upon both interactions and mutual cooperation between Internet access providers. The registry system was believed to have benefited from an ethos that stressed the maintenance of technical continuity and integrity.[15] As a result, many individuals felt that, prior to 1995, this evolutionary process had the cumulative impact of forging 'good' relationships between competing organizations in the UK Internet industry.

By 1995 the rapid increase in demand for Internet-related services began to accentuate procedural and structural weaknesses in the Naming Committee's registration practices.[16] Structurally, there was a growing division between a rapidly evolving commercial Internet-working industry and a technical infrastructure that was operated

---

[13] At this time the organizational representatives to the Naming Committee were primarily technical engineers, most of whom were directly involved to varying degrees in the establishment of LINX.

[14] Former members of the committee generally suggested that the relatively high degree of autonomy they exercised was attributable to the newness of the Internet at the time.

[15] This corresponds to the institutional ethos of consensual adoption of ethics and the propagation of voluntary technical standards, or *rough consensus and running code* that characterized the early history of Internet working (Hafner and Lyon 1996; Rony and Rony 1998; see also Bradner 1996, Hanseth, Monteiro, and Hatling 1996, and Malkin 1994).

[16] When the Naming Committee began processing requests for domain names in 1993, it was receiving, on average, two to three requests per week. By mid-1995 it was receiving in excess of 100 requests per day.

and maintained on a volunteer basis. Procedurally, the Naming Committee's administrative processes were highly contentious, fuelling disputes between full members with voting rights and guest members, as well as between individuals within the voting member constituency. At the root of these controversies were questions about the registry's decision-making authority and its legitimacy. The next two subsections highlight some of the features of the power struggles between these social actors that were associated with the use of specific techniques to allocate domain names prior to the establishment of Nominet UK.

The Naming Committee Voting Structure

By late 1994 the technical, social, and political relationships that evolved around the management and administration of the *co.uk* name space achieved a form of closure. At the technical level, closure was reflected in the stabilization of the technology facilitating the registration of names in the *.uk* domain. Socially, the domain was managed by a relatively small homogenous group of technical engineers with shared perceptions about how to structure and administer addressing needs within the *.uk* name space. At the political level, power relationships within the Naming Committee had been structured in a manner that allowed representatives from a relatively small number of ISPs to determine the acceptability of name requests. Simply put, the closure reached reflected the fact that by late 1994 internetworking in the UK was characterized both by the presence of relatively few actors with a limited variety of needs, and by a relatively limited set of ideas about how best to meet these needs.

This closure manifested itself in two ways. The first was due to the fact that the ability to submit requests for domain names was restricted to members of the Naming Committee. Individual network users and providers of Internet services were not permitted to apply directly to the registry for domain names. From the perspective of non-members, this was problematic because, although the registration of domain names was a service that was provided free, member registrars were levying fees for providing this service.[17]

The second manifestation of closure was attributable to the Naming Committee's bifurcated voting structure. In order to offer domain name registration services in the *co.uk* sub-domain, domain name registrars who were unable to meet the LINX/Naming Committee membership criteria (that is, a permanent, independent, international connection to the Internet and the sale of Internet connection services) had to be introduced electronically to other Committee members by a full/voting member. In most cases, the introduction to the Committee's discussion list was made by the full/voting member that provided bandwidth for the applicant organization. The applicant was granted guest status and permitted to apply for domain names on his or her own organization's behalf and that of its customers. Guest members were not permitted, however, to vote on domain name requests or policy-related matters.

---

[17] Most interviewees claimed that the registration process was very competitive because service registrars were competing on the basis of price.

Some of the founding Naming Committee members claimed that there was no reason for denying voting rights to guest members. Rather, the diversity of the organizations that might wish to offer clients domain name registration services had been underestimated and it had been assumed that, if an organization did not provide Internet access services, it would not require input into how the registry system was managed. Other full voting representatives, however, suggested that the reasons for denying voting rights for guest members were somewhat different. 'The rationale for not allowing guests to vote was that it would have involved too much administration. Counting up just a few votes was much easier to do. Besides if we allowed all the resellers/guests to vote, they could have theoretically voted to dismiss all the rules, and one has to ask where this would have left the system' (interview, 9 Sept. 1998). One consequence of this action was that the Naming Committee structure was perceived by most guest member representatives, and by some full/voting member representatives, as being elitist. One former voting member representative who was sympathetic to this view pointed out that the lack of congruence between LINX membership criteria and eligibility for voting membership in the committee meant that guest members were prevented from exercising the same powers as their access providers (interview, 5 June 1998). This state of affairs underpinned a popular consensus among guest member representatives that this exclusionary voting structure meant that they were being subjected to the tyranny of the larger providers of Internet services.

The fact that all domain name requests were made in a 'public' forum reinforced concerns about the arbitrary and subjective nature of the Naming Committee's decision-making processes. Discussion-list members were able to identify who was registering domain names for various companies simply by looking at the messages posted to the naming discussion list. The discussion list was described as 'a source of free intelligence' that allowed members to keep tabs on their rivals (interview, 9 Sept. 1998). The fact that full/voting members also registered domain names on behalf of their clients further bolstered the perception of the Naming Committee as an elitist organization because voting members had the authority to object to their competitor's requests for names. In the light of these circumstances, at least one company began recommending that its clients register domain names under the generic *.com* top-level domain, since registrations in this domain were not public knowledge and the process in *.com* was believed to be far less arbitrary than in the *co.uk* domain (interview, 29 Apr. 1998).

The Subjective Implementation of Arbitrary Rules

The absence of a comprehensive and unambiguous rule set indicating the criteria and policies for domain name registration in the *co.uk* domain perpetuated numerous disputes amongst the actors. The guidelines for suitable names were originally expected to function on the basis of the 'common-sense interpretations' that manifest themselves 'in a gentlemanly environment' (interview, 9 Sept. 1998). In essence, there were no formal rules for registering domain names. Instead, committee members used guidelines outlining what might be regarded as 'good' or 'bad' names. The voting members were supposed to base their decisions about the acceptability of requested names on these guidelines (see Box 10.1).

Box 10.1. Summary of Naming Commitee *co.uk* registration rules

Restriction	Rationale
1. No two- or three-character domain	Two-letter names may clash with two-letter country name designation[a] Not sufficiently informative
2. No abbreviations of company names	Not sufficiently informative
3. Only one domain name per company	Use of hierarchical structure of the DNS and sub-domains of common domain more appropriate
4. No registration of brand names	Domain names should not represent products. Facilitate the prevention of cyber-squattting.
5. No offensive names (i.e. *rudeword.co.uk*)	Such names convey the wrong image and are immoral

[a] Country code top-level domains are based on two character codes detailed in ISO-3166, an international standards agreement establishing two and three character abbreviation country codes for sovereign nations (see ITU 1999).

Source: Naming Committee Information Sheet 1999.

Examination of this box suggests that, although some of the guidelines were based on technical requirements (that is, two-character names may conflict with top-level and second-level domains), others focused on non-technical requirements that some voting representatives believed domain names should represent (for example, only one domain name per trading entity; no registration of acronyms; and restrictions on registration of potentially offensive names). Efforts to enforce this highly subjective rule set led to a situation where requested names were often objected to, and/or rejected, on stylistic grounds. In numerous instances, voting representatives objected to requests for names and suggested alternative names that were in line with their personal interpretations of the registration guidelines. For instance, law firms often had their name requests rejected on the grounds that the *law.co.uk* sub-domain was more informative for people seeking to find law firms on the Internet than *co.uk*.

Initially, the Naming Committee approved registration requests in the *co.uk* sub-domain only for limited companies, as it was argued that this implied commercial legitimacy. This policy prevented organizations that were not registered with Company's House in the UK—for example, trading companies, partnerships, and sole traders—from registering domain names matching the character string under which they had established a commercial reputation. It also meant that companies trading under a name that was different from their registered name were prevented from registering the name(s) with which they were commonly identified in the physical realm. To overcome this, a decision was made by the voting representatives to register names for entities that could provide 'proof of existence'. However, the meaning of 'proof' was never clearly defined. As a result, the kind of information solicited upon submission of requests for names tended to be arbitrary and dependent upon the whims of representatives from voting member organizations.[18] In the face of difficulties in enforcing this ambiguous rule set, some representatives argued that the process of voting on domain name requests was an impediment to the commercial provision of registration services.

Between 1995 and 1996 the problem of subjective implementation of arbitrary rules and guidelines was aggravated by the rapid growth in the number of full/voting and guest members of the Naming Committee. By July 1995 the number of voting members had increased from five to eleven and the number of guest members had increased to thirty-seven. By August 1996 the number of voting members had more than doubled over the previous year to twenty-four organizations. Figures for the numbers of guest members are not available, but it is estimated that their ranks grew dramatically during this period to approximately 100 organizations. As the numbers of full/voting members with a major interest in the registration process decreased relative to guest members, the earlier goodwill and camaraderie that had characterized member interrelationships began to wane. An 'us-and-them' environment developed between those representatives seeking to exercise ill-defined control over the registry process and the advocates of a more open registry system.

Proponents of a 'closed' process tended to be technical engineers serving as representatives of voting members. Their primary concerns tended to focus on the technical and moral dimensions of the registry system.[19] Advocates of a more 'open' registry system sought the establishment of a neutral body that would process domain name registrations on the basis of a non-restrictive formal rule set that would operate on the principle of first come, first served. There was a consensus among advocates of this approach that proponents of a 'closed' process often lacked an understanding of the commercial realities and were concerned instead with how the Internet as a whole should evolve.

The Emergence of Nominet UK

A meeting convened to address the problems associated with the Naming Committee's operations was held in Cambridge in September 1995. This meeting was the product of a posting on the committee's discussion list soliciting interest in an informal gathering. The meeting was the first opportunity for the majority of member representatives to meet in person. It was agreed that the meeting would incorporate a discussion of procedural issues with respect to the *.uk* domain and its future management, and that votes should be taken on the issues raised. The meeting was attended by thirty-five people, including representatives from seven full/voting organizations and eighteen guest organizations.

---

[18] The evidence demanded ranged from requests for company letterheads to demonstrating the existence of a bank account in a company's name.

[19] 'Moral' refers here to concerns about possible misuses of the DNS that might cast a negative light on Internet working—for example, allowing the registration of names that might be seen as offensive, such as *rudeword.co.uk*.

The meeting achieved very little in terms of tangible outcomes. The attendees agreed that the registry system needed to be changed, but how it should be changed, remained elusive. Agreement was reached on the need to formalize rules for registration procedures and membership, but no immediate action was taken. It was decided that new full/voting members would be accorded guest status during their first three months on the committee. In addition, it was agreed that the ranks of the voting membership would be enlarged to include two members co-opted from, and elected by, guest members as representatives of their interests. And, finally, because of growing dissatisfaction with the operation and maintenance of the registry database, EUnet GB was required to commit to a service level agreement.[20]

In the light of the problems facing the Naming Committee, representatives from UnipalmPIPEX and BTnet submitted proposals reflecting their personal views about the establishment of alternative registry procedures to UKERNA in late 1995. Their documents closely paralleled one another.[21] Both called for the abolition of the committee structure and a shift of the registration process into the commercial realm. The creation of a more centralized registry structure incorporating other neutral subdomains—that is, *ltd.uk*, *org.uk*—was also recommended. Both documents also called upon UKERNA to take a greater role in overseeing the registry to reduce the likelihood that committee members would exploit the registration process to their advantage. The proposal from the BTnet representative went further than that of his counterpart in this regard, suggesting that, if UKERNA was not prepared to perform registration services, the rights of administration and operation of the *co.uk* name space should be sold to an organization that would be willing to run the UK Internet Naming Service in a manner that would guarantee high-quality service.

In January 1996 a second meeting to remedy the problems facing naming administrators was held. It was attended by twenty representatives of full/voting and guest members of the Naming Committee. The meeting was chaired by Dr William Black, the then director of UKERNA and the person responsible for the *.uk* domain.[22] Building on the ideas in the BTnet and UnipalmPIPEX proposals, he called for the creation of a neutral legal body to run the name spaces under the *.uk* domain. By this time similar approaches to the management of national domains had been adopted in The Netherlands, Germany, and Japan. Three working groups charged with developing more detailed proposals were created. Dr Black took responsibility for developing a

[20] A representative from UnipalmPIPEX proposed that his organization should assume control of the registry database because it was better equipped to meet the technical needs of the registry, but this motion was not passed.

[21] This was somewhat ironic, because these two individuals were generally perceived by the other representatives as antagonists, and their exchanges on the discussion list were among the most heated debates concerning the registry's procedures and structure.

[22] Dr Black's authority over the *.uk* domain was rooted in RFC-920 and was independent of his membership of UKERNA. RFCs are official Internet documents that provide information about Internet standards, specifications, protocols, organization notices, and individual points of view. RFC-920 outlined the norms for assigning responsibility for two-letter top-level domains. It was also in this document that the association of domains with specific organizations, and the concept of domain registration entailing a hierarchy of delegation among organizations, were initially outlined.

business plan for the proposed registry organization, a representative from BTnet volunteered to establish a financial working group to create a funding model, and a representative from UnipalmPIPEX volunteered to establish a working group to develop an operational model dealing with the technical requirements of the naming functions to be performed by the new entity.[23] These working groups were composed of several volunteer representatives from a range of UK-based providers of Internet services. A formal proposal for the development of a new registry organization was presented to the UK Internet service provider community at a public meeting in April 1996. The proposal consisted of three components:

- the establishment of a not-for-profit management company providing legal protection, limited liability, and a professional full-time organization to carry out the necessary tasks;
- the creation of a Steering Committee open to all organizations that were prepared to pay membership fees; and
- the establishment of a charging regime for sub-domains registered under the neutral domains.

This became the proposal for the establishment of Nominet UK.

Although being the 'responsible person' appears to have given Black considerable *de jure* power, his *de facto* power depended upon the support of the interested parties in the UK Internet industry. Consequently, views about this initiative were sought from a broad array of participants in the UK Internet industry. However, uncertainty about this proposal led to further disputes between actors, mainly regarding the extent to which a new organization would ameliorate the structural and administrative weaknesses of the Naming Committee.

According to the business plan, the new organization would operate on the basis of a Shared Registry System where all members would manage the registry. The proposed organization would have two executive directors, two non-executive directors, and a steering committee composed of all the organizations that were willing to pay nominal annual subscription fees. The function of members would be to decide on a naming policy for the *.uk* domain and on the appointment of non-executive directors. A particularly contentious issue was the voting structure of this new organization. Many smaller service providers were critical of this aspect of the proposal, suggesting that it would simply graft the elitist structure of the Naming Committee onto the new registry organization.

Subscription fees were initially to be based on the annual turnover of members and voting entitlements were to be proportional to the fees paid up to a maximum of £5,000. However, at the first annual general meeting of the new organization, Nominet UK, the number of votes for members was linked to their respective rates of name registration and provisions were made allowing all members to purchase votes up to a

---

[23] The BTnet and UnipalmPIPEX representatives who assumed responsibility for creating these working groups were the same two individuals who had submitted proposals to UKERNA for reforming the *.uk* registry structure.

defined limit.[24] A voting structure where the number of votes was proportional to respective registration volumes was seen as a means of ensuring that the steering committee would be representative of the relative commercial strengths of its members, thus minimizing the risk of unrepresentative groups exerting undue influence on the registry.

The business plan for Nominet indicated that it would introduce a charge for each name registered in neutral sub-domains of *.uk*. Prior to the start of Nominet's operation in August 1996, the registration of domain names had been free to Naming Committee members. When Nominet began administering the *.uk* domain, the registration of names in the *.uk* sub-domains became subject to a fee of £100 per name for the first two years of registration, followed by a renewal fee of £50 for every subsequent year that the registration was maintained. However, member organizations were entitled to a 40 per cent discount on the cost of registering domain names. Despite this discounted rate, there was some initial concern that these charges would be levied before names had been entered into the registry database and that the prices might be beyond the expenditure capabilities of small entities. This might benefit large providers of Internet services, who would be able to afford larger capital expenditures up front. Such fears were allayed by an agreement to levy charges on domain name registrars on a monthly per registration basis for registrations during the preceding month.

Another key area of controversy focused on Dr Black and how best to limit the power he could exert personally over the registry process. He had offered to leave UKERNA to become the managing director of Nominet UK, and some feared that he might seek to take advantage of his position for personal gain.[25] The majority of actors regarded the new registry as being responsible for coordinating and administering a 'public good', and there was a consensus that this entity should not be run as a for-profit endeavour. This concern was circumvented by recommending the creation of a not-for-profit company limited by guarantee.[26]

There was also uncertainty regarding Dr Black's suitability for the position of managing director of Nominet. He had an academic background and some industry participants asserted that the new registry should be managed by an individual with a commercial background. Dr Black served as managing director in a voluntary capacity from Nominet's incorporation in May 1996 until it commenced operations in August 1996. At the first annual general meeting in August 1996, he was formally elected to the position.

By 1999 Nominet had come to be perceived within the UK Internet industry as having successfully remedied the structural and administrative weaknesses plaguing the

[24] The upper limit on the number of votes allowed per member was set at ten. Members with lower registration volumes were permitted to purchase additional voting entitlements for a fee of £500 per vote.

[25] This would have been in contradiction with RFC-1591, a document focusing on the structure of top-level domain names and the administration of domains, which states that, for responsible persons, 'concerns about "rights" and "ownership" of domains are inappropriate. It is appropriate to be concerned about "responsibilities" and "service" to the community' (Postel 1994: 4).

[26] In the UK, companies limited by guarantee have no shares or shareholders; and those who control the company have no financial interest in the company's assets.

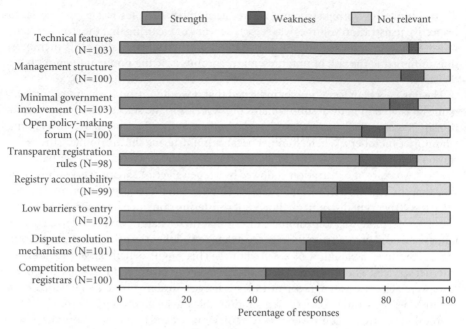

Figure 10.1. *UK Internet industry perceptions of Nominet operations*

old registry system. A survey of UK-based providers of Internet services in November 1998 indicates that Nominet's operations were regarded positively at that time.[27] Respondents were asked to indicate the strengths, weaknesses, and relevance of nine factors pertaining to Nominet's operations (see Fig. 10.1).

The results shown in Fig. 10.1 suggest that the technical features of the registry system were held in high regard by UK-based domain name registrars. Approximately three-quarters of the respondents indicated that the transparency of Nominet's rule set was one of its strengths. Nominet's open policy-making forum was seen by 73 per cent of respondents as another of its strengths. By establishing formal registration rules and practices Nominet appears to have succeeded in removing the ambiguities that had been associated previously with the registration of domain names in the UK. In other words, domain name registrars and registrants were no longer believed to be subject to the *tyranny* of any particular group of actors after the establishment of Nominet.

---

[27] The sample population for this survey encompassed the diverse UK-based companies that potential registrants could approach to register domain names in the autumn of 1998. The sample comprised service providers of various sizes, specializing in the provision of a variety of Internet working based services, targeted to an assortment of potential clients. Potential participants were selected on the basis of whether their respective organization offered domain name registration services. The questionnaire was sent to 408 potential respondents; or approximately one half of the organizations offering this service in the UK at the time. The majority of companies who participated in the survey viewed the provision of domain name registration services as an important aspect of their competitive strategies.

3. THE POLITICS OF INTERNET GOVERNANCE

There has been a dramatic surge of interest in Internet governance in recent years. Prescriptive accounts of developments in Internet governance reflect bipolar perceptions of how Internet working should evolve. Some authors argue that the Internet encompasses numerous technical and non-technical elements that, when taken together, constitute a conceptual whole. They tend to advocate the implementation of top-down governance frameworks to ensure the well-being of the conceptual whole (Foster 1996; Gould 1996*a*; Mathiason and Kuhlman 1998*a,b*). In contrast, others assert that there is no conceptual 'whole' and that the only policy required is one of *laissez-faire* (Gillett and Kapor 1996; Mueller 1997, 1998; Rutkowski 1998*a,b*). These authors do not account for emergent or changing structures of Internet-addressing regimes and they offer relatively little insight into the dynamic relationship between regulatory change and the technical characteristics of networks.

Another line of work on Internet governance has pursued a process-oriented approach. In this case, the focus is on procedures that give rise to various outcomes in the cyber-realm. Adopting a libertarian perspective, Johnson and Post(1996*a,b*) maintain that Internet working, including Internet addressing, can be governed by 'decentralized, emergent law',[28] where privately produced rules would be fashioned through decentralized collective action, leading to the emergence of common standards for mutual coordination. They suggest that these processes might lead to a redefinition of notions of civic virtue within online environments (Johnson and Post 1996*a,b*).[29] In contrast, Reidenberg (1996, 1998) suggests that the hardware and software that constitute the virtual realm impose a set of default rules or *Lex Informatica* on communication networks. Reidenberg argues that, whereas political governance processes usually establish the substantive laws of nation states, in the cyber realm the primary sources of default rule making are the technology developer(s) and the social processes through which customary uses of the technology evolve.[30] These default rules define possible behaviours and the values that are upheld within the cyber-realm.[31] Similarly, Lawrence Lessig argues that architecture is the most significant constraint on behaviour within physical and cyber domains.[32] It establishes conditions of entry and participation and reflects a distinct philosophy of social ordering (Lessig 1998*b*, 1999).[33]

[28] This is similar to the concept of polycentric or non-statist law (see T. W. Bell 1992, 1998).
[29] For instance, the ability of network users to enter or exit online spaces may increase the probability that the democratic tradition of rational debate among elected representatives would be replaced by more dispersed and complex interactions at local levels.
[30] See also Quintas (1996) on the role of software designers in default rule making.
[31] Adopting a perspective based on Foucault's work, Boyle (1997: n.p.) expands on this theme to demonstrate how, rather than enabling users to circumvent state rules, certain information technologies may provide the state with 'a different arsenal of methods to regulate content materially rather than juridically, by everyday softwired routing practices, rather than threats of eventual sanction'.
[32] He labels the architecture of cyberspace *code*. It refers to the 'software and hardware that constitutes cyberspace as it is—the set of protocols, the set of rules, implemented, or codified, in cyberspace itself, that determine how people interact, or exist in this space ... It [code], like architecture, is not optional' (Lessig 1998*b*: 4).
[33] Lessig's approach to Internet governance is based on an assumption that behaviour in the physical and virtual realms is regulated by four constraints. Law that regulates by the imposition of sanctions *ex post*; social

Because values underpin the architecture, there is the potential both for government encroachment into the virtual realm and for the private sector to embed its values within the code. Therefore, according to Lessig there is a need to ensure that the architectures of cyberspace protect values such as liberty, free speech, privacy, and access (Lessig 1998*a*, 1999).

Internet governance entails a mixture of regulations embedded within the architecture and other forms of regulation that apply to various aspects of the cyber-realm. Unfortunately, the perspectives outlined above are highly speculative and seem to underestimate the significance of the social dynamics of governance processes. They yield little insight into how regulatory initiatives establish social, political, and cognitive legitimacy within cyberspace. This highlights the need for empirical analysis of what social actors actually do rather than prescriptive, ideologically laden speculations about what they should do. In the context of implementing changes to the *.uk* domain and the global Internet addressing regime, the catalysts for change were embedded in the techniques employed by social actors to administer the DNS.

In so far as specific design features of a system establish the patterns and boundaries of interaction, the choices and decisions that are implemented by social actors may be regarded as a political phenomenon (Barley and Tolbert 1997; Easton 1965; Wendt and Duvall 1989; Wendt 1987). These interactive processes involve multifaceted power relations that give rise to the authoritative allocation of values. These relationships can be understood as reflecting the instrumental use of resources to attain specific goals such as motivating an organization, or group of organizations, to do something that they would not otherwise do (Dahl 1957; Lasswell and Kaplan 1950). However, the analysis of power relationships cannot focus solely upon overt conflict because power may also be exercised in a manner that limits the scope of, or even restricts, decision making (Bachrach and Baratz 1962, 1963). Power may also operate in a more sublime manner in the form of institutional norms and procedures that bind and constrain the arenas of conflict and its resolution (Lukes 1974). This multidimensional view of power offers a framework for examining how organizational politics influence innovation in the present case of changes in the Internet domain naming system.[34]

Recognizing that organizational politics consist of power in action as well as power in conception, this framework is premised upon a notion of dynamic interactions between 'surface-level' and 'deep-structure' politics. The former are the day-to-day contests and struggles for collaboration between actors—for example, 'attempts by one or more parties to exploit (bend, resist, implement) the rules of the situation that are to their own advantage' (Frost and Egri 1991: 236). Less easily observable deep-structure politics refer to power that influences, 'usually in hard to detect ways, not only the way the rules

---

norms that regulate by enforcing expectations within particular communities; markets that regulate through price and availability; and architecture, or the constraints imposed upon behaviour by the world as we find it. Regulation is the sum of the interactions of these constraints (see Lessig 1999). For an alternative view, see Post and Johnson (1997*a,b*).

[34] This framework was developed by Frost and Egri (1991) and has been used in the industrial networks literature to interpret decision-making processes within asymmetrical exchange networks of interdependent firms (see Elg and Johansson 1997).

of a situation are played but the very way the rules are framed in the first place' (Frost and Egri 1991: 236). Power relations in the deep structure covertly, and/or overtly, shape and influence actions at the surface level. Surface-level actions shape the articulation of power in the deep structure, thereby influencing future surface-level politics.

Frost and Egri postulate that conflict over the introduction of a proposed organizational innovation involves conflict between the status quo and the implementation of change. However, in the case of the restructuring of the Internet-addressing regime at both the .uk and international levels, there was agreement on the part of all the relevant actors that a change was required. Thus, conflicts, and their resolution by social actors, are likely to be manifest in the process leading to the definition of the appropriate parameters for change.

To understand the dynamics of innovation in cyberspace, the architectural issues that influence both the deep-structure and the surface-level politics need to be taken into account. The architecture defines the parameters of the deep-structure relationships through which actors seek to influence the outcomes of the value allocation process.[35] The architecture is not deterministic and perceptions about default rule sets may also be influenced by the history of internetworking and the value structures coinciding with that process. Given this interpretative diversity, we should expect that deep-structure power relations will be manifest in the relevant actors' perceptions about the Internet's architecture.[36] In negotiating new domain-naming regimes, efforts to establish social, political, and cognitive legitimacy are likely to manifest themselves in the form of conflicts between those who advocate governance regimes in accordance with the inclusive values traditionally associated with internetworking and those who advocate governance regimes that are more exclusive.[37] The concept of surface level politics that is used here is similar to that suggested by Frost and Egri (1991). It refers to the strategies and tactics employed by actors to manipulate and influence outcomes to benefit themselves or others.

In the following discussion a three-tier model of the interactions between surface-level and deep-structure politics and the Internet's architecture is used to understand the processes leading to the reconfiguration of the .uk addressing regime. The objective is to show how the actors involved in these processes sought to identify and establish a consensus on the attributes of intermediaries for managing Internet addressing.[38]

[35] This perspective draws on two interpretations: (1) the manner in which specific design features or arrangements of artefacts and systems influence power relations and patterns of authority; (2) the linking of technological properties with institutionalized relations of power and authority. See Winner (1986: 3 58) for a comparative analysis of these interpretative traditions.

[36] Although this level is embedded and implicit, it has been suggested that some actors can recognize and harness deep-structure power relations to their advantage (Frost and Egri 1991).

[37] The degree of community dependence on new organizational forms correlates positively with legitimacy problems because stakeholders may not fully understand the nature of new ventures (Hunt and Aldrich 1998). Moreover, pioneering organizations cannot base trust-building strategies only on technological efficiency. Uncertainties must be framed in such a way that the proposed innovation becomes credible (Aldrich and Fiol 1994).

[38] The intensity of participation in the Naming Committee's activities and interactions was fluid and occurred on a volunteer basis. Variations in participation levels may have influenced the constellations of power relations that emerged and some of the value allocation decisions that were made (Denis, Langley, and Cazale 1996; Wallis and Dollery 1997).

In the period preceding the reconfiguration of the *.uk* addressing regime, appeals for change were often presented as loose collections of ideas arising from a variety of ill-defined and frequently inconsistent individual preferences. Those actors who perceived the architecture—that is, the Naming Committee registry structure—as an impediment to commercial success emphasized the need to eliminate restrictions on the registration process. Those actors who viewed the architecture as a necessary extension of the hardware and software comprising the domain name system stressed the importance of maintaining technical continuity and integrity. These divergent perspectives can be understood as a reflection of differences in deep-structure interests. Some actors were able to prevent changes in the existing architecture by employing deep-structure tactics. Such tactics included treating the architecture as inviolate, presenting positions as being unbiased, and adopting higher values—for example, the need to convey a positive image of internetworking. In contrast, others employed surface-level tactics such as deliberate testing of the allocation process to highlight discrepancies and contradictions in the Naming Committee's operating procedures. These surface-level actions, combined with the rapid increase in both the volume of registration requests and the number of committee members, altered the power relationships within the deep structure. The interactions between the social agents, along with the technical shortcomings of the registry, undermined the Naming Committee's authority and legitimacy.

A significant aspect of this process of change in the power relationships was the interactions between two individuals who were perceived to be advocating diametrically opposed values. The interactions between these two individuals, and their outcomes, served as a catalyst for transforming the administration and management of the *.uk* domain. The notion of establishing a neutral legal body to manage the *.uk* name space emerged as a product of these exchanges and, ultimately, from their proposals for the establishment of alternative registry procedures.

The unanimous dissatisfaction with the existing registry system makes it unlikely that the proposal to establish a new registry architecture in line with principles of neutral service provision represented a threat to any deep-structure power interests. Therefore, most of the political activity occurred at the surface level. Much of the controversy focused on finding optimal strategies for ensuring the impartiality of a new organization. Although the legitimacy of Dr Black's authority over the *.uk* domain was broadly accepted, this did not hold for the proposed Nominet organization. Given the widely held perception of Nominet's predecessor as a closed structure, the social, political, and cognitive legitimacy of the new registry architecture hinged on its openness to participation by all interested actors and on its technical merit.[39]

A number of supportive political tactics were deployed to ensure the success of the Nominet proposal. No interested parties were restricted from participating in the development of a coherent proposal and participants were able to define the parameters of the features of the proposed architecture—that is, structure, technical base, and

---

[39] Suitable degrees of openness have become a focal point of debate about reconfiguring the technical management of Internet names and addresses at the global level.

funding model. When concerns were raised that the voting arrangements might graft the elitist voting structure of the previous registry onto the new architecture, the allocation of votes was modified. Once this was implemented, membership in Nominet remained open to all, thus empowering all interested parties to contribute to domain name policy for the *.uk* name space. The initiative acquired legitimacy, in part, because its detractors were also incorporated in the planning processes and, later, within the company itself.[40] By the time Nominet commenced operation in August 1996 virtually all opposition had been diffused.

This initiative was entirely industry based with little or no government involvement. This was perceived by participants in the process as giving social, political, and cognitive legitimacy to the new organization. The UK government's interest had focused on the extent to which user interests, business and individual, would be adequately represented in the restructuring process. According to the former head of Infrastructure and Convergence Policy at the Department of Trade and Industry, 'the key issue is that the registry system be open, transparent, and objective; which Nominet is' (interview, 13 May 1998). The Nominet approach was regarded as a good model reflecting a pragmatic industry decision.

The perception of the parties concerned with Nominet was that it was inclusive; it prevented the interests of any entity from dominating the process; and it was independent and therefore not in competition with industry players. The perceived degree of openness of the process of change appears to have been the primary deep-structure consideration of the social actors involved. Consistent with historically embedded values of internetworking, the openness of the processes associated with the establishment of a new regulatory regime allowed interested parties to define the types of checks and balances that would be incorporated in the new architecture. The initiative acquired legitimacy because it was perceived as a bottom-up endeavour that permitted the actors to define the constraints that would be placed on their behaviour. Once established, the continuing legitimacy of Nominet was related to its technical efficacy and to the maintenance of its neutrality.

4. CONCLUSION

The Internet is comprised of the hardware and software that make internetworking possible, and of the formal and informal organizational structures that are evolving around the technical infrastructure. The structures responsible for coordinating and administering core functions, such as addressing, are important dimensions of the architecture. Because of its prescriptive overtones and its tendency to underestimate the significance of the social dynamics of the infrastructure, much of the Internet-related governance literature does not address the importance of social dynamics in the establishment of new architectural configurations.

---

[40] For example, at Nominet's first annual general meeting, the members elected one of the most vehement opponents of this organization as a non-executive director. According to him, getting 'on board' offered a means of improving the system from within (interview, 1 May 1998).

The catalyst for establishing a new governance regime for the *.uk* domain was embedded in the power struggles over the techniques for domain name allocation and the Naming Committee's administrative procedures. Change did not occur as a result of a specific plan. Instead, it was shaped by dynamic processes of cooperation and competition between participating social actors. The analysis of the dynamics of Internet politics illustrates how the characteristics of internetworking influenced the actors responsible for managing one of the Internet's core functions. These dynamics have been revealed by examining the processes of interest mediation associated with the changes in the coordination structures for managing domain name registrations and allocations.

A focus on changing power relationships offers a means of interpreting the way that the social, political, and cognitive legitimacy of regulatory organizations is being established in the cyber-realm. Although the 'responsible person' for the *.uk* domain had authority over this segment of the Internet, maintenance of the legitimacy of this authority depended upon the extent to which interested parties perceived this individual as a neutral actor. The success of the Nominet initiative can be attributed to the inclusive strategies adopted to develop a proposal for change. The bottom-up manner in which collective decision making was conducted was congruent with the values of internetworking. The outcomes of these processes influenced the legitimacy bestowed on the new intermediary organization—that is, Nominet—by all interested parties.

This analysis of the particular case of the transformation of the governance regime for the *.uk* name space suggests that other features of the evolution of Internet governance can best be understood by examining the manner in which new configurations of power constellations are influencing the emergence of the structures responsible for managing and administering the Internet's core functions. The benefit of this approach is that it does not fall prey to ideologically motivated positions with respect to the appropriate roles of the private sector or governments in the evolution of governance regimes for the Internet. It also offers a means of coupling investigations of the determinants of the technical architecture of the Internet and the way the social and political interests of its designers and users become embedded in that architecture.

11

Missing Concepts in the 'Missing Links' for Brazilian Telecommunications

ANA ARROIO

The figure of progress. She holds in one hand an aeroplane, in the other some small object symbolic of improved education. I will give you the detail of that later. The idea will come to me . . . a telephone might do . . . I will see.

(Waugh 1962: 18)

1. INTRODUCTION

Since the first uses of satellite technology for international communication in the late 1960s, it has been suggested that this technology would be deployed to provide the missing telecommunication infrastructure links that might help to improve the social and economic conditions in developing countries by closing the telecommunication 'access gap'.[1] By the late 1990s a new generation of satellites, low earth orbiting satellites (LEOS), were being promoted by industrial players and by many governments. Although the promise of the earlier generation of geostationary satellites had yet to be met, this time it was claimed that LEOS would be implemented in a way that would bring new opportunities for developing countries to meet their communication policy goals. A key goal expressed by many of the proponents of LEOS was to extend afford- able telecommunications services to rural and disadvantaged areas. The analysis of one country's experience of the introduction of LEOS is presented in this chapter. It aims to demonstrate that, in so far as the implementation of this ambitious policy goal is a realistic expectation, it is much more likely to occur when specific technological and other capabilities are present within the decision-making communities in the domes- tic environment than when they are absent or relatively weak.

The chapter is based on a detailed study of the types of negotiating capabilities that accumulated in Brazil to support decisions that led to the development and imple- mentation of satellite technology over an extended period (Arroio 2000). The objective of the research was to determine whether and how these capabilities had been deployed

[1] The telecommunication's 'access gap' generally refers to worldwide disparities in telecommunication main line penetration rates (see IEEE Transactions on Communications 1976, Schramm 1968, and Smythe 1962).

in the negotiations between indigenous and external players that occurred prior to the launch of LEOS services in Brazil. An additional objective was to assess the extent to which these capabilities influenced how and for whom telecommunications services have been developed. A key issue in the analysis is whether the capabilities that were embodied within actors in key institutions in Brazil were favourable or unfavourable to measures to encourage the deployment of an innovative satellite system that would provide services in line with the stated policy goals.

The negotiation process that led to satellite deployment is considered here as a special form of institutional intermediation—that is, the coordination of institutional actors within the economy with the aim of achieving an explicit set of social and economic objectives for satellite deployment. Intermediation is treated as an important feature of the negotiation processes that involved a substantial number of policy, regulatory, and other institutions in Brazil. Numerous individuals became involved in the decisions that established the framework within which it would be possible for private-sector firms to introduce LEOS-based telecommunications services. The changing strengths and weaknesses of the capabilities of key actors are examined through an intensive round of semi-structured interviews conducted in Brazil in 1997 and 1998. This evidence is considered in the light of the bargaining positions adopted by these actors in their negotiations with the mostly foreign suppliers of satellite technology and services. With one notable exception, the supplier firms were all US based.

An overview of the technical characteristics of LEOS systems and of the promises for their performance that were made by their suppliers is provided in Section 2. These promises focused on the way LEOS services would provide the 'missing link' in infrastructure provision to achieve the goal of universal telecommunications service provision that has been set by many developing countries.[2] The negotiations leading to the deployment of satellite services in Brazil are highlighted in Sections 3 and 4. The discussion provides insights into the changing strengths and weaknesses of the Brazilian actors' negotiating capabilities during the negotiation stages. The evidence is organized using indicators that enable an assessment of the capabilities of various actors to negotiate towards the goals that had been established within the policy-making process for telecommunication service provision. In Section 5 the contributions of the main concepts used in this analysis are considered in terms of how they provide a basis for generating improved understanding of the interrelationships between technological changes and the actual deployment of new technologies. The main conclusion from the analysis is presented in Section 6. The analysis suggests that, in some instances, strong negotiating capabilities embedded within Brazilian institutions contributed to the opening of 'windows of opportunity' that increased the likelihood that certain social policy goals, including the extension of services to rural and disadvantaged areas, would be met when the new satellite services were deployed in Brazil. Thus, the accumulation of strong technological and other capabilities by the actors within national

---

[2] The 'missing link' is a reference to a report prepared by the Independent Commission for Worldwide Telecommunication Development (1984).

institutions is shown to have contributed positively towards the goal of harnessing the potential benefits of an innovative technology to extend the telecommunications infrastructure.

2. SATELLITES FOR DEVELOPMENT?

There is ample evidence that satellites have not been used very effectively to diminish the telecommunication 'access gap' in developing countries. Although there has been considerable investment in the telecommunications infrastructure in these countries, the data for the late 1990s indicate that the access gap between the industrialized and most developing countries persists. Even in the middle-income developing countries, it is being closed only relatively slowly.[3] A recent analysis has suggested that it will be many years before much of the developing world will 'catch up' with industrialized countries in terms of penetration of telecommunications service (Mansell and Wehn 1998: ch. 2).

In Brazil telecommunications infrastructure investment increased significantly after the mid-1990s, but the access gap also remained significant. In 1999, for example, teledensity—that is, main telecommunication lines per 100 inhabitants—had reached approximately 14 per cent (Anatel Agência Nacional de Telecomunicações 1999) as compared to the average of 54.72 per cent for high-income countries in the same year (ITU 1999a). Although satellite technology has been deployed for domestic telecommunications services in Brazil and these services are heavily used for voice, data, and television transmission, the available satellite capacity has not been extensively utilized to provide services such as tele-medicine or tele-education or to extend telephone services to rural or remote low-income areas.

In the 1990s a new generation of satellite technology was being championed as the means of accelerating the rate of catch up for countries with relatively low telecommunication penetration rates. Constellations of LEOS were being designed to support global mobile telecommunications services and the first of the LEOS systems was launched in 1997. The technology developers claimed that the new technology could be used to bring telephone services to areas where the terrestrial infrastructure was underdeveloped and to provide advanced telecommunications services to the most undeveloped parts of the world 'without the large investment that otherwise would be required to develop a local infrastructure' (Iridium 1996: 11).

In contrast to the older, geostationary, satellite communication technology, the new LEOS systems were designed to support global mobile voice telephone service in addition to global data, email, and positioning services. Some of the business plans of the initial LEOS operators included a fixed service component to support solar-powered telephone booths for use in remote villages and low population density areas (Globalstar 1996; Nourozi and Blonz 1998). Table 11.1 provides an overview of some of the technical features of LEOS systems.

---

[3] See Independent Commission for Worldwide Telecommunication Development (1984), ITU (1997), and Mansell and Wehn (1998: ch. 2).

Table 11.1. *LEOS system overview*

Characteristics	ECCO	Globalstar	ICO	Iridium	Teledesic
Number of satellites	46	48	10	66	288
Altitude (km)	2,000	1,414	10,355	780	1,375
System cost (US$ billion)	2.8	2.6	4.6	3.4	9.0
Start-up date	2002	1999	2000	1998	2002
Circuits per satellite	1,000	3,000	4,500	1,100	125,000
Inter-satellite links	No	No	No	Yes	Yes
No. of gateways	n.a.	150–210	Many	25	Many
Cost of mobile phone ($)	1,000	750	n.a.	3,000	1,000[a]
Airtime charge (min. $)[b]	0.50	0.35–0.55	0.50	3.00	0.04

Note: n.a. = not available.
[a] Price for computer equipment with capacity for 64 kbps.
[b] All prices refer to wholesale local rates except Iridium charges (price for international calls).

Source: compiled from Evans (1998), Gilder (1994), Nourozi and Blonz (1998), and Telebras (1995).

In the mid-1990s middle-income countries were expected to account for a large part of the LEOS operators' market and the technology developers and service operators had launched a well-orchestrated campaign to gain support at the national level in these countries for their new services. The new services were to be subject to national policy and regulatory provisions and national governments would need to give permission for service provision within their territories. In many of these countries, policy at the time did not permit competitive service providers to operate alongside the state-owned (or privatized) monopoly telecommunications operator. The LEOS operators' campaign to be allowed to supply services within developing countries was conducted within international radio frequency allocation forums to gain access to the required frequency bands, and through marketing statements that were designed to attract the interest of government officials. Corporate and political actors on the national and international stages began to argue that the new technology for satellite communication would offer a major opportunity to assist developing countries to reduce the gaps with industrialized countries in their telecommunications infrastructure.[4] The International Telecommunication Union (ITU), for instance, suggested that 'the global or near global character of [LEOS] should permit the provision of basic telecommunication services particularly in those rural and remote areas which may not be reached in an economical way by other means' (ITU 1996). Many corporate and political actors linked this achievement with the potential for major improvements in the capacities of developing countries to meet their social and economic development goals. Access to LEOS services would, it was argued, bring 'improvement in the quality of life of the population, an influx of investment and participation in the ownership and operation of [LEOS] facilities and services' (Wright 1997: 74, citing ITU).

The Brazilian government was not immune to such hyperbole. In 1995 the government recognized the potential of LEOS for developing countries and asserted that

---

[4] See e.g. ITU (1996) and MacLean (1996).

LEOS services will be vastly relevant because they will enable quick coverage in locations where normally this would be very difficult, or would mean extremely high costs. This is the case, for instance, of a great number of rural areas and isolated localities in the country, mainly in the northern, mid-western, and part of the north-eastern regions (Ministério das Comunicações 1995: 69).

Minicom, the government department responsible for communication policy, went even further saying that LEOS deployment would lead to 'faster access by users not presently covered' and 'access opportunity for low income users' (Ministério das Comunicações 1995: 69). Was there a real basis for these high expectations? Earlier generations of satellite technology had been used primarily to meet business demand for new services or had been priced in ways that attracted mainly wealthy users in the urban areas. What conditions would be favourable to the deployment of the new generation of satellites to contribute to the social and commercial goals of Brazilian telecommunication policy? In the next section, the conditions that might be expected to give rise to a more favourable outcome from deployment are considered.

3. BUILDING TECHNOLOGICAL AND INSTITUTIONAL CAPABILITIES

Brazilian activities in satellite communication have been organized and conducted mainly by two public institutions: the Brazilian National Institute for Space Research (INPE) and the organizations associated with the domestic telecommunication holding company, Telebras. To assess the nature of the changes in the technological and related capabilities of these organizations, a variety of indicators was constructed and analysed along two dimensions.[5] The first dimension is the evolution of domestic negotiating capabilities for influencing the development and use of satellite technology. Negotiating capabilities are defined in terms of both technological and institutional skills. As shown in Box 11.1, indicators of negotiating capability, a concept that is developed in Section 5, include expenditure on research and development (R&D), investment in education and training, spending on skills accumulation in key areas including management, and the nature of collaborative linkages between institutions that would be expected to influence the deployment of LEOS services. The second dimension is the discrete intervals in time during which capabilities for negotiation would be expected to have accumulated. Three time periods, chosen on the basis of key events in Brazilian policy making, are examined: from 1965 to 1975, from 1976 to 1985, and from 1986 to 1998.[6]

[5] See e.g. Chiesa, Coughlan, and Voss (1996), who develop an audit methodology to benchmark the strengths and weaknesses of firms' innovation capabilities. Rush et al. (1995) propose criteria to benchmark 'best practice' in government-funded research and technology institutes.

[6] This analysis is based on the results of in-depth interviews with policy-makers and engineers at the Ministry of Communications, Telebras, Embratel, and the Brazilian Space Agency (AEB). Interviews were also conducted with engineers, scientists, and project managers at the National Institute for Space Research (INPE) and with representatives of Iridium, Globalstar ICO, and Teledesic. Most of the interviews were conducted in 1997 and 1998 by the author. Secondary sources including government directives, policy documents, and the trade literature were also used (see Arroio 2000).

Box 11.1. Indicators of negotiating capability

Indicator	Definition
Technological capabilities	
R&D investment	Overall expenditure specified as R&D investment
Nature of R&D investment	Features of R&D investment: equipment and projects
Training investment	Expenditure on raising the technical skills and knowledge of human resources
Nature of investment	Training activities and specific courses
Institutional capabilities	
Institutional initiatives	Formal social, economic, commercial, and technology policy goals
Programme management	Management features of policy or project initiatives
Policy implementation	Outcome of policy or project introduction
Domestic linkages	Nature of linkages between technical and non-technical institutes

Source: adapted from Bell and Pavitt (1993), Dalum, Johnson, and Lundvall (1992), and Freeman and Soete (1997).

The negotiations associated with the development and implementation of policies intended to strengthen Brazilian capabilities for developing space-sector technologies and systems are examined. These policies involved attempts to implement a satellite-based distance-education project, domestic efforts to design satellite technologies, and an initiative to introduce an indigenous LEOS system called ECO-8.

3.1. *The Organization of Space R&D Activities*

Brazil was one of the first developing countries to establish an institutional framework for the administration and development of space-related R&D. In 1961 the National Commission for Space Activities (GOCNAE) was created as a civilian unit that would report to the National Research Council (CNPq).[7] A training programme was established with the objective of producing a cadre of Brazilian scientists who would be capable of directing R&D projects in the field, and of leading domestic laboratories. In 1968 M.Sc. courses were introduced and implemented and two scientists—from India and from the United Kingdom—were attracted to come to Brazil to serve as thesis advisors. This measure was backed up by the arrival of twenty doctoral graduates from India to provide initial training in the field of space and atmospheric science.

[7] In 1971 GOCNAE was officially redesignated the National Institute for Space Research (INPE).

The main areas of R&D activity and personnel training that were established in the late 1960s and early 1970s were meteorology, remote sensing, electronics and telecommunication, applied computer science, systems analysis and application, and systems engineering. All of these fields of expertise could be expected to strengthen Brazilian capabilities in the satellite technology field. By the end of 1973 some 150 researchers had received Master's degrees and fifty doctoral students had been trained abroad, mainly in the United States, in these research areas (F. Oliveira 1991).

From the late 1960s, there was intense cooperation between the US and the Brazilian governments on a diverse range of space R&D activities. Of particular note was the development of satellite technology to support distance education in the Advanced Interdisciplinary Communication Satellite Project (SACI/EXERN). This project involved the use of the American ATS-6 satellite for the distribution of educational materials that were produced at INPE. Broadcasts were received by televisions and radios installed in primary and secondary schools in the economically disadvantaged north-east region of Brazil. INPE argued that 'a key element in the project was its potential for improving educational facilities in remote regions of the interior of the country, encouraging the permanence of the population in these regions, as opposed to their migration to the major cities' (F. Oliveira 1991: 51).

An interdisciplinary group of 108 specialists, mainly in the fields of educational technology and satellite systems engineering, was formed in addition to the introduction of the Master's training programme at INPE. By 1975 education programmes were reaching about 500 primary schools covering around 20,000 pupils, and approximately 1,500 teachers had been trained to work with tele-education programmes (Terracine 1997). Unfortunately, the project ran for only one year from 1974 to 1975. Although the Ministry of Education had stated that satellite education would be extended progressively throughout Brazil, the programme was cancelled. The reasons for the cancellation are not very clear. According to Nettleton and McAnany (1989: 165), the 'difficulty of arranging interministerial co-operation between social and technical agencies in Brazil' was a major factor, which made it difficult to gain support for the continuation of the project. From the available evidence, it seems that the Ministries of Education and Planning did not have access to information about how the new technologies could be used to support their goals. There were few, if any, incentives for them to alter their traditional approaches to implementing policy.

It also appears that there was little interinstitutional cooperation in the development and implementation of policies for satellite education (Terracine 1997). For example, there were disagreements between INPE and the Ministry of Communications (Minicom) over the appropriate deployment of domestic satellites. Minicom did not support INPE's education project, arguing that the Institute should focus primarily on space R&D rather than on the development of applications for the new satellite technologies.

3.2. *Defining Directions for Space Research Activities*

As illustrated in Fig. 11.1, from 1977 there was increasing public investment in INPE. This investment was linked to a space policy programme called the Brazilian Complete

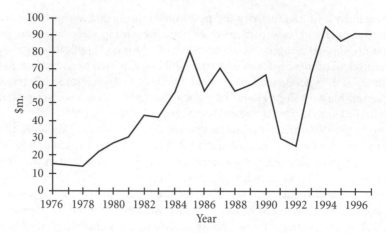

Figure 11.1. *INPE annual investment, 1976–1997*

Source: compiled from Brisolla (1994) and INPE (1976–97).

Space Mission (MECB). Its objectives were to develop four scientific satellites by 1988: two for data collection (SCD-1 and SCD-2) and two satellites for remote sensing (SSR1 and SSR2). A launch vehicle (VLS) and a launch base were also to be developed. The satellites were to be designed at INPE, and the Brazilian Commission for Space Activities (COBAE) took responsibility for overall project management.

During the MECB, INPE began to develop space technology—that is, the payloads and equipment for building scientific satellites. These technologies included satellite batteries, liquid and ionic propulsion material, technology and materials for the mechanical structure, thermal control, on-board cameras, and equipment for telecommunications services using the 'S' radio frequency band. Training was provided in such areas as on-board computer control engineering for satellite systems and software systems for satellite orbital and altitude control (telemetry and tracking).

The decision to pursue the development of domestic technological capabilities in the satellite and other space technology areas suggests that the researchers involved had accumulated considerable bargaining strengths. The INPE research team promoted a strategy of fostering a national capability through the indigenous development of essential equipment and subcomponents. This strategy was opposed by other scientists who argued that 'specific technologies should be bought off-the-shelf in major markets'.[8] This group of scientists were in favour of buying in the necessary components from the external market as the preferred means of 'leapfrogging' towards the modernization of Brazilian space technology-related R&D and deployment activities.

During the implementation of the MECB, a mixed strategy was employed such that some components were developed locally while others were purchased abroad. In the

---

[8] Interview with H. Cardoso, engineer at INPE, 25 Apr. 1997. The components for implementing the laboratory for satellite testing and integration, for example, were purchased externally, mainly from France.

case of this initiative, the director of INPE, Nelson Parada, was able to manage and reconcile diverging views about how to proceed towards a common space technology development objective, at least in the early stages. Parada was the director of INPE for ten years. During this period substantial federal funds were made available for R&D activities in a number of areas, including plasma physics, computation, applied mathematics, combustion, and propulsion. Research activities were grouped together to form Associated Laboratories that were intended to provide scientific support for satellite development programmes. As shown in Fig. 11.2, the number of scientists employed at INPE increased significantly in this period and the total number of staff engaged in R&D activities more than doubled from 300 in 1980 to over 700 in 1985. Around 8 per cent of the annual budget was invested in internal human resource training, according to INPE's records.

The MECB provided a basis for establishing the long-term policy directions and goals for space R&D activities. During the mission there was strong leadership within INPE and financial resources for investment in appropriate skills and training. This enabled a large number of research projects to be conducted, which gave rise to considerable innovative capabilities in the development of satellite equipment and subcomponents.

3.3. *The Search for Alliances: The Development of ECO-8*

Unfortunately, INPE faced severe financial constraints from 1985. Fig. 11.1 shows the decreasing levels of investment in space R&D, particularly from 1990, and Fig. 11.2 indicates that from 1988 research staff employment declined. Diminishing financial and human resources and weaknesses in the linkages between public institutions that

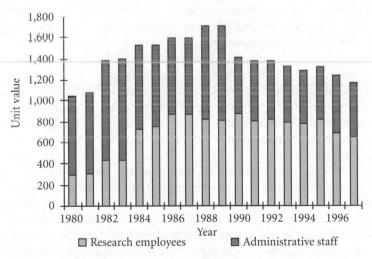

Figure 11.2. *INPE research and administrative employees, 1980–1997*

Source: compiled from INPE (1980–97).

might have encouraged innovative activities meant that the MECB was delayed and the objectives for satellite development were substantially revised.

Tapia (1995) has suggested, for instance, that there were major weaknesses in the management of the activities related to the MECB. The links between the research groups at INPE and the agency responsible for coordinating overall policy direction, COBAE, were not very strong. The technical knowledge was concentrated in INPE, and COBAE does not appear to have had the capability to plan and organize space R&D activities or to evaluate the progress of projects. COBAE's lack of technical knowledge was a stumbling block when it became necessary to negotiate increases in financial resources and for alterations to the scope and objectives of MECB with the military establishment and various funding agencies. INPE proposed the establishment of a team to assist COBAE personnel with technical issues but this was rejected mainly because COBAE staff believed that, if the limitations and difficulties that were experienced in the course of technological innovation projects were made explicit, funding levels would be reduced.

The objectives for the satellite development programme remained stable throughout the 1980s. However, by 1986 it was obvious that neither the Brazilian satellite launch vehicle (VLS) nor the four scientific satellites would be ready by the planned date. Researchers at INPE decided to prioritize the development of one data collection satellite and to search for partners for the development of remote sensing satellites (Tapia 1995). By July 1988 a partnership agreement calling for a joint programme with the Chinese Academy for Space Technology (CAST) was established to develop two remote sensing satellites. The various subsystems, such as the mechanical structure and energy supply technology as well as satellite telecommunications services and equipment (telemetry), were to be the responsibility of the Brazilian participants in the initiative.

The first Brazilian data collection satellite, SCD-1, was not launched until 1993— that is, seven years after the expected date. By this time, INPE no longer had sole responsibility for the development and launch of the satellites. Further changes from 1993 occurred as investment increased (see Fig. 11.1) and COBAE was replaced by the Brazilian Space Agency (AEB) in 1994, a civilian organization that was empowered to establish space R&D objectives. The Space Agency began to emphasize the policy goals of strengthening industrial capabilities in the supply of satellite telecommunications equipment and subsystems and also of 'opening' Brazilian space R&D activities by procuring components abroad and commercializing equipment developed in Brazil for the international markets (AEB 1994).

By 1994 INPE had accomplished its initial objectives for satellite development— that is, one satellite had been launched successfully and a second was being assembled. As a means of ensuring continuity in the acquisition of technological capabilities for space R&D, a new satellite project to be called Equatorial Communications (ECO-8) was proposed. The project called for the launch of eight satellites in low earth equatorial orbit, an orbit that would be particularly suitable for establishing telecommunications services for countries in the southern hemisphere. The satellites were to employ a relatively simple design to encourage the provision of cost-effective telecommunications services for the developing countries in the region (Bambace and Bastos 1996; Ceballos 1994). The ECO-8 project was expected to enable Brazil to enter the lucrative commercial international telecommunication market. In 1994, Telebras, the holding company responsible for the supply of public telecommunications services in Brazil, was given responsibility for implementing the ECO-8 project and for the commercial provision of services following the launch of the new satellites.

Thus, by the mid-1990s, thirty years of R&D activity in space technology-related fields had resulted in the accumulation of robust technological capabilities especially in fields such as meteorology and the subsystems and payloads for smaller satellites. The scientific expertise at INPE was being developed specifically with regard to the design and deployment of small satellites that could operate in low earth orbit, rather than with regard to the capabilities required for developing equipment and components more closely associated with larger geostationary satellites.

4. PROMOTING TELECOMMUNICATIONS NETWORK EXTENSION

As noted above, the ECO-8 project plan brought Telebras, the Brazilian telecommunications service supplier, into the picture. The direct involvement of Telebras introduced additional actors into the process of designing a new technological system and of negotiating to secure the necessary financial, technological, and human resources for the eventual deployment of an indigenous LEOS system.

4.1. *The Organization of Telecommunications Activities*

Until the late 1960s, telecommunications services in Brazil were provided by over 900 small operating companies. The authority to grant operational licences and set prices was distributed across many institutions, including central government, the federal state governments, and the local municipal authorities. The decentralized process of granting concessions for network operation and service supply was regarded as being responsible for the fact that in 1965 the quality of telecommunications service provision was very poor. It was also considered to be the main reason for the extremely low telephone penetration rates of around 1 per 100 inhabitants (Ministério das Comunicações, 1996). In addition, weak operational and administrative coordination between the various concessions resulted in a high degree of fragmentation of the telephone network. As a result, the provision of telephone services between rural areas and small or medium towns was practically non-existent (Vianna 1993).

In 1957 a congressional commission was established to propose new policies that would lead to the establishment of a unified national telecommunications system and in 1962 a National Telecommunication Code (CNT) was agreed by the federal Congress. The National Council for Telecommunication (Contel) was created and given responsibility for regulation and oversight of network planning, the integration of the fragmented network infrastructure, and the pricing of telecommunications services. The National Telecommunication Code included recommendations for the extension of public telecommunications services throughout the territory of Brazil and for the promotion of technical and professional training.[9]

In 1965 Brazil became a founding member of the International Telecommunication Satellite Organization (INTELSAT). The discussions leading to the decision to become a member of INTELSAT were characterized by intense debate concerning the appropriate arrangements for ownership of the satellite earth stations and for the provision of international telecommunications services in Brazil. The ministries with responsibility for the Brazilian economy opposed public ownership and supply of the services delivered using the INTELSAT satellites, and the military institutions in Brazil actively promoted nationalization of the existing network infrastructure and public ownership and supply of telecommunications services (Arroio 1995; E. Q. Oliveira 1992). The views of the military establishment were promoted mainly by the then Head of the Military Cabinet, General Ernesto Geisel, and these views gained ascendancy. As a result, the newly created public telecommunications operator, Embratel, which had responsibility for interstate and international telecommunications services, was given responsibility also for the implementation of satellite telecommunications operations.

The Brazilian government concluded negotiations with Hughes Communication International and other companies in 1967 for the construction of a satellite earth station for international telecommunications service supply within the INTELSAT agreement. This earth station, Tanguá 1, became the Brazilian gateway for international telecommunications services. The contracts with Hughes, RCA, and Ericsson for the supply of earth station equipment specified that Brazilian technicians would receive training in the operation and maintenance of the equipment. Eleven engineers were trained and these individuals 'became responsible for training new technicians who eventually took charge of the operation and maintenance of Tanguá 1' (Ferreira Silva 1997: 110).

A Ministry of Communications (Minicom) was created in 1967 with a remit embracing the regulation of the telecommunications and postal services. A few years later, in 1972, Telecomunicações Brasileira SA, Telebras, an open capital and mixed economy enterprise, was created to extend and integrate local, interstate, and international telecommunications service supply into a coherent national system. By 1976 Telebras had acquired and absorbed 914 local telecommunications companies (Ministério das Comunicações 1995). These were combined into twenty-seven local and regional operating companies, which were responsible for service operations and network extension in specific regions.

According to Minicom (Ministério das Comunicações 1996: 3), the creation of Telebras was an important step in the development of the local telephone service, since 'the measures following the CNT led to improvements in international and interstate services, but the same could not be said for local services'. However, others saw the network integration strategy of Minicom as a political manœuvre. A representative of Contel, for example, suggested that the actors within Minicom had insisted on

---

[9] By 1993 there were fifteen universities with highly qualified faculty members running undergraduate and postgraduate courses in telecommunications engineering (Pessini 1993).

the creation of a strong telecommunication holding company in order to 'control' Embratel. The creation of Telebras would divest Embratel of some of its authority and simultaneously provide Minicom with greater bargaining power to achieve its policy goals.[10]

During this period there was a series of innovations in the organization of the institutions with responsibility for telecommunication policy and service supply. The Ministry of Communications and Telebras were created and policies were established to encourage learning in key technological areas. This included the establishment of telecommunications engineering courses and training in satellite ground segment operations.

4.2. *Telecommunications Build-up*

As shown in Fig. 11.3, overall investment in the Telebras system increased significantly between 1974 and 1997. These are considered to be the 'golden years' of the Telebras system (Wohlers 1996). The telecommunications network expanded considerably and by 1997 both the number of main lines and the number of public telephones had increased fourfold in relation to the previous period.

Consolidation of the Telebras telecommunications system was promoted actively by the federal government and the company was allowed considerable decision-making and investment autonomy. The expansion of the network infrastructure was promoted

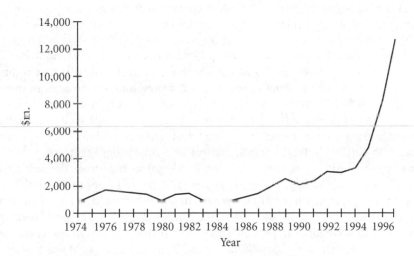

Figure 11.3. *Telebras system annual investment, 1974–1997*

Source: compiled from Telecomunicações Brasileiras SA (1974–97).

---

[10] Maculan (1981) observes that, in addition to Embratel, the local and municipal telecommunication concessions and the equipment industry were strongly opposed to the concentration of network planning and purchasing power in a single institution.

by the strong and continuous leadership of General Alencastro, who was director of Telebras for more than ten years. During this period merit awards for selecting staff were introduced, particularly for senior employees in the regional operating company executive management positions. Performance standards, including regional targets for telephone service extension and quality, were also implemented.

Another important development in 1976 was the creation of the Centre for Telecommunication Research and Development (CPqD) to foster R&D in telecommunications transmission and switching technology. The centre was expected to establish effective links between universities, equipment suppliers, and the telecommunications operating companies, and it also had responsibility for developing new programmes and adapting foreign technologies (Göranson 1993). A Satellite Communication Programme was established at CPqD in close cooperation with Embratel and more than 250 satellite-related projects were undertaken during the 1980s. These included work on time division multiple access (TDMA) systems used in satellite communication and on equipment used in ground stations such as off-set C Band antennae,[11] amplifiers,[12] satellite link measurement and simulation, satellite system interference, and signal processing and transmission equipment (Erber and Amaral 1994). University researchers were involved in the Satellite Communication Programme as well as in related telecommunications equipment development projects.[13] By the end of the 1970s, domestic suppliers of telecommunications equipment were meeting some 85 per cent of demand in the Brazilian market (Ferreira Silva 1997; Pessini 1993).

From 1980, although the rate of investment by Telebras in the telecommunications infrastructure began to decline, Embratel continued to finance a significant R&D effort aimed at strengthening Brazilian capabilities in the design and operation of satellite technologies (see Fig. 11.4). Funding for research activities jumped from 0.67 per cent of Embratel's operating reserves in 1980, to 7.6 per cent in 1985, and the funding levels continued to be maintained at a high level in 1986 and 1987. In addition, about 1.5 per cent of Embratel's annual operating revenue was invested in internal training activities. From 1980 to 1990, thirty-two engineers and technicians from INPE and Embratel received training at Spar Aerospace Ltd in Canada, a company with substantial experience in the design and launch of satellites for domestic use. Training focused on satellite orbital control, control and operation engineering, and the management of space programmes. Embratel sought to encourage the diffusion of the technological knowledge acquired by externally trained personnel throughout the company by implementing training programmes including courses on multiplex facilities control, data communication traffic control, and other technical operational activities. The substantial importance accorded to training within the company is illustrated by the fact that responsibility for the Training Department was transferred from General Administration to the Vice-President's office.

[11] Antennae of 6.0 and 4.5 metres.
[12] Power Amplifiers (PA) and 100 watt High Power Amplifiers (HPA).
[13] These include the Pontifica Universidade Catolica of Rio de Janeiro; the University of Sao Paulo (Microelectronics Laboratory); the Engineering Department of Unicamp; the National School of Engineering; the Aeronautics Technological Institute, and three universities in the south (Pessini 1993).

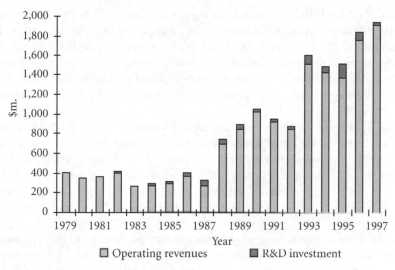

Figure 11.4. *Embratel operating revenues and R&D investment, 1979–1997*
Source: compiled from Embratel (1979–97).

The availability of increased R&D and training resources was closely tied to a proposal for the introduction of a Brazilian Satellite Telecommunication System (SBTS), a domestic telecommunication satellite operating in geostationary orbit. Investment in a new domestic satellite system was expected to contribute to 'the extension of telecommunication to all Brazilian municipalities and the creation of a robust and competitive industrial and R&D sector investment' (Mattos 1984: 30). There was opposition from the Planning and the Finance ministries for the proposed acquisition of a domestic satellite system. The actors in these ministries argued that the cost of the proposed Brasilsat system should be financed by a levy on domestic telecommunication revenues (Bensimon 1988). In addition, an influential group within Embratel favoured investment in the integration of the domestic telecommunications network using the terrestrial microwave network that was under construction at the time, rather than in a satellite system (Terracine 1997). Despite substantial opposition, however, the domestic satellites, Brasilsat A1 and A2, were launched in 1985 and 1986.

In summary, during this period many policies and measures were implemented to strengthen the skills and capabilities needed for developing and operating satellite systems. The CPqD was created and the Satellite Communication Programme was established. Embratel R&D investment increased significantly over the earlier period and there is evidence that relatively strong linkages were forged between CPqD, Embratel, the universities, and local firms.

4.3. *The Search for Partners: Liberalizing Telecommunications Markets*

The Brasilsat system operated with spare capacity from 1986 to 1990—that is, well below initial demand forecasts. Relatively little attention was given by the system operator, Embratel, to the commercial use of satellite services through initiatives aimed at stimulating demand for satellite service applications in the domestic market. In addition, the necessary earth station network was not expanded rapidly enough to provide a network for satellite uplinks and downlinks for telecommunications service provision. More than half of the planned service life of the satellites was wasted while Embratel developed the earth station network (Ferreira Silva 1997; Nettleton and McAnany 1989).

By 1992, although Embratel had begun to promote the use of satellite telecommunications services for businesses, the services that were available remained costly and in small localities there were very few users. In 1995, for example, despite the deployment of the Brasilsat system, less than 2 per cent of rural dwellings had access to telecommunications services and the system was not being exploited to support the delivery of tele-education or tele-medicine (Ministério das Comunicações 1996; Terracine 1997).

During the early 1990s there was intense pressure for changes to the structure and organization of the supply of telecommunications services in Brazil. Business users established the Brazilian Institute for the Defence of Telecommunication (IBDT) to lobby the government executive and the legislative assembly for the launch of measures to introduce competition in the supply of telecommunications services.[14] The government responded by liberalizing the telecommunications equipment market and by initiating procedures to allow competition in the supply of cellular telephone and satellite-based telecommunications services. Telecommunication equipment manufacturers and service providers that had exited the Brazilian market in the 1970s established joint ventures with the major telecommunications operating companies and with Embratel. The result was dramatic. In 1995 the balance of trade for the Brazilian telecommunications sector registered a deficit of $1,230 million, a substantial increase over the 1993 deficit of $300 million. By 1997 an unprecedented level of $2,500 million deficit was reached (Cassiolato, Szapiro, and Andrade 1998; Gazeta Mercantil 1998).

Embratel began to introduce measures to improve its competitiveness in the newly liberalized marketplace. These measures included a strategy to increase the use of satellite technology by expanding the range of data communications services available both domestically and in neighbouring Latin America countries. Embratel also introduced a training programme in Management of Total Quality Control using an executive television service to provide incentives for exchanges of management and other information within the company.

Although the implementation of larger geostationary satellite technology through the introduction of the Brasilsat system was successful, little progress had been made towards the planned development of the LEOS project, ECO-8. In the mid-1990s, this project was an important component of Minicom's strategy to improve the reach of

---

[14] IBDT was backed by fourteen powerful industrial and financial groups, including construction companies, banks, and telecommunications equipment and service suppliers (Hollanda 1993).

the domestic telecommunications network beyond that which had been accomplished by Embratel (Ministério das Comunicações 1995). Extending telecommunications access to low traffic density areas by offering services at low prices was the main objective of investment in the project. Telebras negotiated a memorandum of understanding with the American companies, Constellation Communications Inc. (CCI) and Bell Atlantic Enterprises International Corporation, in 1995 for the construction and operation of a low orbit satellite system. The agreement stated that: 'The system called ECCO—Equatorial Constellation Communications—combined the independent Telebras projects for an equatorial system and the CCI project for a global system. This joint effort . . . will enable development of the project from a commercial and business perspective' (Telecomunicações Brasileiras SA 1994: 1).

The original ECO-8 proposal was expanded to include eleven satellites in low equatorial orbit and a further thirty-five satellites in low, inclined orbit. This constellation of forty-six satellites in low earth orbit would comprise the Equatorial Constellation Communications system (ECCO) designed to provide global mobile telecommunications service coverage (see Table 11.1). The ECO-8/ECCO project was considered by the Brazilian government to be 'the jewel in the Brazilian space programme's crown' and expenditure of some $280 million was approved by the government for the project (Freitas 1995: 47).

From 1996, however, Brazilian participation in the project came to a standstill. Representatives of Telebras suggested that 'the decision was a political one' in the sense that continuation of the project required legislative (congressional) authorization and the procedure to obtain such authorization had to be initiated by Minicom.[15] Minicom did not initiate the necessary authorization after 1996 for a variety of reasons.

One of the main reasons for the stalemate in the domestic negotiations concerning the further development of the ECO-8/ECCO project was the pending privatization of Telebras. Representatives of Telebras and Minicom argued that it would be easier to privatize the monopoly telecommunications operator if it were unencumbered by this costly project. The competing LEOS telecommunications service providers who were seeking to obtain an operating licence in Brazil—that is, Globalstar, ICO, and Iridium—also supported this viewpoint. The potential new entrants argued that the proposed ECO-8/ECCO system was unlikely to be profitable and that, therefore, it would require public 'subsidy' for Telebras. Opponents also suggested that the government should not finance a high-risk domestic venture at public expense and that a Brazilian organization should become a member of an international LEOS consortium (*Exame* 1994).

In 1997 the Brazilian National Congress approved the General Telecommunications Act (LGT), which established guidelines for structural reforms in the telecommunications sector. One year later, the twenty-seven local operating companies in the Telebras system were privatized, creating three new regional operating companies.

---

[15] Interviews with Ronaldo Sá, Secretary for Radio frequency allocation, Ministry of Communications, and João Mello, Manager of the Division for International Affairs, Telebras, 25 Mar. 1997 and 9 Apr. 1997, respectively.

Embratel was sold off separately to MCI.[16] A new regulatory office, the National Telecommunications Agency (Anatel), was created and granted authority for the allocation of radio frequencies and the regulation of satellite communications services as well as other terrestrial telecommunications services. It was given the authority to grant radio frequencies, with or without public auctions, and to determine whether a licence fee should be levied. Considerable pressure was exerted by prospective LEOS service suppliers on the new regulatory agency to move quickly to introduce regulations for the supply of LEOS services. The LEOS operators sought to ensure that any provisions for licence fees and public auctions would be waived because 'if auctions were held it would render most LEOS projects impracticable'.[17]

Provisions for the regulation of LEOS services were established through a legislative decree No. 560 in 1998. This decree stated that permission to operate LEOS services should be granted to licensees for periods of fifteen years and that individual LEOS service providers should pay approximately $100,000 to the government for the use of radio frequencies to provide services within Brazil. Apart from a payment to cover inspection costs incurred by the regulator during the licensing process (approximately $9,000), no additional charges were to be levied on operators seeking to exploit the non-geostationary satellite market. The reason given for this by the government was that 'the size of the market is not known and therefore, these charges could not be based on reasonable criteria' (Valente 1998: 3). LEOS operators' reservations about the potential demand for their services in the Brazilian market do not appear to have been shared by the prospective operators. Iridium, for example, expected to have a subscriber base of 250,000 users in Brazil and to generate annual revenues of $180 million by the year 2000 (Uchôa 1997: 22). Forecasts such as that by Nourozi and Blonz (1998) suggested that Brazil would be the eighth largest market worldwide for LEOS services.

Despite the existence of a potentially large market given the still relatively low penetration of affordable telecommunications services (that is, a 14 per cent teledensity), the regulatory provisions did not establish infrastructure extension targets for LEOS service operators or targets for promoting service use in rural and remote areas. The option of levying licensing fees to support the development of services in less economically advantaged regions or for the provision of village telephones was not pursued. The government also opted not to study the economic viability of linking the sale of Embratel to commitments by potential investors to the further development of the ECO-8/ECCO project. Another option might have been to allocate some portion of

---

[16] The Telebras system was sold for a total of $R22,057 bn. (approximately, $US22 bn. at the exchange rate applicable at the time) (Furiati 1998). The regional companies were purchased by consortia comprising Spanish, Portuguese, US, Italian, and Brazilian interests. Embratel was sold to the US-based telecommunications carrier MCI for $R2.6 bn. (approximately $US2.6 bn. in 1998) (Furiati 1998: 13). The privatization strategy included the sale of 'mirror licences'. These were to be sold to different consortia to promote competition against fixed and cellular incumbents. In early 1999 Bonari Holding, a consortium comprising the National Grid for electricity supply in the United Kingdom, France Telecom, and the American telecommunications company Sprint, bought the 'mirror licence' to compete in the long-distance and international telephone service market against Embratel/MCI (Decol 1999: 8).

[17] Interview with José Moraes, representative of ICO, on 15 Apr. 1997.

the proceeds from the privatization of Telebras to the further development of the LEOS ECO-8/ECCO project. It also seems that the domestic experience with LEOS service development was not exploited to assess the conditions for licensing foreign LEOS service operators that would be most favourable to the Brazilian government. In other words, the accumulated learning from the domestic LEOS project did not provide the basis for negotiating arrangements with the new service providers that would have encouraged the provision of services for low-income users.

In summary, from the mid-1990s Brazil experienced significant changes with respect to telecommunication and space R&D policy formulation and implementation as a result of market liberalization measures. Although there was overall growth in R&D investment and in INPE R&D investment between 1993 to 1998, and Telebras succeeded in expanding the reach of telecommunications services to some extent, these developments were not accompanied by programmes to promote the continued strengthening of the domestic capacity for developing and implementing space and satellite technology systems. In fact, there was a sharp decline in the number of researchers at INPE and the ECO-8/ECCO was suspended despite its objective of increasing domestic industry participation in the space and satellite equipment market.[18] Increased telecommunication investment by the government and the purchase of an additional domestic geostationary satellite (Brasilsat B4) to support television and data applications were part of the Cardoso administration's strategy designed to improve the commercial value of Telebras prior to its privatization. However, this strategy of using public resources to attract investors did not extend to enhancing the likelihood that the new LEOS services would be deployed in a way that would be directly responsive to the goals of extending telecommunications services to rural and remote economically disadvantaged areas.

5. MEDIATING THE RULES OF THE GAME

The terms and conditions that are established by national governments for satellite service operator licensees are the result of intense negotiations between a large number of private and public stakeholders. Regulators, satellite system operators, members of the scientific and engineering communities, and representatives of various government ministries engage in negotiations that establish the 'rules of the game' for investment and service provision. Such rules must be responsive to objectives with respect to investment in telecommunications networks to meet business demand, but there are other objectives that may be addressed through the establishment of the 'rules of the game'. For instance, it may be feasible to establish terms and conditions for telecommunications operators such that they are obliged to contribute to the development of the local skills base through arrangements for the accumulation of technical or management-related knowledge. As a condition of their licences, operators may be

---

[18] Although the US–Brazil ECCO joint venture was abandoned in the late 1990s, the company CCI proceeded with plans for a commercial LEOS project that was similar to the original ECO-8/ECCO project (see CCI 1998).

required to meet targets for the extension of the telecommunications infrastructure to provide affordable access to services for users in rural remote areas, or they may be charged with supporting the distribution of tele-health or education programmes.

In order to establish national regulations for telecommunications service supply in a liberalized market environment that are responsive to a wide range of economic and social policy goals, the stakeholders in the domestic environment need to be able to bargain effectively with new entrants in the marketplace. A failure to bargain effectively is more likely to enable prospective service providers to introduce arrangements that favour their short-term interests in achieving a return on their investment in the telecommunications network. Effective mediation—that is, the coordination of a variety of processes involving negotiations between domestic and foreign stakeholders—is likely to result in greater opportunities to find innovative means of implementing social or commercial policy objectives that are linked to the extension of the telecommunications network and the provision of affordable services.

There are two sets of capabilities that are essential for the representatives of developing countries to accumulate if they are to acquire strong capabilities to pursue negotiations in ways that are responsive to domestic policy objectives and priorities: these are technological and institutional capabilities. It is important to analyse how and by whom these capabilities are accumulated over time.[19] Technological capabilities are defined as the resources needed to generate and manage technical change and they are manifested in intangible social resources that include the human skills, knowledge, and experience required for initiating and implementing technological change (Bell and Pavitt 1993: 163–5). A dynamic process of learning is required to develop and strengthen technological capabilities and a key aspect of the learning process is fostered by continuous participation in the generation and improvement of technologies. As the sections of this chapter have shown, analysis of specific aspects of the development of these capabilities helps to assess the extent to which actors can respond flexibly and adapt to the opportunities created by the introduction of innovative technologies in a complex and competitive market.

The technological capabilities that contribute to and support the introduction of innovative technologies are only one side of the coin that needs to be considered in understanding the relative strengths and weaknesses of actors in a national setting to negotiate effectively for outcomes that are in line with policy objectives. 'Institutional capabilities' also need to be examined. These capabilities relate to the role of public-sector institutions and actors that become involved in financing, using, or regulating the deployment of new technological developments. As illustrated in the analysis of the developments leading to the deployment of satellite technologies, public-sector institutions and actors play an important role in coordinating negotiations leading to the deployment of new technological systems and services. In some cases, interactions between public and private participants in these negotiations lead to enhanced learning opportunities for domestic players and to the formation of common expectations and goals for the introduction of new technologies in line with a broad range of social

---

[19] See e.g. Dosi (1991), Freeman and Soete (1997), Nelson (1993), and Perez and Soete (1988).

and economic policy goals. In other instances, especially where technological and institutional capabilities are rather weak, the outcomes of negotiations are more likely to support only some aspects of those goals.[20]

Jessop (1997: 111) has suggested that effective learning is most likely to occur when conditions are sufficiently stable to facilitate the selection of responses by parties in a negotiation process in such a way that their goals are jointly achieved. However, in periods of rapid technological change, discontinuities emerge between the technological and institutional capabilities that are necessary to engage in effective negotiation and the capabilities of actors who are involved in establishing the conditions for the deployment of the new technologies (Mansell 1996: 27–8). Developing appropriate responses to new technical opportunities involves substantial investments in the acquisition of new skills and knowledge through various forms of learning. It also requires that domestic actors acquire the capabilities necessary to exercise the power to modify and adapt existing institutional arrangements. Technological and institutional capabilities are manifested not only in intangible social resources (Bell and Pavitt 1993), but in the power to act that is embodied in the individual members of a variety of institutions (Mansell 1996).

The capabilities that became embedded in the authorities responsible for policy and regulation of telecommunications services in Brazil, in the personnel employed by companies involved in the development of equipment and deployment of satellite systems, and in the representatives of telecommunications operating companies have been examined in this chapter. The analysis of some of the interactions between these actors yields insights into the major factors that have influenced the outcomes of negotiations leading to the deployment of satellite technologies. A variety of intricate power relationships was instrumental in establishing the extent to which domestic policy goals would be met and these relationships represent a form of mediation — or intermediation—between actors within their respective institutional settings.

The indicators used to assess specific types of negotiating capabilities were presented in Box 11.1. Substantial and sustained investment in the acquisition of these capabilities may be expected to provide a strong basis for the formation of an informed community of domestic stakeholders who can influence the introduction of a new technology. When both technological and institutional capabilities are rather strong, it is more likely that negotiations between domestic and foreign stakeholders will produce outcomes that favour a range of domestic policy goals. Jessop (1997: 95) defines governance as the 'complex art of steering multiple agencies, institutions and systems which are both operationally autonomous from one another and structurally coupled through various forms of reciprocal interdependence'. If reciprocal interdependence is to be achieved, there is a need for the coordination of action on the part of numerous stakeholders to achieve strategic or policy goals. Coordination, in turn, requires communication between public and private actors to establish mutual understandings and expectations about alternative ways of implementing the new technologies. Negotiation processes, such as those described in the preceding sections,

---

[20] This argument has been developed by Freeman and Soete (1997), Lundvall (1992*b*), and others.

offer a basis for communication that is more likely to mediate the relationships in ways that lead to the joint articulation of common objectives. In effect, social actors embedded in a variety of institutional contexts accumulate the technological and institutional capabilities that are necessary to act to establish new 'rules of the game' for the deployment of technological innovations.

The experience of the Brazilian actors illustrates that the combination of negotiating capabilities at different times in the history of the debate around satellite telecommunications service provision had a decisive impact on how the new technologies would be deployed and used. In the case of INPE's failure to promote the Advanced Interdisciplinary Communication Satellite project (SACI) after 1975 and the long delay in the launch of the remote sensing satellite (SCD-1), there were major problems in the interactions between those involved in the negotiations. There were clear signs of problems in the mediation of the relationships between key actors and their institutions. These difficulties contributed to INPE's inability to gain support for the SACI project from the Ministry of Communications and resulted in decreased financial support after 1985 because of weak linkages with the Commission for Space Activities.

In contrast, the development of the telecommunications network from the mid-1970s, and the promotion of R&D in satellite technologies in the early 1980s, suggests that during this period the outcomes of various negotiations were a reflection of the accumulation of stronger negotiating capabilities. This period was characterized by increasing levels of investment in the telecommunications infrastructure and in space-related R&D aimed specifically at developing a domestic satellite system to provide more extensive and affordable telecommunications services within Brazil. There is substantial evidence during this period of skills development in the fields of satellite equipment manufacturing and the design of satellite components. Strong leadership at INPE was also important for the implementation of R&D projects and for encouraging the accumulation of technological and human resources. Good management by Telebras also appears to have contributed to the successful integration of private local operating companies into a national telecommunications network and to the creation of CPqD.

By the mid-1990s considerable progress had been made in establishing a coordinated and interconnected public switched telecommunications network in Brazil. All cities and 71 per cent of villages were being served by the public switched telecommunications network with at least one point for public access (Ministério das Comunicações 1995). However, despite these significant achievements, telephone penetration rates remained low compared to the levels being achieved by neighbouring countries, including Argentina, Chile, and Uruguay, and were significantly lower than the penetration rates achieved in high-income countries (ITU 1999*a*). Representatives of Telebras and INPE proposed the ECO-8 LEOS project as a means of increasing telephone penetration in low teledensity areas and argued the need for a satellite system that would be cost effective and therefore able to support service provision at affordable prices.

Initially, the AEB was successful in coordinating policies relating to the ECO-8 project. However, from 1996 the agency was unable to negotiate successfully for continued federal government support for the project, even in the face of the strong projected demand for LEOS services. This suggests that the bargaining strengths of the actors in INPE, Telebras, and the AEB had become relatively weak in the face of the bargaining strengths of those who were lobbying for the introduction of LEOS services by foreign suppliers.

Three key capabilities and strengths in each of these areas seem to have been important at times when satellite technologies have been deployed in line with a broad range of national economic and social priorities. First, the capability to establish and maintain the coordination of institutional linkages played a major role, for example, in the partnership between Embratel and CPqD for the development of satellite equipment and software applications and in the development of a coordinated approach to telecommunications network expansion by Telebras in the 1970s. The second important capability lies in the skilled leadership present, for example, during the 1980–5 period at INPE and in the 1974–85 period at Telebras. A third important capability is the capacity to mobilize investment in R&D aimed at developing new technologies, as in the case of investment in satellite equipment at CPqD in the late 1970s.

6. CONCLUSION

All too frequently there are high expectations that the fruits of technological innovation such as the LEOS systems will be deployed in ways that are fully responsive to both commercial and social policy goals that have been formally established by governments.[21] Unless there is substantial investment in various capabilities, these expectations are not fulfilled. It was initially expected that telecommunications service users in both the industrialized and developing countries would reap many social and economic benefits as a result of the implementation of the new LEOS technology. However, little attention was given to the fact that the application of this new technology would incur substantial costs related to the build-up of technological and institutional capabilities. The foregoing analysis has illustrated the importance of these capabilities to support learning and to encourage negotiated outcomes that are in line with policy expectations. There has been very little detailed investigation of the specific capabilities that are needed by domestic stakeholders to enable them to establish the 'rules of the game' for the supply of new technological systems and services that encourage their use in ways that are consistent with established economic and social goals.

The Brazilian experience presented in this chapter offers three main explanations for the dynamics of negotiation that have been at play over an extended time period. First, in order to grasp the opportunities created by innovations in technologies, it is crucial that domestic actors be able to negotiate effectively to achieve the desired policy outcomes. The accumulation of strong negotiating capabilities depends on the build-up of specific technological and institutional capabilities. In the case of the Brazilian government's efforts to take advantage of innovations in satellite technology

---

[21] See e.g. Hudson (1997) and Jussawalla and Tehranian (1993).

during the SACI experiment and during the implementation of the domestic satellite communication programme—that is, Brasilsat—the necessary capabilities were either absent or weak. The domestic stakeholders were unable to negotiate with external actors to ensure that a wide range of social and economic policy goals would be met.

Secondly, when governments seek to secure both social and commercial policy objectives as a result of satellite technology deployment, it is necessary to ensure that effective processes are in place to enable the relevant actors to negotiate in ways that establish common expectations for meeting jointly agreed objectives. When this mediation process is successful, it is more likely that these objectives will influence and become embedded in the 'rules of the game' for the introduction of the new technologies. The analysis of Brazil's experience suggests that little attention was given to encouraging the formation of common objectives for harnessing the social as well as the business potential of satellite technologies. As a result, the regulatory authorities have been concerned mainly with other issues such as creating the appropriate ground rules for the extension of telecommunications services to meet business demand for data and mobile telephony services. The introduction of the domestic Brasilsat system was not accompanied by regulatory measures to promote the extension of affordable telecommunications services in remote and rural areas, and the opportunity to exploit the new technological infrastructure to support tele-education and other social services was not pursued. Following the round of negotiation to establish the 'rules of the game' for LEOS services, there were few signs that the regulatory provisions would ensure that LEOS service operators would promote improved access for rural and remote telecommunications service users.

Thirdly, the Brazilian experience of the implementation of satellite technologies suggests that the liberalization of the telecommunications market to permit the entry of competing LEOS service operators has not been accompanied by efforts to take advantage of the potential scale of demand in the domestic market or the accumulated technological and institutional capabilities. The potential scale of market demand for the new services and the knowledge base of the domestic stakeholders could have provided a strong bargaining point for policy-makers to establish terms and conditions for the new entrants that would have strengthened the technological capabilities in the satellite industry in Brazil. The lessons and experience of the early development of the ECO-8 project were not integrated with policies for the privatization of Telebras. In the absence of a focused R&D programme on LEOS technology (or related satellite technologies), it is likely to be difficult for INPE to encourage the development of management and technical skills in this area.

In the light of the Brazilian experience with LEOS technologies, what realistic measures might be taken to provide a greater incentive for the deployment of LEOS and other new generations of innovative technologies in a way that is responsive to both business and social goals? Given the existing legislative framework for the regulation of LEOS service providers, there will be a need to find innovative ways of responding to the requirements of low-income users. A favourable environment could be created by strengthening the linkages between INPE, AEB, Anatel, local equipment suppliers, and groups representing low-income users. Forums in which different stakeholder interests can be acknowledged and discussed would enable representatives of low-income users to accumulate greater knowledge about how to lobby for policy measures that would be responsive to their needs.

If economically disadvantaged user communities are to be encouraged to accumulate better negotiating capabilities, the processes that lead to the formation and implementation of policies and regulation for telecommunications service supply must be made more transparent. The creation of the AEB and the regulatory authority, Anatel, in 1994 and 1998, respectively, has begun this process of transparency, and decisions are no longer the preserve of the representatives of government ministries or the military establishment.[22] However, it continues to be difficult for civilians to participate in the discussions of these institutions. In order to facilitate learning and provide effective governance that embraces the concerns of both social and commercial policy, new incentives will be needed to encourage informed common consensus on the advantages of providing new services in response to social as well as commercial goals.

Of particular importance in the case of satellite telecommunication policy in Brazil is the need to strengthen the R&D capabilities of the domestic space technology developers. INPE, despite a limited budget, was able to support the development of domestic capabilities for design and deployment of satellite technology. Given the need for public investment in other priority areas in Brazil, including health and education, it seems unlikely that INPE's resources for R&D will increase substantially in the short or medium term. Nevertheless, there are relatively low-cost options that could be pursued that would help to strengthen technological capabilities in the industrial and scientific communities—for example, encouraging the involvement of local firms in the development and production of LEOS-related equipment such as satellite gateways and hand-held terminals. This could be achieved if domestic policy-makers are able to bargain more effectively with foreign LEOS operators in the future to establish conditions conducive to the participation of domestic firms in the design of equipment and components.

The government could also offer incentives to attract investment from private, foreign, or domestic sources to continue the development of the ECO-8 project. The potential benefits would include the continued strengthening of local technological capabilities and improved relationships between government decision-makers and local firms. The successful development of the project would open new opportunities for exploiting the LEOS system to support the provision of affordable telecommunications services for users in low-income areas.

The analysis in this chapter is important in any consideration of the ways that technological innovations become embedded in countries where their use is expected to yield widespread social and economic benefits. LEOS systems represent a small subset of the many advanced ICTs that are providing a foundation for improved communication and access to information. Despite clear signs that the telecommunications access 'gap' is unlikely to be closed rapidly, there continue to be expectations amongst the

---

[22] Anatel has implemented procedures to promote greater public participation in decision-making processes, including public consultations and extensive use of the Internet.

policy communities that new telecommunications service providers, such as those operating global mobile services, will offer services that will help to bring telecommunications access to those areas that currently are excluded.

This case study of the Brazilian experience suggests that expectations will not be fulfilled unless national governments and other stakeholders make coordinated efforts to strengthen their capabilities for effective negotiation with the new service suppliers. This is not simply a matter of investing in skills accumulation. It involves a complex process of investing in specific sets of technological and institutional capabilities over extended periods of time. Many middle-income countries do have the resources to make sustained investment when the need to do so is clearly recognized by policymakers, the industrial and scientific communities, and other stakeholders. Although such investment cannot guarantee that the outcomes, in the form of the accumulation of stronger negotiating capabilities, will lead to the deployment of innovative technologies in line with both social and commercial objectives, it clearly offers a greater likelihood that domestic actors will introduce 'rules of the game' that are more, rather than less, supportive of these objectives. A major factor contributing to the implementation of social and economic objectives is the creation of decision-making processes that are transparent and that favour the mediation of shared perceptions of the value of commonly agreed objectives.

12

Conclusion: Social Relations, Mediating Power, and Technologies

ROBIN MANSELL

1. INTRODUCTION

Many observers are predicting profound and wide-ranging social change as a result of a global transition towards what is variously called informational capitalism (Castells 1996), the information economy (Shapiro and Leone 1999), the digital economy (Margherio *et al.* 1998), the weightless economy (Quah 1997), or simply the 'new economy' (Kelly 1998). Frances Cairncross (1998) has used the phrase the 'death of distance' to express the idea of fundamental and pervasive changes in the way that individuals and social groups relate to one another. Manuel Castells (1996: 1) writes of a major 'restructuring of capitalism' itself. In each of these characterizations of global transformation, digital information and communication technologies (ICTs), the technological networks they support, and the digitized and often interactive services they provide, are implicated as major contributors to potentially massive economic and social reorganization of society.

In the introductory chapter to this volume (Chapter 1), we presented a conceptual framework developed initially by Christopher Freeman and Carlota Perez. This framework puts the analysis of the dynamics of paradigmatic techno-economic change at the top of the research agenda for improving understanding of how innovations in ICTs influence social and economic organizations and development. In that chapter, we noted that their framework encourages analysis of emergent 'guiding principles', which facilitate the establishment of dominant economic, social, political, or technological regimes.

Research on the determinants of paradigmatic social and technological transformations has shown the potential for major disruptions in the economy to coincide with the pervasive deployment of new digital technologies and services (Freeman 1994*b*; Freeman and Soete 1997). Research has focused on the capacity for human adjustment to new technological regimes and it has examined mismatches between the supply of

This chapter benefited substantially from David Neice's contribution and also integrates and extends the arguments developed by the other contributing authors. The final text in this chapter is the author's responsibility.

new technologies and market demand (Mansell and Steinmueller 2000) for the new ICTs. However, there has been too little research on the specific dynamics of the social and technical interactions that are influencing change in social organization and on whether social actors (individually and collectively) are able to ensure that the new technical systems are consistent with their values and aspirations. The research presented in the preceding chapters of this book reveals several important aspects of these dynamics. In order to do so, the chapter authors have had to penetrate many widely held views about the dynamics 'inside the communication revolution'. They have had to select sites for research where we might expect to find evidence of the key social processes that are encouraging or discouraging the features of a new dominant technological regime.

The development and use of digital ICTs offer the potential for changes in patterns of face-to-face and 'virtual' interaction, just as earlier generations of ICTs offered new opportunities (Levinson 1997; Marvin 1988). The pervasive reach of the new digital technologies, the extent to which their use reduces the salience of time differences and geographical distance, and the potential for a wide variety of modes of communication and types of information and media services mean, however, that we cannot look only to history as a guide to the implications of current trends in social and technological developments. New empirical studies are needed to examine how people are interacting with the new generations of technology and to assess the social consequences. We elected to focus on these interactions as instances of mediation which we define as a dynamic relational process that binds (or unbinds) networks of individual actors or institutions.[1] To meet this research challenge, we have sought to illustrate how each of the instances of mediated experience that we have investigated sheds light on the ways that society is likely to change through human shaping, accommodation, and resistance to the emerging 'techno-economic paradigm' as it is manifested in specific systems and technological artefacts.

Our empirical studies are all concerned, either directly or indirectly, with an aspect of the sites of interaction between social and technological processes. In some instances, we are principally concerned with the social processes informing the use of a given constellation of ICTs (for example, computer-mediated communication and the Internet in Chapters 2, 3, and 4). In other instances, our concern is focused on social processes that shape institutions of governance that are emerging alongside the growth of networking via the Internet (for example, certification authorities for electronic commerce in Chapter 5, frameworks for sharing the data used in geographic

---

[1] See Ch. 1, Sect. 3. Our use of mediation is distinct from 'mediation theory', which was proposed by the philosopher and statistician Karl Pearson. His work referred to any approach to learning that assumes that there are mediating processes between a stimulus and a response and used statistical techniques to model the presence and effect of such mediating processes (Bothamly 1993: 337). Re-intermediation is a term frequently employed in the financial sector of the economy (e.g. Coleman 1994; Cooley and Nam 1998; Easterwood and Morgan 1991; Yanelle 1989). It is also employed in the library sciences (White 1995), in economics and rational choice theory (Cosimano 1996), and in studies of electronic markets (Kraut *et al.* 1998; Sarkar, Butler, and Steinfield 1998). The term 'remediation' is used mainly in psychology in reference to learning disorders, but Morris and Hopper (1980) have discussed the concept in the context of communication in the public sphere.

information systems in Chapter 9, and the Internet domain name organization in Chapter 10). In still others, our research aims to elucidate the social processes involved in the early design of digital technological systems (including information systems within firms in Chapters 6, 7, and 8 as well as large-scale hardware and software systems in Chapter 11). We do not provide detailed histories of these ICT systems, because our focus is on the social and technical interactions that structure, enable, or constrain possibilities for human action in a world where digital technologies are becoming more pervasive in people's lives.[2]

Our research has embraced two major themes (which are reflected in the two parts of the book). The *first* is an investigation into how new patterns of social and technical interaction can yield new understandings and power relationships between social actors and new parameters for social inclusion and exclusion. The *second* is an investigation of the institutional structures and processes that are evolving and the capabilities that facilitate the development of the new technological systems and artefacts that may facilitate knowledge exchange.

This chapter sets out the main insights arising from our research programme by drawing on the results reported in the preceding chapters and considering these within the wider framework of questions about how we can further develop a research agenda that is informed both by a sensitivity to the crucial importance of social and technical mediation and by a desire to provide insights for policy-makers, technology designers, and users. In Section 2 we explain the role of 'guiding principles' in the emergent social and technological regime. This discussion is followed in Section 3 by a consideration of objectification processes that give rise to the ambiguity of social and technological systems in terms of their implications for those whose lives are mediated by them. Section 4 provides an extended discussion of our use of the concept of mediation and explains how this concept is applied to explore the two major themes of our research agenda. In Section 5, the way in which our research focus on mediation complements the more familiar information-processing view of the 'communication revolution' is demonstrated by drawing out the key insights in the preceding chapters of the book. In Section 6, we conclude with observations about the need to strengthen policy dialogue about the values and guiding principles of the emerging digital technology paradigm.

2. GUIDING PRINCIPLES AND EMERGENT SOCIAL AND TECHNICAL REGIMES

The reader may conclude that, because human behaviour is always characterized by social and technical interactions of many kinds, there is nothing particularly new about mediated action in the context of the emergent digital ICT paradigm. Manuel Castells, for example, is careful to emphasize that 'information' exchange (understood as the communication of knowledge) is, and always has been, crucial for all societies.[3]

---

[2] For a comprehensive examination of the innovation processes that are giving rise to the pervasive deployment of digital ICTs, see Mansell and Steinmueller (2000).

[3] This view was also central to the work of Canadian economic historian Harold Innis (1950, 1951).

To highlight the important characteristics of today's society he stresses 'informational' activities that, he argues, are associated with 'a specific form of social organization in which information generation, processing and transmission become the fundamental sources of productivity and power, because of the new technological conditions emerging in this [current] historical period' (Castells 1996: 21n. 33).

This distinction is an important one, because it emphasizes the importance of the inheritance of patterns of social and technical interaction and their guiding principles, even as a new paradigm is being shaped. As Freeman (1992a: 165) argues, we should expect 'a new set of guiding principles' or common-sense practices to be interwoven with a new techno-economic paradigm. These principles or practices become embedded in institutions and in society. In so far as they create new asymmetries of power, they can be expected to introduce a new dynamic of social inclusion and exclusion and they may exacerbate existing divisions and distinctions within society. This means that it is essential to explain how and where the new principles are operating, how they alter the operation of existing principles, and what they mean for socially valued human action.

There is no generic model of a global information society that is likely to be universally desired by all interested social actors. The so-called new economy and its relationship with the predominant social order that is associated with the earlier techno-economic paradigm is unlikely to give rise to an immaterial future in which all people prosper equally. As Christopher Freeman and Carlota Perez have consistently suggested, a new paradigm does not simply replace an older one. The principles of the old regime and the new one become entangled. They may give rise to entirely new modes of social and technical interaction, the dynamics of which may produce improved or worsened social outcomes. The studies in this book suggest clearly that the values, structures, and processes of mediation—that is, the guiding principles of the digital technology paradigm—are in a state of flux.

Paradigmatic technological shifts and their associated guiding principles are not only technical and economic. Shifts in beliefs and perceptions and in the social and political order are also intertwined with the dynamics of the unfolding technological regime. We have examined how certain forms of digital technology come 'between people' and become implicated in a reconstitution of the social order. The press of computers and other digital technologies into our lives is an ongoing mediation process.

As Freeman (1974) has argued in his book *The Economics of Industrial Innovation*, the previous techno-economic paradigm was dependent upon the mediation of social and technical interaction created by the availability of relatively inexpensive petroleum products. Technical innovations associated with this development, and, arguably, changes in the technologies for one-to-one and one-to-many mass communication, were sufficiently powerful to encourage a new way of organizing urban and rural life in many parts of the world. This way of life progressively isolated people and their affairs, reduced the spheres for public participation, carved through local communities, and melted social bonds, replacing them with cosmopolitan yearnings and affiliations.

The mediation of digital technologies similarly inserts the digital (tools and media content) into various forms of social interaction. The reweaving of the technical, the personal, and the social is a ceaseless process. Nevertheless, today's digital technologies and services offer an opportunity to examine some of the new forms of mediation while they are crystallizing. They can provide opportunities to reformulate the 'guiding principles' of social and technical interaction in the 'new economy'.

Many analyses focus on these social processes or guiding principles, but, for the most part, these analyses have persisted in ignoring the intricacies of mediation processes. With some exceptions, such as MacKenzie's work (1996*b*) and some research within the 'social shaping' or constructivist traditions (e.g. Williams and Edge 1992; Williams and Slack 1999; Wyatt *et al.* 2000) social analysis continues to treat digital technical systems as if they are neutral with respect to their implications for 'users'. Alternatively, they concentrate mainly on the generation of innovative technical systems and give little attention to the interfaces between their generation and appropriation by users (Mansell 1996).

In this book, we have focused on those 'interfaces' in order to detect the dynamics of these interactions and their implications for the emergence of the guiding principles that shape the contours of the emerging digital techno-economic paradigm. We have found evidence both of accommodation to, and acceptance of, the new technical systems as they are 'received' from their developers. But we have also found considerable latitude for social actors to make alternative choices about the design of these technical systems and about their use (and in some cases, non-use). Mediation is not simply a structural linking between people and the technical interfaces that enable them to connect to new virtual spaces. Mediation is a dialectical, relational process. The formation of relationships always requires mediations that alleviate the tensions in any exchange, whether physical or 'virtual'. Such processes are interactive and they are infused with asymmetries of power at both the individual and collective levels.

Computer-mediated communication, for example, is a technical intermediation that influences the meanings of access to electronic services for the individual.[4] This, in turn, involves various forms of institutional (collective) intermediation in order to sustain participation in the online world. There are substantial variations in the resources and attention that are accorded by social actors to the new digital technologies. These depend on how they become institutionally embedded and on the contexts of their use. In the case of the Internet, for example, mediation involves a continuous dynamic whereby the technical features of the Internet shape people's activities through its unique requirements and architecture, but, at the same time, this architecture and its governance are, themselves, outcomes of new patterns of mediation between the social and technical worlds.

3. THE AMBIGUITY OF SOCIAL AND TECHNICAL SYSTEMS

As suggested in the preceding section, examining social and technical interaction through the lens of mediation processes introduces a relational and dialectical view of

---

[4] We recognize that 'meaning' is an elusive concept that can be theorized from many perspectives (Bothamly 1993: 336). Our use of the concept is in line with definitions concerned with signification and the significance of human experience (Flexner and Hauck 1996: 1191).

the changes underway 'within the communication revolution'. Technology is humanly designed and constructed. However, in most people's minds, one of the implications of the ascendancy of the dominant 'guiding principles' associated with the new digital technologies is that the technology itself comes to be perceived as an autonomous force and as a force that acts on people. Technology is not unique in being subject to processes of objectification. Social institutions such as slavery, apartheid, capitalism, or 'the family' also become objectified such that, at various times, they display a certain force by virtue of social and cultural expectations and values. They appear to become 'facts'. These 'facts' lull us into forgetting that they are socially constructed and they can be altered, abandoned, or subordinated to other 'facts' or values. Thus, as social meanings are constructed around digital technologies, they become internalized collectively. Analysis of mediation processes is important for understanding, not merely the physical presence of a technology that comes between people, but also the phenomena whereby computers, the Internet, software, and information and media services come to be regarded as being autonomous and beyond the force of individual or collective action.[5]

As a dialectic, mediation processes leading to the formation of guiding principles within the communication revolution are not without their contradictory dynamics and ambiguities. For instance, we are often informed that the communication revolution means the end of our governments' capacities to shape the spread of the Internet (Kelly 1998; Leadbeater 1999; Post and Johnson 1997b). We are informed that new skills must be acquired to participate in a more knowledge intensive society (DTI 1998b; World Bank 1998). The new digital technologies are becoming essential for participating in many kinds of valued work and leisure activities and 'e' technologies and services can be used meaningfully only by those with a particular profile of resources in terms of skill, literacy, and, often, finance. However, if the new 'e' systems are not simply to be regarded as being autonomous or, as Mumford (1964) put it, as the latest 'megamachines' with a life of their own, we need to examine the nature of the mediation processes that are giving rise to the values and social processes that are embedded in the new technical systems and the new asymmetries of power that influence participation in the new technologically mediated social order.[6]

For some technologists, the new digital technologies are simply the latest 'megamachines'. They are designed to empower people only to the extent that the user may interact with the technological systems in specific and prefigured ways. For instance, Cochrane (1999: 227) argues that 'digital television is exciting, not because the picture is better, but because it is going to be a point of transaction. Most people will never *master* the perverted interface of the PC, but they will be able to *control* the television

---

[5] The notion of 'autonomous technology' has been developed by Winner (1977).

[6] Mumford (1964: 263) discusses the creation of the 'new megamachine'—i.e. the atom bomb. He also argued that 'the problem of integrating the machine in society is not merely a matter of making social institutions keep step with the machine. The problem is also one of altering the nature and rhythm of machines to fit the actual needs of the community' (Mumford 1934: 367). This highlights the importance of (re)emphasizing the critical importance of social and technological interactions in the context of the spread of the new digital technologies (Mumford 1934, 1964).

set. Government, however, does not have a hope in hell of controlling any of this'
(emphasis added).[7] This argument is consistent with the view that people must adjust
to the new technologies. Various public policies, commercial practices, and social
values cannot be expected to influence either the design or the use of the new technologies, because the paradigm shift to the digital world is conceived of as an inevitable
process. However, our observations of the dynamics of social and technical mediation
processes lead to a very different conclusion. Society operates with reference to cultural,
social, and political processes and practices and these and the emerging technical
systems are modified through their reciprocal mediation. The result of this dynamic
gives rise to the potential to modify existing power relationships with *ambiguous* implications for social and economic welfare.[8]

One means of revealing the dynamics of mediation in the context of the new digital
technology paradigm is to develop a distinction between the technology systems
'behind the wall' and those 'in front of the wall' (Taylor 1998).[9] The former include
'information utilities' or digital technologies and the technologists and entrepreneurs
who are marketing the new interactive products and services. The latter include the
digital appliances (the personal computer (PC), the mobile handset, the television,
etc.) and those who use them. The latter also include the World Wide Web browsers,
Internet banking services, software, and digital content of all kinds. Social action is
technically mediated when people connect to networks delivering broadcasts or when
they connect to one another using fixed and mobile telephones.

The technical systems behind the wall are being built by software developers and
technicians (Mansell and Steinmueller 2000; Negroponte 1995). The design of these
systems affects how the digital appliances are used. With a few exceptions, the systems
behind the digital wall are not readily understood by most users. This aspect of our
mediated experience is 'taken for granted' except when it fails to work in the way we
have come to expect. For example, in general, people give very little thought to the way
the Internet's intelligent agents or 'bots' communicate with one another about people's
habits, with, and sometimes without, their knowledge (Dutton 1999; Lyon 1994). Yet,
these and many other developments mediate our social lives, just as our social actions
ultimately adjust to, or resist, such developments.

Attempts to breach the wall between the technology designers and the users of the
new digital technologies generally encounter an emotional battleground that makes it
difficult to comprehend the underlying dynamics of emergent forms of social and
technical interaction. On the one hand, fear, scepticism, and loathing fuel a dystopian
vision of the ascendant ICT paradigm (Angell 2000; B. Metcalfe 2000; Rochlin 1997;
Wyatt *et al.* 2000). Evidence of digital system breakdowns and global financial systems

---

[7] Cochrane was Chief Technologist at BT Laboratories (now BTexaCT, Adastral Park) at Martlesham; his
1999 article was based on a speech at a conference on the digital economy in 1999.

[8] Lessig (1999) makes this point with respect to the Internet architecture particularly and in the context
of the United States.

[9] Taylor (1998) defines 'behind-the-wall' systems as information utilities and entrepreneurs, and
'in-front-of-the-wall' systems as information appliances and individuals.

that are 'out of control' is often presented, together with the threats of surveillance and intrusions into people's privacy (Agre and Schuler 1997; Perri 6, Lasky, and Pletcher 1998; Raab and Bennett 1998). This vision is held by those who argue that people are disempowered by the new technologies (Braverman 1974; Rifkin 1995).

The other vision is utopian (or allows that the new technological tools can be used to benefit people). In this case, empirical evidence is used to point to the way the new technical systems enable globally dispersed production facilities that create employment in both the industrialized countries and the entrepreneurial corners of developing countries (KPMG 2000; Mitter and Bastos 1999). Efficiencies derived from the use of electronic commerce, for example, are shown to deliver major reductions in the costs of producing services and manufactured goods, thereby advantaging producers in developing countries (UNCTAD 2000). Proponents of this view foresee vast opportunities for greater inclusion of the poor through distance learning and improved health care delivered using digital networks (Mansell and Wehn 1998; OECD 1997, 1999, 2000).

The dystopian and utopian viewpoints will continue to coexist. Breaching the digital wall in ways that reveal the dynamics of the mediation processes that give rise to its lack of permeability requires examination of developments in technologies and in the social organization of the means for governing and using them. This observation is no less important for the current developments in digital technologies than it has been for other major technological developments. For example, the 'Atomic Age' gave rise to dreams of electrical power that would be 'too cheap to meter'. Atomic Age technologies were contained in vast sealed facilities tended by acolytes, but the Three Mile Island and Chernobyl nuclear disasters showed that such dreams can become nightmares and that these technologies could be subject to specific policies and industrial practices to enhance their safety.

Computer technologies were initially designed mainly for military purposes, but PCs and many other digital appliances have been 'domesticated' (Silverstone 1994)— that is, they have been refashioned by a new set of guiding principles. The design of the PC has transformed it into a digital appliance that is associated with a fascination with the power that it provides to enable participation in the spaces created by the Internet (Turkle 1996). Noble (1998: 159) argues that enthusiasm for the Internet and virtual experiences may fulfil a human need 'to dwell empowered or enlightened on other, mythic, planes'. These hints about the nature of the way social and technical interactions are mediating human experience and the potential for individuals and groups to exercise an element of power in their lives need to be subject to more detailed and systematic investigation. We have elected to accomplish this through the lens of mediation and the way it is experienced in off- and online worlds.

4. MEDIATION AND SOCIAL AND TECHNOLOGICAL CHANGE

Martin Buber (1970: 69) observed that 'in the beginning is the relation' and his interest was in how our experiences are mediated by dialogue.[10] Participants in dialogues, even

---

[10] As Buber (1965: p. xx) put it, 'people are creatures of the between'.

those mediated by services like email and other forms of computer-mediated communication, establish relations between themselves. They also maintain relationships that are less directly mediated by digital technologies through their face-to-face interactions with others. Relationships and their mediation are central to the *first* theme of our research.

Rather than concern ourselves separately with what might be 'real'—that is, physically present face-to-face interactions—and what might be 'virtual'—that is, experience involving ICTs—we have sought to expose key features and dynamics of interactions that occur in both worlds.[11] Life online and offline is increasingly interwoven for those who have the capacity to access the new digital technology services. By tracking changes that are becoming embedded in the new forms of technical and social interaction, we can highlight the significance of the dynamics of change. As Kraut *et al.* (1998) and Nardi (1997) have found, research that examines the relationships that motivate work and play in virtual and offline places is most revealing when it considers how these relationships are modified through their interaction.

Our view differs from that of some contributors on the way the new digital technologies are becoming embedded in people's lives. For instance, notions of mediated communication appear in studies of the impact of computer-mediated communication. A computer network, for example, may be seen simply as a configuration of hardware and software that provides a means for people separated by space and time to exchange messages (Emmott 1995; Zorkoczy and Heap 1995). Some researchers suggest that non-electronic communication is also 'non-mediated' interaction (Lombard and Ditton 1997). We argue instead that all communicative action employs some mediational means (Wertsch 1985). Those means include language, culture, technologies, and institutions. Our ability to act, including our ability to think and speak, follows from and depends upon our participation in society, rather than preceding it.[12]

A widely held view of electronically mediated environments is that they are relatively cold and cueless compared to offline environments (Borsook 2000). The former are perceived as environments where human beings withdraw from the 'normal' social order so that they can conduct their activities as 'rational' or 'quasi-rational' information processing actors (Roche 2000). Another similarly widely held view is that cognitive tasks and capabilities are functions that occur 'in the head' and that they should

---

[11] We are not the first to observe that the pursuit of distinctions between the 'real' and the 'virtual' is a rather unproductive line of enquiry. For instance, Miller and Slater (2000: 4) argue as follows: 'But by focusing on "virtuality" as the defining feature of the many Internet media and then moving on to notions such as "cyberspace", we start from an *assumption* that it is opposed to and disembedded from the real.'

[12] As Silverstone (forthcoming) argues in the context of his discussion of mediation, communication, and, specifically, the new interactive media: 'Mediation is a fundamentally dialectical notion which requires us to address the processes of communication as both institutionally and technologically driven and embedded. Mediation, as a result, requires us to understand how processes of communication change the social and cultural environments that support them as well as the relationships that participants, both individual and institutional, have to that environment and to each other. At the same time it requires a consideration of the social as, in turn, a mediator: institutions and technologies as well as the meanings that are delivered by them are mediated in the social processes of reception and consumption.' Silverstone draws particularly on Thompson (1995) and Martin-Barbero (1993).

transfer smoothly from offline environments into the digital environment (Kurzweil 1999). If there is any change, cognitive capabilities should be enhanced by a shift to online environments as other social contexts are removed. Studies of computer-mediated communication have begun to illustrate that electronically mediated environments may be very rich in social or culturally important cues and that a multiplicity of mediations within variegated, sometimes separate and sometimes overlapping, on- and offline contexts are what differentiate today's interactions of social and technical systems from those of the past (Bolter and Grusin 1999; Rice 1999).

Although we employ the concept of mediation in our research, it is a troublesome concept. It may be difficult to comprehend how the use of a passive technical 'interface' can influence the outcome of social interactions. However, information may be held in many forms, including Internet web pages, CD-ROMs, or online databases. These offer a degree of interactivity through a technical 'interface'. They contain much more digitized information than earlier generations of information systems and they are distributed over many sites. For many digital technologies, mediation 'involves literally putting a message into media, or encoding a message into electronic, magnetic, or optical patterns of storage and transmittal' (December 1995: 7). There are many ways to accomplish this. The choices lead to different outcomes in terms of the mediated experience of individuals.[13]

Our focus on mediation is also a departure from the emphasis on information processing that characterizes much of the work on the attributes of the digital technology paradigm. The information-processing view is highlighted, for instance, in Castells's observation (1996: 21) that 'information generation, processing and transmission become the fundamental sources of productivity and power'. An emphasis on information processing is also present in much of the research in the 'knowledge-management' field (Barnatt 1995; Neef 1998). In the information-processing tradition of analysis, information exchange is generally represented as the transmission of multi-layered messages where information is channelled through a medium that acts as conduit but interferes with its accurate transmission.[14] Mediation in this context is simply a filter that allows a given amount of information to be transmitted. Recent studies of 'info-mediaries' (Armstrong and Hagel III 1996; Hagel III and Rayport 1997), still within the knowledge-management field, are more concerned with how social and technical mediation processes operate to forge, for example, trusted relationships, but there has been little effort to understand how the dynamics of such relationships depend on distinct mediation processes. Information processing and its efficiency for problem solving remain the principal concern.

In contrast, we take the view that technically mediated environments are created and then experienced in specific ways. Steuer (1992) and Sheridan (1992), among others, have introduced a concept of 'telepresence' that they employ in their studies of the specific relationships that are created in the online environments within which

---

[13] As Lessig (1999: 59) puts it in a somewhat different context referring to the architecture of the Internet: 'Architecture is a kind of law: it determines what people can and cannot do.'

[14] This view is reminiscent of Shannon and Weaver's theory of communication (1949).

individuals interact. In this work, online environments have been examined to assess whether they create a sense of 'presence'—that is, 'a sense of being there' (Lombard and Ditton 1997; Reeve 1998). This perspective suggests that it is important to take the social processes associated with offline environments into account when considering the potential of new online environments to support new forms of interaction. Åkesson and Ljungberg (1998) also emphasize the importance of coupling these environments especially in contexts where information systems are being developed. Mediation, therefore, is best understood as the process of making sense of the world through a variety of means of expression and interaction.[15] Thus, mediation or intermediation may be understood as an interactive process that bridges between human activity (including mental activity) and the cultural, historical, institutional, and technical contexts in which it is situated. It has the potential to shape and transform social activity. Our experience is mediated through interactivity or joint social (and sometimes technical) activity. In some instances, this process is reciprocal and symmetrical, but in others it is not.

This view also provides a basis for a departure from studies of how digital technologies *impact* on organizations.[16] Brown and Duguid (1998: 97) suggest, for instance:

It is a mistake to equate knowledge and information and to assume that difficulties can be overcome with information technologies. New knowledge is continuously being produced and developed in the different communities of practice throughout an organization. The challenge occurs in evaluating it and moving it. New knowledge is not capable of the sorts of friction-free movement usually attributed to information. Moreover, because moving knowledge between communities and synthesizing it takes a great deal of work, deciding what to invest time and effort in as well as determining what to act upon is a critical task for management.

They emphasize the importance of social institutions in their examination of the evolution of complex adaptive systems and they focus on the dynamics of social organization. Thus, 'most champions of complex adaptive systems . . . overlook the importance to human behavior of deliberate social organization. It is well known that humans distinguish themselves from most other life forms by the increasingly sophisticated technologies they design. It is less often noted that they also distinguish themselves by designing sophisticated social institutions' (Brown and Duguid 1998: 92). It is people who design digital technologies and appropriate them for use or reject them. It is people who establish social institutions and the guiding principles that govern social interactions on- and offline.

Our departure from a predominantly information-processing view of the communication revolution also implies an important distinction between the nature of information and knowledge. For example, following Boisot (1998: 26), we take the

---

[15] This view is informed by, but distinct from, Latour's widely acclaimed view (1999) that mediations and translations between actors and actants dissolve erroneous distinctions between, for instance, 'the Internet' and 'society', or, in our case, ICTs and people or social processes. We do not wish to impute intention or agency to technical objects.

[16] Studies of technological 'impact' are pervasive in the literature; see Wyatt *et al.* (2000) for a critique of this research tradition. Work by Lave and Wenger (1991) and Brown and Duguid (1991) on communities of practice provides a useful stepping stone for examining social and technical systems within which organizations interact.

view that 'information is something that is extracted from data in order to modify knowledge structures, taken as dispositions to act'. Boisot argues that the growth of information-processing activities in all types of organizations has led to a predominant focus on the available means of achieving data-processing economies and on improving transactional efficiency. This bias, he suggests, needs to be 'counterbalanced by a greater organizational investment in uncodified forms of knowledge, in exploratory as well as exploitative learning, and in the cultural values and beliefs that make these possible' (Boisot 1998: 266). Greater emphasis on mediation processes leads to insights into the technology design choices that become dominant. This perspective on mediation of both the social and the technical is the foundation for our research addressing the *first* main theme in this book—that is, how new patterns of social and technical interaction produce new understandings on the part of social actors and potentially new dynamics of power relationships.

Our *second* main theme concerns the forms of institutional structures and processes that are evolving with the digital technical systems and the capabilities needed to facilitate the development of the new systems. In this context, we have explored a range of institutionalized learning processes at several levels including the micro-level experiences of individuals who interact with information and communication systems and the social dynamics of relationships that are created across organizational boundaries when these systems are in use. We also have examined instances of learning aimed at creating institutions for managing the governance of the new digital technologies.[17] Institutional or collective responses to selecting the components of the dominant digital technology regime arise through mediation processes. For example, Ruggie (1975) and Haas (1975) argue that the management of technical change is not restricted to responses to technology. Instead, it is the result of conflict resolution between social actors and the potential of technical change, and between the desire of social actors for flexibility and autonomy in their utilization of a technical system and collective responses to problems and opportunities posed by technical change.

5. COMPLEMENTING THE INFORMATION-PROCESSING VIEW

Our main themes—that is, how new patterns of social and technical interaction produce new understandings on the part of social actors and potentially new dynamics of power relationships, and the nature of evolving forms of institutional structures and processes to facilitate the development of new technological systems—have been addressed in the chapters of this book. Box 12.1 summarizes the key issues that are addressed in Part One, Mediating Social and Technical Relationships.

In Part One, there are detailed investigations of technical and social systems and of the nature of the relationships between the social dynamics of off- and online activities. The investigation of the mediation of these developments is in its infancy and, as

---

[17] Institutions are understood, in the first instance, as norms and practices, some of which may be formal. Others may be informal and only tacitly understood (Edquist 1997; Senker 1995). In the second case, institutions are understood as collective organizations such as a particular firm or public-sector organization.

Box 12.1. Mediating social and technical relationships

Chapter	Technical system	Social system	On- and offline relationship
2	Internet and computer-mediated communication	Procedural and institutional authority	Virtual communities find innovative means of reconstituting some form of authority to sustain growth
3	Internet and computer-mediated communication	Reciprocity and social interaction	Experienced Internet-users engage in interactions that value conspicuous contributions in ways that parallel traditional science communities
4	Old and new technologies supporting 'virtual' working	Knowledge exchange	Organizational change is related to how knowledge exchange occurs across space, time, and organizational communities
5	Electronic commerce, mainly Internet based	Trust services and authentication of identity	Digital certificates rely on a wide variety of technical developments as well as non-technical procedures and practices
6	Electronic commerce system development (LIMNET)	Insurance industry risk management	Resistance to electronic trading and a search for shared meaning
7	Electronic information systems development; range of information and communication technologies	Banking industry credit assessment	Selecting new technologies that enhance enhance capacities to 'repersonalize data'

Box 12.1 suggests, we have focused on only a few of the guiding principles of these social and technical interactions. We have examined the nature of procedural and institutional authority, reciprocity, systems of knowledge exchange, relationship building to created trusted services, and the use of ICT systems within the social organization of the insurance and banking industries.

We began by examining social actors who may be expected to have considerable imagination about the alternatives available for constructing new technically mediated environments—that is, those who spend significant amounts of time interacting with others through computer-mediated communication based on the Internet. In Chapter 2 Steinmueller asks whether virtual communities (defined as groups of individuals who meet and interact with each other in 'cyberspace') offer more than a novel means for mediating social interaction—that is, do such communities have embedded within them innovative guiding principles for organizational change? The social system for generating procedural and institutional authority sustains organizations in the offline world, but little is known about the extent to which the technical and social interactions within online worlds gives rise to the authority that is needed to sustain the growth of virtual communities. Steinmueller concludes that there is considerable potential for the growth of such communities, but that there are constraints to the expansion of these communities because it is unclear how they can encourage social practices that generate sustained procedural and institutional authority. Virtual communities are voluntary associations and, as such, they do not have embedded within them the conventional norms of hierarchy that give rise to such authority in the offline world.

Neice observes in Chapter 3 that members of communities in the offline world construct their social order based on identifiable categories of social status and esteem. He demonstrates that, despite the fact that the social norms in the online world are in a state of flux, patterns of social and technical mediation are emerging that are reminiscent of the values of reciprocity that characterize scientific communities. Neice's study of intensive Internet-users shows that the principles for online acquisition of social esteem are located precisely at the point where the 'old' and the 'new' social orders intersect. His research confirms the presence of 'gifting' behaviour in social interactions mediated by the Internet, and goes further to suggest that this form of mediation does not necessarily imply a transformation of the commercial or market values that predominate in the offline world. Neice suggests that the specific form of technical mediation that is supported by the current architecture of the Internet is an amplification of social tendencies that are subject to change as the network evolves.[18]

Changes in the social practices that mediate team-based work in physically present and spatially distributed environments are examined in Gristock's study of the use of ICTs to support new ways of working in the newspaper industry (Chapter 4). She specifically considers how the 'virtual' modes of working are coupled with other modes of mediated experience. She shows that the dynamics of organizational change are influenced by the ways that knowledge exchange occurs across space, time, and organizational communities. Gristock's study illustrates that both the old and the new technologies and social practices need to be taken into account to understand the potential for changes in mediation that gives rise to the production of new meanings or knowledge.

The next three chapters (Chapters 5, 6, and 7) offer detailed studies of the interaction of social and technical systems that are central to the development of electronic commerce. Many studies of electronic commerce treat this form of technical mediation of business transactions as if it is simply an adjunct to the organization. Consistent with

---

[18] This observation builds on Agre (1999).

the information-processing viewpoint, the presumption is that firms must adapt to changes in ICTs to gain the benefits of efficiencies generated by automated information and communication systems. Much of the debate about the benefits of electronic commerce is couched in terms of firms' 'readiness' to accommodate to a new technological system that can relocate commercial activity from the physical world of shops and offices into computer-mediated environments.

A digital certificate is a technical device that is expected to provide a means of creating a trusted environment for the conduct of electronic commerce (Mansell, Schenk, and Steinmueller 2000).[19] Ingrid Schenk's study (Chapter 5) examines how trust service providers are developing these certificates as a means of mediating online interaction in ways that enable the identity of social actors to be verified or authenticated. She identifies a gap between the technical characteristics of certification and the social dynamics associated with different types and levels of trust that facilitate commercial interactions in offline environments. She shows that, just as in offline environments, certification procedures are needed to create reputation. Digital certificates govern the conduct of exchanges but they also structure and organize relationships in online trading environments. Trust cannot be created simply as a result of technical mediation and currently available techniques embed specific social values with respect to business culture and practices and the extent to which the privacy of individuals is protected.

In Chapter 6, Rae's examination of the LIMNET information system, which was intended to support electronic commerce in the insurance industry, illustrates clearly the intersection of social and technical mediation. His study highlights the strength of social actors' resistance to the use of digital technologies to mediate risk management. Rae's study of the London Insurance Market shows the preference of social actors for face-to-face negotiation over the use of an information system that initially was expected to provide a more efficient means of mediating risk decision making. His work emphasizes the importance of recognizing that the development of digital information systems is informed by the local experiences of social actors in their search for shared meanings and that some of these experiences cannot easily be transferred to a technologically mediated environment.

Finally, in Chapter 7, Credé's work suggests a similar dynamic of resistance to certain forms of technically mediated decision making—this time, in the context of credit risk analysis in the banking industry. His examination of whether ICTs provide a basis for the online provision of the services offered by the banking industry shows that intermediation involves complex exchanges of information, not all of which are likely to benefit from the availability of digital technologies. He argues that social actors within the banks engage in a 'repersonalizing' of data—that is, they add trust to information. Therefore, only those technical designs that enhance the capacity to repersonalize information will be accommodated by bank personnel.

Overall, our examination of the social and technical mediation processes in the seven distinct areas considered here confirms the importance of analysing the innovations in social and technical mediation in the off- and online worlds. In all instances,

---

[19] For a review of these arrangements, see e.g. Mansell (2001).

we have found attempts to replicate features of the offline social environment, but we have also found considerable creativity in the ways that this is accomplished and sustained in the online world. There appear to be different emphases and there are clearly many alternatives for the 'guiding principles' that can be embedded in the design of the technological systems. We have also found in two industries—insurance and banking—that are widely regarded as leaders in the development and use of digital technologies that social actors resist those technical systems that simply enhance information-processing capabilities and efficiencies. The capacity of social actors to resist the introduction of electronic commerce systems that do not favour the priority that they give to trust, privacy, and confidentiality is clearly in evidence in our research. There are, of course, questions about whether the social values of these actors can ultimately be embedded in the design of electronic commerce systems and whether the social values of the actors themselves will change as a result of the pervasive mediation of ICTs in other aspects of their lives.

In Part Two of this book, we shifted the focus of our research to the second main theme—that is, the nature of evolving forms of institutional structures and processes that are intended to facilitate the development of digital technological systems. Box 12.2 summarizes the levels of analysis, the types of learning processes, and the forms of

Box 12.2. Building capabilities for knowledge exchange

Chapter	Level of analysis	Learning	Institutions
8	Individual within firm	Situated and local	Norms and practices of co-design take precedence over IT system development tools
9	Interorganizational	Situated and local/collective	Norms and practices govern spatial data-sharing practices across organizational boundaries
10	Organizational	Collective	New norms and practices embedded in new institutional formation for Internet governance
11	Interorganizational	Collective	New capabilities evolve for negotiating technological outcomes, but political resistance is strong, leading to failure to develop a new satellite system

institutions (norms and practices) that give rise to specific capabilities that clearly shape the ways in which the technical systems are permitted to mediate social experience and to support various forms of knowledge exchange. In each of the four chapters in this part, learning occurs either in a local or situated context or in a collective context at the organizational level. Learning involves negotiations and conflict resolution and, in each case, the resolution is linked to social actors' choices and preferences. In all instances, the historical power relationships and asymmetries of the offline world are implicated in the decisions concerning the development or use of digital technologies.

In Chapter 8, Millar examines negotiations between social actors within two firms who are engaged in a knowledge-management system 'co-design' process that is mediated partly by the use of software tools. She illustrates the complexity of the learning processes and the extent to which organizational norms and practices in the offline world take precedence over those that could be developed using the digital tools. Millar's study confirms the importance of jointly analysing the processes of social and technical mediation. In these firms, despite their use of identical technical tools to support the knowledge-management system design process, the social dynamics in one case yield acceptance of a new system while in the other, rejection is the outcome.

Uta Wehn de Montalvo's study of the inclinations of social actors to share spatial data-sets across organizational boundaries again foregrounds the way that the situated local experiences of users of geographic information systems (GIS) can take precedence over both the potential for data sharing created by the digital technology system and the incentives created by the collective learning process. She shows the importance of the attitudes of social actors, which influence whether the values promoting sharing that are embedded in the technical features of these systems will be resisted. She argues that the sharing of computerized information resources is difficult to promote because of the social actors' perceptions about their power to act within a given organizational setting. The members of the community of potential data-sharers were strongly influenced by asymmetries in the control they believed they had over whether they would gain or lose by accommodating the data-sharing features of the technical system. Thus, perceptions and the learning experiences created by the organizational context and interorganizational rivalries mediate social interactions in this case, despite the values embedded in the technical standards, and norms and practices established by policy.

The last two chapters in Part Two are specifically directed to analysing the learning processes that are involved in the formation of new capabilities at the organizational and inter-organizational levels of analysis. In the first of these chapters, capabilities and new norms and practices for governing the Internet are considered. In the second of the chapters, we examine the evolution of new capabilities that are intended to create a new technical system—that is, a satellite system.

In the first of these two chapters (Chapter 10), Daniel Paré shows that conflicts between the social actors involved in establishing a governance framework for the allocation of Internet domain names were informed by the extent to which the actors valued the economic potential of the technical system itself. He illustrates how intermediary organizations, domain name registries that coordinate and administer the linking of identities in the physical realm with a virtual identification—that is, a domain name—engaged in specific power relationships, the dynamics of which led to a new technical configuration of the domain name system. The focus on power relationships and asymmetries, in this instance, reveals how choices are made that establish the legitimacy of regulatory organizations for the online world. The political interests of the social actors became embedded in the Internet's technical architecture and in the organization of the governance system.

In the last chapter in Part Two, we show that building up capabilities for developing the new digital technologies involves a very broad range of capabilities, including those that enable social actors to negotiate outcomes that are favourable to their interests. In Chapter 11, Arroio offers a detailed study of the learning processes within the many organizations charged with responsibility for the development of a new low earth orbiting satellite (LEOS) system, in this case, in Brazil. Capabilities for negotiation are crucial to achieve the coordination of social actors and to ensure that key policy objectives are reflected in the design of the technical system. In the face of resistance from some quarters to the deployment of a locally designed satellite system, efforts to develop it failed and negotiations with external suppliers also failed to ensure that social values with respect to the extension of new services to rural areas at low cost were embedded in the design of the new satellite system, which was supplied by external technology system providers. This failure occurred in a context in which most of the relevant technological and regulatory capabilities had been acquired over an extended period of time, but where the political and economic momentum of the 'offline' world stalled efforts to create a new 'online' system.

Overall, in the studies in Part Two, as in Part One, we have considerable evidence, not only of the complexity of the learning processes that yield new developments in digital technology systems, but also of the extent to which social actors individually and collectively shape whether new systems are designed in ways that accommodate their interests. In the case of the design of a new knowledge-management system and in the case of the use of a GIS social norms and practices at the local level within the organization take precedence over the technological potential offered by the technology designers. Norms and practices that are inconsistent with the individuals' perceptions of their relative power within the organization govern what behaviours they are willing to engage in with respect to the potential of technical mediation. In the case of the development of the Internet governance system, we clearly see that the new regime's guiding principles were constrained, not primarily by the features of the technical architecture of the Internet, but by the political and economic aspirations of the social actors. Similarly, in the case of the development of a new satellite system, political and economic interests prevailed, so that the technological potential of a satellite system designed to respond to policy priorities could not be exploited.

6. CONCLUSION

Changes in mediation processes that coincide with the development of digital technologies are all outcomes of public or private deliberations and dialogues. These dialogues result in the selection of alternative values that become embedded in the technical systems. When this is acknowledged, we can begin to understand how the accumulation of these choices affects the way social and technical systems interact. The results of research of this kind can help us to imagine alternatives to the guiding principles that are shaping our mediated experience and that otherwise may be regarded as 'facts'—that is, unalterable features of the digital technological regime.

Policy action is needed to broaden participation in this dialogue. The dialogue must be aimed at achieving a deeper process of enquiry into the dynamics that are unfolding inside the communication revolution. Heidegger (1962: 57) describes such processes of enquiry as involving 'unconcealment'. In this context, we need to 'unconceal' the choices and interests that are contributing to the shape of the new digital technological systems and, in turn, to influence how these systems mediate human choices. Without an improved dialogue, some people will obtain benefits from the development of digital ICTs. They will feel comfortable with the changes in mediation processes that these technologies enable. Some people will be excluded because of temperament, belief, or preference. Yet others will be excluded as a result of social inequalities that become embedded in the guiding principles of the emerging digital technology paradigm. The quality of modern life depends upon active involvement with, and commitment to, interactions with the digital systems. The design of these technical systems and artefacts favours specific outcomes and particular social practices and procedures. Some designs may perpetuate historical patterns of economic growth and social development, and others may favour new patterns and departures.

Although policy action is needed to facilitate wider participation in such dialogues, it is often difficult to see how the majority of people who are not directly involved in active policy-making or implementation might influence these choices. One strategy is available for exercising some power over these choices. Investigations of how changes in the guiding principles favour specific forms of social and technical mediation provide a means of exposing the values that are becoming embedded in the digital technology systems. Smythe (1957, 1977, 1981) and Melody (1973, 1985, 1987, 1997) have argued that distinctive social and technical interactions are characteristic of all societies and that these interactions are not neutral in their effects. By examining the structural or political economy features of the transformations involved in the shift to a digital technology paradigm, we can gain some insight into the dynamics of these changes and their implications (Garnham 2000). But this mode of enquiry cannot offer insights into the way people actually experience these changes and elect to accommodate or resist them. It cannot uncover their hopes, their fears, and their capacities for action that may reconstitute the guiding principles of the dominant technological regime.

Our emphasis on mediation, its structures, its processes, and its patterns, is an attempt to move on from the notion that there is a dichotomy between people and their technologies. When we elide this dichotomy, the dynamics that are at the centre of the social and technological transformations that so many analysts are struggling to comprehend come more clearly into view. These views of the dynamics within the communication revolution provide a basis for more informed dialogues about the emerging values and principles guiding our social order.

For example, strengthening online relationships that are not exclusively oriented towards commercial gain is a policy priority for many governments that are concerned about the potential for social exclusion as a result of the spread of the communication revolution. As Steinmueller has shown (Chapter 2), virtual communities attract people with common interests who engage in voluntary association. These members of communities distribute messages, share their knowledge, and offer mutual support and cooperation for scientific and other purposes. However, in this work, he focuses mainly on the social processes that give rise to procedural and institutional authority in these communities. This approach does not include an analysis of the political economy of the type of market that might render the open Internet spaces in which virtual communities flourish economically sustainable in the longer term. Similarly, while Neice has explored (Chapter 3) how social esteem is generated by reciprocity in the dynamics of the social and technical interactions of intensive Internet-users, we have not linked this analysis to the wider politics that are leading to the stronger enforcement of intellectual property rights in the digital spaces that the intensive Internet-users currently inhabit.

The overriding policy implication that arises from our emphasis on improving the processes of dialogue about the values and mediations of our social and technical interactions within the context of the digital communication revolution is that the policy-making processes at all levels should be used to create more opportunities to provoke discussions among key policy actors. There should be remits that require wider stakeholder representation in these dialogues and that require policy-makers to encourage the formation of specific interests groups (on- and offline). There should be sponsorship that facilitates ongoing participation. In addition, greater efforts can be made to mobilize all levels of the education sector to provide a foundation that enables learners actively to participate in more informed and explicit dialogues about the nature of the guiding principles that are becoming embedded in the social and technical dynamics of the digital technology paradigm. Finally, considerably more could be done within the existing policy apparatus to ensure that those who seek to promote a socially inclusive society through the application of digital technologies engage with other parts of public service organizations with responsibilities for health care, education, social welfare, and so on, to encourage these actors to have a voice in dialogues about the technological design and the guiding principles that are embedded in the digital technologies that are used to support the services they provide. This would give greater opportunities to ensure that the practice of 'outsourcing' information and communication system development reduces the instances where the new systems simply do not offer appropriate solutions for their users.

A goal for furthering the research agenda is to couple analysis of the mediation of social and technical processes with studies of the political economy structures of information and ICT systems. These structures are the context for social action inside the communication revolution and they are only indirectly revealed by studies of mediation processes.

References

Abbott, J. (1996). *The Alignment of GIS in the Public Service in SA*. Report produced for the RDP Ministry in the Office of the President. Rondebosch, South Africa: University of Cape Town.

AEB (1994): Agência Espacial Brasileira, *Política Nacional de Desenvolvimento das Atividades Espaciais (PNDAE)*. Brasília, DF: Agência Espacial Brasileira, Dec.

Ad Resource (2000). 'In Web Advertising, the Rich Get Richer: Web Playing Field Uneven for Advertisers, Publishers Alike'. *Ad Resource: Internet Advertising and Promotion Resources* adres.internet.com/stories/article/0,1401,7561_182451,00.html (accessed 20 Dec. 2000).

Agre, P. (1999). 'The Self-Limiting Internet: Problems of Change in Networks and Institutions'. Paper presented at the Telecommunications Policy Research Conference, Alexandria, VA, Sept.

—— and Schuler, D. (1997) (eds.), *Reinventing Technology, Rediscovering Community: Critical Studies in Computing as a Social Practice*. Norwood, NJ: Ablex.

Ainley, R. (1998). *New Frontiers of Space, Bodies and Gender*. New York: Routledge.

Ajzen, I. (1991). 'The Theory of Planned Behaviour'. *Organizational Behaviour and Human Decision Processes*, 50: 179–211.

Åkesson, K. P., and Ljungberg, F. (1998). 'Coupling Real and Virtual Environments', Ipswich. Paper prepared for the workshop 'Presence in Shared Virtual Environments', BT Laboratories, Martlesham Heath, Ipswich, 10–11 June, www.sics.se/~kalle/Projects/BT-submission.html (accessed 16 June 1999).

Alasuutari, P. (1995). *Researching Culture: Qualitative Methods and Cultural Studies*. London: Sage Publications.

Albitz, P., and Liu, C. (1997). *DNS and BIND*. Sebastopol: O'Reilly & Associates.

Aldrich, H. E., and Fiol, C. M. (1994). 'Fools Rush In? The Institutional Context of Industry Creation'. *Academy of Management Review*, 19/4: 645–70.

Allen, F., and Santomero, A. M. (1997). 'The Theory of Financial Intermediation'. *Journal of Banking & Finance*, 2/11–12: 1461–85.

Alter, C., and Hage, J. (1993). *Organizations Working Together*. Newbury Park, CA: Sage Publications.

Anatel (1999): Agência Nacional de Telecomunicações. *Um Ano depois da Privatização Linhas Telefônicas no Brasil chegam a 24,5 milhões*, www.anatel.gov.br/biblioteca/releases/release.asp?id_noticia=68 (accessed 20 June 1999).

Ancori, A., Bureth, A., and Cohendet, P. (2000). 'The Economics of Knowledge: The Debate about Codification and Tacit Knowledge'. *Industrial and Corporate Change*, 9/2: 255–88.

Anderson, C. (1997). 'A Survey of Electronic Commerce: In Search of the Perfect Market'. *The Economist*, 10 May, 1–26.

Angell, I. (2000). *The New Barbarian Manifesto*. New York: Kogan Page.

Antonelli, C. (1992). *The Economics of Information Networks*. Amsterdam: North-Holland.

Arafat, Y. (1997). 'On Benjamin Netanyahu's Trustworthiness'. *Newsweek*, 19 June, n.p.

Armstrong, A., and Hagel III, J. (1996). 'The Real Value of On-line Communities'. *Harvard Business Review* (May–June), 134–41.

Arroio, A. C. (1995). 'A Politica Externa e o Sistema Brasileiro de Telecomunicações por Satélite'. *Contexto Internacional*, 17/1: 61–88.

Arroio, A. C. (2000). 'Technological Opportunities for Brazilian Social Development: An Examination of Low Earth Orbit Satellite Deployment'. Unpublished D.Phil. thesis, SPRU, University of Sussex, Brighton.

Arrow, K. (1984). *The Economics of Information*. Oxford: Blackwell.

Associated New Media (1999). 'This Is Britain.com—the Definitive UK Portal Website from the UK's leading Regional Publishers', ANM Press Release, available at **www.thisislondon.com/html/britain_press_release.html** (accessed Apr. 1999).

Bachrach, P., and Baratz, M. S. (1962). 'Two Faces of Power'. *American Political Science Review*, 56/3: 947–52.

—— —— (1963). 'Decisions and Nondecisions: An Analytical Framework'. *American Political Science Review*, 57/3: 641–51.

Baier, A. (1995). *Moral Prejudices: Essays on Ethics*. Cambridge, MA: Harvard University Press.

Bambace, L. A., and Bastos, C. A. (1996). 'The Brazilian and Equatorial Countries Communications Needs in Sustainable Development Support and Environmental Monitoring, Control and Enforcement'. Paper presented at International Symposium on the Role of Telecommunication and Information Technologies in the Protection of the Environment, International Telecommunication Union, Tunis, Tunisia, 17–19 April.

Barbour, R. S., and Kitzinger, J. (1999). *Developing Focus Group Research: Politics, Theory and Practice*. London: Sage Publications.

Barley, S. R., and Tolbert, P. S. (1997). 'Institutionalization and Structuration: Studying the Links between Action and Institution'. *Organization Studies*, 18/1: 93–117.

Barnatt, C. (1995). 'Office Space, Cyberspace and Virtual Organization'. *Journal of General Management*, 30/4: 78–91.

Barras, R. (1990). 'Interactive Innovation in Financial and Business Services: The Vanguard of the Service Revolution'. *Research Policy*, 19/3: 215–37.

Barrett, M. I. (1999). 'Challenges of EDI Adoption for Electronic Trading in the London Insurance Market'. *European Journal of Information Systems*, 8 (Mar.): 1–15.

—— and Walsham, G. (1999). 'Electronic Trading and Work Transformation in the London Insurance Market'. *Information Systems Research*, 10/1: 1–23.

Bashe, C. J., Johnson, L. R., Palmer, J. H., and Pugh, E. W. (1986). *IBM's Early Computers*. Cambridge, MA: MIT Press.

Baudrillard, J. (1988). *Jean Baudrillard: Selected Writing*. Stanford, CA: Stanford University Press.

Beamish, R. (1998). 'The Local Newspaper in the Age of Multimedia', in B. Franklin and D. Murphy (eds.), *Making the Local News: Local Journalism in Context*. London: Routledge, 140–53.

Bell, M., and Pavitt, P. (1993). 'Technological Accumulation and Industrial Growth: Contrasts between Developed and Developing Countries'. *Industrial and Corporate Change*, 2/2: 157–209.

Bell, T. W. (1992). 'Polycentric Law'. *Humane Studies Review*, 7/1, **osf1.gmu.edu/~ihs/w91issues.html** (accessed 30 May 1999).

—— (1998). 'Polycentric Law in the New Millennium'. The Mont Pelerin Society, 1998 Golden Anniversary Meeting, Alexandria, VA, **user.aol.com/tomwbell/papers/FAH.html** (accessed 30 May 1999).

Beniger, J. R. (1986). *The Control Revolution: Technological and Economic Origins of the Information Society*. Cambridge, MA: Harvard University Press.

Benjamin, R. I., and Levinson, E. (1993). 'A Framework for Managing IT Enabled Change'. *Sloan Management Review*, Summer: 23–33.

Bensimon, C. (1988). 'Uma Historia de Polêmicas'. *Dados & Ideias*, 13/116: 35–48.

Berkowitz, D. (1997) (ed.). *Social Meanings of News: A Text Reader*. Thousand Oaks, CA: Sage Publications.

Bewley, T. F. (2000). *Why Wages don't Fall during a Recession*. Cambridge, MA: Harvard University Press.

Biddle, C. B. (1997). 'Legislating Market Winners: Digital Signature Laws and the Electronic Commerce Marketplace'. *World Wide Web Journal*, 2/3: www.w3journal.com/ 7/s3.biddle.wrap.html (accessed 5 July 2000).

Bijker, W. E. (1993). 'Do Not Despair: There Is Life After Constructivism'. *Science, Technology and Human Values*, 18: 113–38.

—— and Law, J. (1992) (eds.). *Shaping Technology, Building Society: Studies in Sociotechnical Change*. Cambridge, MA: MIT Press.

—— Hughes, T. P., and Pinch, T. J. (1987) (eds.). *The Social Construction of Technological Systems: New Directions in the Sociology and History of Technology*. Cambridge, MA: MIT Press.

Biocca, F., Kim, T., and Levy, M. R. (1995). 'The Vision of Virtual Reality', in F. Biocca and M. R. Levy (eds.), *Communication in the Age of Virtual Reality*. Hillsdale, NJ: Lawrence Erlbaum Associates, 3–14.

Blackler, F. (1993). 'Knowledge and the Theory of Organizations: Organizations as Activity Systems and the Reframing of Management'. *Journal of Management Studies*, 30/6: 863–84.

Boden, M. (1990). *The Creative Mind: Myths and Mechanisms*. London: Weidenfeld & Nicolson.

Boden, M. and Miles, I. (2000) (eds.). *Services and the Knowledge Based Economy*. London: Continuum International Publishing Group, Pinter.

Boisot, M. H. (1995). *Information Space: A Framework for Learning in Organizations, Institutions and Cultures*. London: Routledge.

—— (1998). *Knowledge Assets: Securing Competitive Advantage in the Information Economy*. Oxford: Oxford University Press.

Bolter, J. D., and Grusin, R. (1999). *Remediation: Understanding New Media*. Cambridge, MA: MIT Press.

Borsook, P. (2000). *Cyberselfish*. New York: Public Affairs, Perseus Book Group.

Bothamly, J. (1993). *Dictionary of Theories*. London: Gale Research International Ltd.

Bourdieu, P. (1984). *Distinction: A Social Critique of the Judgement of Taste*, trans. Richard Nice. London: Routledge & Kegan Paul.

Boyle, J. (1997). 'Foucault in Cyberspace: Surveillance, Sovereignty, and Hard-Wired Censors'. *University of Cincinnati Law Review*, 66, www.wcl.american.edu/pub/faculty/boyle/ foucault.htm (accessed 30 May 1999).

Bradach, J. L., and Eccles, R. G. (1989). 'Price, Authority, and Trust'. *Annual Review of Sociology*, 15: 97–118.

Bradner, S. (1996). 'RFC 2026: The Internet Standards Process—Revision 3', Oct. www.ietf.org/rfc/rfc2026.txt (accessed 22 Nov. 1999).

Braverman, H. (1974). *Labor and Monopoly Capital*. New York: Monthly Review Press.

Brisolla, S. (1994). 'Evolução da Execução Financeira Anual Consolidada de Todas as Unidades do Ministério da Ciência e Tecnologia (MCT) de 1980 a 1992', in S. Schartzman (ed.), *Estudo Atual e Papel Futuro da Ciência e Tecnologia no Brasil*. Brasília, DF: Ministério da Ciência e Tecnologia, Fundação Getúlio Vargas, Programa de Apoio ao Desenvolvimento Científico e Tecnológico, www.mct.gov.br/MCTHome/Estudos/Html/ EAPF.htm, 1–15.

Brown, J. S., and Duguid, P. (1991). 'Organizational Learning and Communities of Practice: Towards a Unified View of Working, Learning, and Innovation'. *Organization Science*, 2: 40–57.

—— —— (1998). 'Organizing Knowledge'. *California Management Review*, 40/3: 90–111.

Bryan, L. (1988). *Breaking up the Bank: Rethinking an Industry under Siege*. New York: Dow Jones, Irvin.

—— (1993). 'The Forces Shaping Global Banking'. *McKinsey Quarterly*, 2: 59–72.

Buber, M. (1965). *Between Man and Man*, trans. R. G. Smith. New York: Macmillan. First published in English in 1948.

—— (1970). *I and Thou*, trans. W. Kaufman. New York: Charles Scribner. First published in English in 1937.

Bultje, R., and van Wijk, J. (1998). 'Taxonomy of Virtual Organizations, Based on Definitions, Characteristics and Typology'. *Virtual-Organization-Net Newsletter*, 2/3: www.virtual-organization.net/news/nl_2.3/table.htm (accessed 19 June 2000).

Bunker, N. (1988). 'Joint Venture Makes Strong Headway, supplement Information Technology in Finance', *Financial Times*, 10 Nov., 15.

Burns, T., and Stalker, G. (1966). *The Management of Innovation*. London: Tavistock.

Burrough, P. A., and McDonnell, R. A. (1998). *Principles of Geographical Information Systems*. Oxford: Oxford University Press.

Byrne, J. A., Brandt, R., and Port, O. (1993). 'The Virtual Corporation: The Company of the Future will be the Ultimate in Adaptability'. *International Business Week*, 8: 36–41.

Cairncross, F. (1998). *The Death of Distance*. London: Orion Business Publications.

Campbell, H. (1991). 'Organizational Issues in Managing Geographic Information', in I. Masser and M. Blakemore (eds.), *Handling Geographical Information: Methodology and Potential Applications*. Harlow: Longman Scientific & Technical, 259–82.

Cappelli, P. (1999). *The New Deal at Work: Managing the Market Driven Workforce*. Boston: Harvard Business School Press.

Carter, R. L., and Falush, P. (1998). *The London Insurance Market*. London: London Insurance Market Strategy Committee.

Cassiolato, J., Szapiro, M., and Andrade, M. (1998). 'Política Industrial no Brasil'. Relatório de Projeto de Pesquisa Apoiado pelo Instituto de Estudos para o Desenvolvimento Industrial, versão prelimnar, Universidade Federal do Rio de Janeiro, Instituto de Economia, Rio de Janeiro, June.

Castells, M. (1980). *The Informational City: Information Technology, Economic Restructuring, and the Urban-Regional Process*. Oxford: Basil Blackwell.

—— (1996). *The Information Age: Economy, Society and Culture*, I. *The Rise of the Network Society*. Oxford: Blackwell.

—— (1997). *The Information Age: Economy, Society and Culture*, II. *The Power of Identity*. Oxford: Blackwell.

—— (1998). *The Information Age: Economy, Society and Culture*, III. *End of Millennium*. Oxford: Blackwell.

—— (2000). 'Materials for an Exploratory Theory of the Network Society'. *British Journal of Sociology*, 51/1: 5–24.

Cawson, A., Haddon, L., and Miles, I. (1995). *The Shape of Things to Consume: Delivering Information Technology into the Home*. Aldershot: Avebury.

CCI (1998): Constellation Communications Inc., *Business Plan, Constellation Communications Inc., System Operator Publication*. Fairfax, VA: Constellation Communications Inc.

Ceballos, D. (1994). 'Equatorial Low Orbit Communication. ECO-8 System'. *Acta Astronautica*, 34/2: 47–54.

Chelmsford, J. (1992). *L is for LIMNET: The Vital Link*. London: DYP Insurance and Reinsurance Research Group Ltd.

Chesborough, H. W., and Teece, D. J. (1996). 'When is Virtual Virtuous? Organizing for Innovation'. *Harvard Business Review*, 74/1: 65–73.

Chiesa, V., Coughlan, P., and Voss, C. (1996). 'Development of a Technical Innovation Audit'. *Journal of Product Innovation Management*, 13/2: 105–36.

Christiansen, T., Christ, H., and Hansmann, B. (1997). 'GIS in German Technical Cooperation: The Status Quo in 1996'.*GTZ GIS Newsletter*, 3: 3–6.

Ciborra, C. U. (1996). *Teams, Markets and Systems: Business Innovation and Information Technology*. Cambridge: Cambridge University Press.

—— (2000) (ed.). *From Control to Drift: The Dynamics of Corporate Information Infrastructures*. Oxford: Oxford University Press.

Citicorp (1994). *Annual Report*. New York, Citicorp.

Clarke, D. G. (1997). 'Mapping for Reconstruction of South Africa', in D. Rhind (ed.), *Framework for the World*. Cambridge: GeoInformation International, 48–62.

—— Gavin, E., Honu, W., Krieg, T., Muller, M., Smith, H. J., Smith, T., and Vorster, S. (1998). 'National Spatial Information Framework Workshop'. Paper presented at the NSIF Workshop, Pretoria, 11 Feb.

Clarke, S., and Roome, N. (1995). 'Managing for Environmentally Sensitive Technology: Networks for Collaboration and Learning'. *Technology Analysis and Strategic Management*, 7/2: 191–216.

Cleaver, A. (2000). 'First UK Newspaper on the Web'. Personal email communication to J. Gristock.

Clemente, P. C. (1998). *State of the Net: The New Frontier*. New York: McGraw-Hill.

Cochrane, P. (1999). 'If You Turn your Back for a Moment, You're Dead'. *Information Technology and Public Policy*, 18/1: 227.

Cohendet, P., and Steinmueller, W. E. (2000). 'The Codification of Knowledge: A Conceptual and Empirical Exploration'. *Industrial and Corporate Change*, 9/2: 195–209.

Cole, M. (1985). 'The Zone of Proximal Development: Where Culture and Cognition Create Each Other', in J. V. Wertsch (ed.), *Culture, Communication and Cognition: Vygotskian Perspectives*. Cambridge: Cambridge University Press, 146–61.

Coleman, W. E. (1994). 'Banking, Interest Intermediation and Political Power: A Framework for Comparative Analysis'. *European Journal of Political Research*, 26/1: 31–58.

Cooley, T. F., and Nam, K. (1998). 'Asymmetric Information, Financial Intermediation and Business Cycles'. *Economic Theory*, 12/3: 599–620.

Coombs, R., Saviotti, P., and Walsh, V. (1987). *Economics and Technological Change*. London: Macmillan Education.

Coppock, J. T., and Rhind, D. W. (1991). 'The History of GIS', in D. J. Maguire, M. F. Goodchild, and D. W. Rhind (eds.), *Geographical Information Systems—Principles and Applications*. Harlow: Longman Scientific & Technical, 21–43.

Corporation of Lloyd's (1998). *Corporation of Lloyd's Annual Report*. London: Corporation of Lloyd's.

Cosimano, T. F. (1996). 'Intermediation'. *Economica*, 63/249: 131–43.

Cowan, R., and Foray, D. (1997). 'The Economics of Codification and the Diffusion of Knowledge'. *Industrial and Corporate Change*, 6/3: 595–622.

—— David, P. A., and Foray, D. (2000). 'The Explicit Economics of Knowledge Codification and Tacitness'. *Industrial and Corporate Change*, 9/2: 211–54.

Coyle, D. (1998). *The Weightless World: Strategies for Managing the Digital Economy*. Cambridge, MA: MIT Press.

Credé, A. (1997). 'Technological Change and the Information Society: An Examination of Credit Risk Assessment and Cash Handling Procedures in Commercial Banks'. Unpublished D.Phil. thesis, SPRU, University of Sussex, Brighton.

Crompton, R. (1993). *Class and Stratification*. Cambridge: Polity Press.

Dahl, R. A. (1957). 'The Concept of Power'. *Behavioural Science*, 2: 201–15.

Daley, W. M. (1999). 'Remarks by Secretary of Commerce, Conference on Understanding the Digital Economy: Data, Tools and Research', Washington, as prepared for delivery at **204.193.246.62/public.nsf/docs/remarks-by-secretary-digital-economy-conference-052599** (accessed 17 Sept. 2000).

Dalum, B., Johnson, B., and Lundvall, B.-Å. (1992). 'Public Policy in the Learning Society', in B.-Å. Lundvall (ed.), *National Systems of Innovation: Towards a Theory of Innovation and Interactive Learning*. London: Pinter Publishers, 296–317.

Dasgupta, P., and David, P. A. (1994). 'Toward a New Economics of Science'. *Research Policy*, 97/387: 487–521.

Dataquest (1996). 'GIS Applications Move into the "Mainstream"', **gartner5.gartnerweb.com** (accessed 19 June 2000).

—— (1999). 'Pushing into the Mainstream', **gartner5.gartnerweb.com** (accessed 19 June 2000).

Davenport, T. H. (1993). *Process Innovation: Reengineering Work through Information Technology*. Boston: Harvard Business School Press.

—— and Prusak, L. (1998). *Working Knowledge: How Organizations Manage What They Know*. Boston: Harvard Business School Press.

David, P. A. (1995). 'Rethinking Technology Transfers: Incentives, Institutions and Knowledge-Based Industrial Development'. Paper prepared for presentation at the British Academy/Chinese Academy of Social Sciences Joint Seminar on Technology Transfer, Beijing, 5–6 Apr.

—— and Foray, D. (1995). 'Accessing and Expanding the Science and Technology Knowledge Base'. *OECD STI Review*, 9/10: 13–68.

Davidow, W. H., and Malone, M. S. (1992). *The Virtual Corporation: Structuring and Revitalizing the Corporation for the 21st Century*. New York: Harper Collins.

de Bony, E. (1998). 'Europe in Digital Signature Drama'. *Industry Standard*, 30 Nov., 6–7.

December, J. (1995). 'Units of Analysis for Internet Communication'. *Journal of Computer Mediated Communication*, 1/4, **jcmc.huji.ac.il/vol1/issue4/december.html** (accessed 1 Sept. 2000).

Decol, R. (1999). 'Crisis Pays off for Brazil's New Entrant'. *Communications Week International*, 218: 8.

DeLisi, P. S. (1990). 'Lessons from the Steel Axe: Culture, Technology and Organizational Change'. *Sloan Management Review*, 32/1: 83–93.

Denis, J.-L., Langley, A., and Cazale, L. (1996). 'Leadership and Strategic Change under Ambiguity'. *Organization Studies*, 17/4: 673–99.

Department of Commerce (1998). *Falling through the Net: A Survey of 'Have Nots' in Rural and Urban America*. Washington: US Department of Commerce.

—— (1999). *The Emerging Digital Economy II*. Washington: Department of Commerce, June.

—— (2000). *Digital Economy 2000*. Washington: Department of Commerce, June.

Dess, G. G., Rasheed, A. M. A., McLaughlin, K. J., and Priem, R. L. (1996). 'The New Corporate Architecture'. *IEEE Engineering Management Review*, 24/2: 21–8.

Diamond, D. (1984). 'Financial Intermediation and Delegated Monitoring'. *Review of Economic Studies*, 51: 393–414.

Dibona, C., Stone, M., and Ockman, S. (1999) (eds.). *Open Sources: Voices from the Open Source Revolution*. Sebastopol, CA: O'Reilly & Associates.

Dickenson, P., and Sciadas, G. (1996). *Access to the Information Highway, Services Indicators, 1st Quarter*. Ottawa: Statistics Canada.

Dizard Jr., W. (1997). *Meganet: How the Global Communications Network will Connect Everyone on Earth*. Boulder, CO: Westview Press.

Dosi, G. (1982). 'Technological Paradigms and Technological Trajectories'. *Research Policy*, 11/3: 147–62.

—— (1991). 'Perspectives on Evolutionary Theory'. *Science and Public Policy*, 18/6: 353–61.

—— and Malerba, F. (1996). 'Organisational Learning and Institutional Embeddedness', in G. Dosi and F. Malerba (eds.), *Organisation and Strategy in the Evolution of the Enterprise*. London: Macmillan, 1–24.

—— Marsili, O., Orsenigo, L., and Salvatore, R. (1995). 'Learning, Market Selection and the Evolution of Industrial Structures'. *Small Business Economics*, 7/6: 411–36.

Drenth, H., Morris, A., and Tseng, G. (1991). 'Expert Systems as Information Intermediaries'. *Annual Review of Information Science and Technology*, 26: 113–54.

Drucker, P. F. (1964). *Managing for Results*. New York: Harper & Row.

—— (1988). 'The Coming of the New Organization'. *Harvard Business Review*, 66/1: 45–53.

DTI (1989): Department of Trade and Industry, *EDI Standards, A Guide for Existing and Prospective Users*. London: HMSO.

—— (1996). *IT for ALL: A Survey into Public Awareness of Attitudes toward and Access to Information and Communications Technology*. London: Department of Trade and Industry, Information Society Initiative.

—— (1998a). *The Latest Findings Concerning Attitudes towards IT*. London: Department of Trade and Industry, Information Society Initiative.

—— (1998b). *Our Competitive Future: Building the Knowledge Driven Economy, The 1998 Competitiveness White Paper*. London: Department of Trade and Industry, Dec.

Dudman, J. (1994a). 'Staying the Pace'. *Computer Weekly*, 10 Feb., 30.

—— (1994b). 'Tools you can Trust'. *Computer Weekly*, 29 Sept., 44.

Duff, A. S. (2000). *Information Society Studies*. London: Routledge.

Durlach, N., and Slater, M. (1998). 'Presence in Shared Virtual Environments and Virtual Togetherness'. Paper presented at the Workshop 'Presence in Shared Virtual Environments', BT Laboratories, Martlesham Heath, Ipswich, 10–11 June, **www.cs.ucl.ac.uk/staff/m.slater/BTWorkshop/durlach.html** (accessed 4 Aug. 1999).

Dutton, W. H. (1996) (ed.). *Information and Communication Technologies: Visions and Realities*. Oxford: Oxford University Press.

—— (1999). *Society on the Line: Information Politics in the Digital Age*. Oxford: Oxford University Press.

Easterbrook, S. (1991). *Negotiation and the Role of the Requirements Specification*. School of Cognitive and Computing Sciences Research Report No. 197. Brighton: University of Sussex.

Easterwood, J. C., and Morgan, G. E. (1991). 'Eroding Market Imperfections, Reintermediation and Disintermediation'. *Journal of Financial Research*, 14/4: 345–58.

Easton, D. (1965). *A Systems Analysis of Political Life*. New York: John Wiley & Sons.

Edquist, C. (1997). 'Institutions and Organizations in Systems of Innovation: The State of the Art'. Background paper for the Program Planning Workshop on 'Strengthening Innovation Systems' (SI3), Linköping, Linköping University, June.

Elg, U., and Johansson, U. (1997). 'Decision Making in Inter-firm Networks as a Political Process'. *Organization Studies*, 18/3: 361–84.

Eliasson, G. G. (1987). *The Knowledge Base of an Industrial Economy*. Research Report No. 33. Stockholm: Industrial Institute for Economic and Social Research.

—— (1990). *The Knowledge-Based Information Economy*. Stockholm: Almquist & Wiksell International.

Embratel (1979–97): Empresa Brasileira de Telecomunicações, *Relatório Anual*. Rio de Janeiro: Empresa Brasileira de Telecomunicações.

Emmott, S. J. (1995) (ed.). *Information Superhighways: Multimedia Users and Futures*. London: Academic Press.

Erber, F., and Amaral, L. (1994). 'Os Centros de Pesquisa das Empresas Estatais: Um Estudo de Três Casos', in S. Schwartzman (co-ordinator) (ed.), *Estado Atual e Papel Futuro da Ciência e Tecnologia no Brasil*. Brasília, DF: Ministério da Ciência e Tecnologia, Fundação Getúlio Vargas, Programa de Apoio ao Desenvolvimento Científico e Tecnológico, 30–60.

European Commission (1997). *A European Initiative in Electronic Commerce.* Brussels: European Commission COM(97) 157, 30 May, www.cordis.lu/esprit/src/ecomcom.htm (accessed 4 July 2000).

Evans, E. (1998). 'New Satellites for Personal Communications'. *Scientific American*, 278/4: 60–7.

Exame (1994). 'Começa a Guerra nas Estrelas', 26/10: 31–3.

Federal Register (1994). *Coordinating Geographic Data Acquisition and Access: The National Spatial Data Infrastructure.* Washington: Executive Order of the President of the United States (No. 12906).

Feghhi, J., Feghhi, J., and Williams, P. (1999). *Digital Certificates: Applied Internet Security.* Reading, MA: Addison Wesley Longman.

Feigenbaum, J. (1998). 'Towards an Infrastructure for Authorisation, Position Paper'. 3rd USENIX Workshop on Electronic Commerce, Sept.

Ferné, G., Hawkins, R. W., and Foray, D. (1996). *The Economic Dimension of Electronic Data Interchange (EDI).* Geneva/Paris: International Organization for Standardization and Organization for Economic Cooperation and Development.

Ferreira Silva, L. F. (1997). 'US Cold War Foreign Policy and Satellite Communication: The Case of Earth Station Network Build-Up in Brazil and Argentina'. Unpublished D.Phil. thesis, SPRU, University of Sussex, Brighton.

Financial Times (1997). 'FT 500, The World's Top 500 Companies', 24 Jan.

—— (1999). 'Insurers and Brokers to set up E-Commerce Network', 30 Mar.

Finholt, T. A., and Olson, G. M. (1997). 'From Laboratories to Collaboratories: A New Organizational Form for Scientific Collaboration'. *Psychological Science*, 8/1: 28–35.

Flaaten, P. O., McCubbrey, D. J., O'Riordan, P. D., and Burgess, K. (1989). *Foundations of Business Systems.* Chicago: Dryden Press.

Flexner, S. B., and Hauck, L. C. (1996) (eds.). *Random House Compact Unabridged Dictionary.* Special Second Edition. New York: Random House.

Flowerdew, R., and Green, M. (1991). 'Data Integration: Statistical Methods for Transferring Data between Zonal Systems', in I. Masser and M. Blakemore (eds.), *Handling Geographical Information: Methodology and Potential Applications.* New York: Longman Scientific & Technical, 38–54.

Foray, D. (1995). 'Information Access, Knowledge Distribution and the Institutional Infrastructure'. Paper prepared for the Systems of Innovation Research Network Conference, Lanzarote.

—— and Lundvall, B.-Å (1996). 'The Knowledge-Based Economy: From the Economics of Knowledge to the Learning Economy', in OECD (ed.), *Employment and Growth in the Knowledge-based Economy.* Paris: OECD, 11–32.

Ford, W., and Baum, M. (1997). *Secure Electronic Commerce: Building the Infrastructure for Digital Signatures and Encryption.* Upper Saddle River, NJ: Prentice Hall.

Forrester Research (1996). *Will the Web Kill EDI?* New York: Forrester Network Strategy Report No. 10(3), Feb.

Foster, W. A. (1996). 'Registering the Domain Name System: An Exercise in Global Decision Making'. Paper presented at the Coordination and Administration of the Internet Workshop, Kennedy School of Government, Harvard University, Boston, 8–10 Sept., ksgwww.harvard.edu/iip/cai/foster.html (accessed 2 June 1997).

Fourie, H. (1998). 'GI in the SANDF', in D. Clarke, E. Gavin, W. Honu, T. Krieg, M. Muller, H. J. Smith, T. Smith, and S. Vorster (eds.), *National Spatial Information Framework Workshop Conference Proceedings*. Pretoria: Sinodale Sentrum, 30–1.

—— (1999). 'Should We Take Disintermediation Seriously?'. *Electronic Library*, 17/1: 9–16.

Freeman, C. (1974). *The Economics of Industrial Innovation* 1st edn. Harmondsworth: Penguin Modern Economic Texts.

—— (1987). 'The Case for Technological Determinism', in R. Finnegan, G. Salaman, and K. Thompson (eds.), *Information Technology: Social Issues. A Reader*. London: Hodder & Stoughton, 5–18.

—— (1988). 'Information Technology and the New Economic Paradigm', in H. Schutte (ed.), *Strategic Issues in Information Technology: International Implications for Decision Makers*. Maidenhead: Pergamon Infotech, 159–75.

—— (1992a). *The Economics of Hope: Essays on Technical Change, Economic Growth and the Environment*. London: Pinter Publishers.

—— (1992b). 'Technology, Progress and the Quality of Life', in C. Freeman (ed.), *The Economics of Hope: Essays on Technical Change, Economic Growth and the Environment*. London: Pinter Publishers, 212–30.

—— (1994a). 'The Diffusion of Information and Communication Technology in the World Economy in the 1990s', in R. Mansell (ed.), *The Management of Information and Communication Technologies: Emerging Patterns of Control*. London: Aslib, 8–41.

—— (1994b). 'The Economics of Technical Change'. *Cambridge Journal of Economics*, 18: 463–514.

—— (1995). *Information Highways and Social Change*. Ottawa: International Development Research Centre.

—— and Perez, C. (1988). 'Structural Crises of Adjustment, Business Cycles, and Investment Behaviour', in G. Dosi, C. Freeman, R. Nelson, G. Silverberg, and L. Soete (eds.), *Technical Change and Economic Theory*. London: Pinter Publishers, 38–66.

—— and Soete, L. (1994). *Work for All or Mass Unemployment? Computerised Technical Change into the Twenty-First Century*. London: Pinter.

—— —— (1997). *The Economics of Industrial Innovation*. 3rd edn. London: Pinter Publishers.

Freitas, O. (1995). 'O Céu é Brasileiro', Isto É, 9 Aug., 46–51.

Froomkin, M. A. (1996) 'The Essential Role of Trusted Third Parties in Electronic Commerce'. *Oregon Law Review*, 49/75, www.law.miami.edu/~froomkin/articles/trusted.htm (accessed 4 July 2000).

Frost, P. J., and Egri, C. P. (1991). 'The Political Process of Innovation', in L. L. Cummings and B. M. Staw (eds.), *Research in Organizational Behaviour*. Greenwich, CT: JAI Press Inc., viii 229–95.

Furlan, G. (1990). 'Sonho em Realidade: Telebras Privatizada', *Revista Nacional de Telecomunicações*, 20/229: 12–32.

Galaskiewicz, J., and Zaheer, A. (1999). 'Networks of Competitive Advantage', in S. B. Andrews and D. Knoke (eds.), *Network In and Around Organizations*. Stamford, CT: JAI Press Inc., 237–61.

Gane, C., and Sarson, T. (1979). *Structured Systems Analysis*. Englewood Cliffs, NJ: Prentice Hall.

Garcia, D. L. (1995). 'Networking and the Rise of Electronic Commerce: The Challenge for Public Policy'. *Business Economics* (Oct.), 7–14.

Garfinkel, S., and Spafford, G. (1997). *Web Security and Commerce*. Sebastopol: O'Reilly & Associates.

Garnham, N. (1994). 'Whatever Happened to the Information Society?', in R. Mansell (ed.), *The Management of Information and Communication Technologies: Emerging Patterns of Control*. London: Aslib, 42–51.

Garnham, N. (1996). 'Constraints on Multimedia Convergence', in W. H. Dutton (ed.), *Information and Communication Technologies: Visions and Realities*. Oxford: Oxford University Press, 103–20.

—— (2000). *Emancipation, the Media, and Modernity*. Oxford: Oxford University Press.

Gauntlett, D. (2000) (ed.). *Web.Studies: Rewiring Media Studies for the Digital Age*. London: Arnold.

Gavin, E. (1998). 'Introducing the NSIF', in D. Clarke, E. Gavin, W. Honu, T. Krieg, M. Muller, H. J. Smith, T. Smith, and S. Vorster (eds.), *National Spatial Information Framework Workshop*. Pretoria: Sinodale Sentrum, 18–19.

Gazeta Mercantil (1998). 'Relatório da Gazeta Mercantil Latino-Americana, Telecomunicações', 28 Sept.

Gibbons, M., Limoges, C., Nowotny, H., Schwartzman, S., Scott, P., and Trow, M. (1994). *The New Production of Knowledge: The Dynamics of Science and Research in Contemporary Societies*. London: Sage Publications.

Gibson, W. (1984). *Neuromancer*. New York: Ace Books.

Gilder, G. (1994). 'Telecosm Ethersphere'. *Forbes Technology Supplement*, 10 Oct., 1–10.

Gillett, S. E., and Kapor, K. (1996). 'The Self-Governing Internet: Coordination by Design'. Paper presented at the Coordination and Administration of the Internet Workshop, Kennedy School of Government, Harvard University, Boston, 8–10 Sept., **ccs.mit.edu/ccswp197.html** (accessed 2 June 1997).

Gilster, P. (1993). *The Internet Navigator: The Essential Guide to Network Exploration for the Individual Dial-up User*. New York: Wiley.

—— (1997). *Digital Literacy*. New York: Wiley.

Glaser, B. G., and Corbin, J. (1990). *Basics of Qualitative Research: Grounded Theory Procedures and Techniques*. Newbury Park, CA: Sage Publications.

—— and Strauss, A. L. (1967). *The Discovery of Grounded Theory: Strategies for Qualitative Research*. Chicago: Aldine.

Globalstar (1996). *Wireless Communications for the World*. Globalstar System Operator Publication.

Globe and Mail (1997). 'Finland Suddenly Takes Over as World's Most Wired Country'. 1 Feb.

Goffman, E. (1963). *Behavior in Public Places: Notes on the Social Organization of Gatherings*. New York: Free Press.

Goggin, G. (2000). 'Pay Per Browse? The Web's Commercial Futures', in D. Gauntlett (ed.), *Web.Studies: Rewiring Media Studies for the Digital Age*. London: Arnold, 103–12.

Goldhaber, G. M., and Barnett, G. A. (1988) (eds.). *Handbook of Organizational Communication*, Norwood, NJ: Ablex.

Goldstein, A., and O'Connor, D. (2000). *E-Commerce for Development: Prospects and Policy Issues*. Paris: OECD Development Centre, **www.oecd.org/dev** (accessed 5 Feb. 2001).

Göranson, B. (1993). *Catching up in Technology: Case Studies from the Telecommunication Equipment Industry*. London: Taylor Graham.

Gould, M. (1996*a*). 'Governance of the Internet—A UK Perspective'. Paper presented at the Coordination and Administration of the Internet Workshop, Kennedy School of Government, Harvard University, Boston, 8–10 Sept., **aranea.law.bris.ac.uk/HarvardFinal.html** (accessed 29 June 1997).

—— (1996*b*). 'Rules in Virtual Society'. *International Review of Law, Computers and Technology*, 10/2: 199–218.

Greenwood, D. J. (1998). 'Risk and Trust Management Techniques for an "Open But Bounded" Public Key Infrastructure'. *Jurimetrics*, 38/3: 277–94.

Gristock, J. (2001). 'Organisational Virtuality in the UK Newspaper Publishing Industry'. D.Phil. thesis (under examination), SPRU, University of Sussex, Brighton.

Guardian (1999). 'Poll points to Lift Off for the Internet'. 11 Jan.

Guardian Weekly (1999). 'Celebrity Slaughter'. 161/13, 23 Sept.

Guthke, K. S. (1990). *The Last Frontier: Imagining other Worlds, from the Copernican Revolution to Modern Science Fiction*. Ithica, NY: Cornell University Press.

Haas, E. B. (1975). 'Is There a Hole in the Whole? Knowledge, Technology, Interdependence and the Construction of International Regimes'. *International Organization*, 29 (Summer): 827–76.

Hafner, K., and Lyon, M. (1996). *Where Wizards Stay up Late: The Origins of the Internet*. New York: Touchstone.

Hagel III, J., and Armstrong, A. G. (1997). *Net Gain: Expanding Markets through Virtual Communities*. Boston: Harvard Business School Press.

—— and Rayport, J. F. (1997). 'The New Infomediaries'. *McKinsey Quarterly*, 55–70.

Hagstrom, W. O. (1965). *The Scientific Community*. New York: Basic Books.

Hammer, M., and Stanton, S. (1999). 'How Process Enterprises Really Work'. *Harvard Business Review*, 77/6: 108–18.

Handy, C. (1996). *Beyond Certainty: The Changing Worlds of Organizations*. London: Arrow Books.

Hanseth, O., Monteiro, E., and Hatling, M. (1996). 'Developing Information Infrastructure: The Tension between Standardization and Flexibility'. *Science, Technology, and Human Values*, 21/4: 407–26.

Hanson, H. (2000). 'Advertising Networks: What About the Little Guy?', *Ad Resource: Internet Advertising and Promotion Resources*, **adres.internet.com/feature/article/ 0,1401,8961_442761,00.html** (accessed 20 Dec. 2000).

Hardin, R. (1993). 'The Street-Level Epistemology of Trust'. *Politics & Society*, 21/4: 505–29.

Harrigan, K. R., and Newman, W. H. (1990). 'Bases of Interorganization Co-operation: Propensity, Power, Persistence'. *Journal of Management Studies*, 27/4: 417–34.

Harris, R. C., Insinga, R. C., Morone, J., and Werle, M. J. (1996). 'The Virtual R&D Laboratory'. *Research Technology Management*, 39/2: 32–6.

Hart, P., and Saunders, C. (1997). 'Power and Trust: Critical Factors in the Adoption and Use of Electronic Data Interchange'. *Organization Science*, 8/1: 23–41.

Hawkins, R. W., Mansell, R., and Steinmueller, W. E. (1999). 'Toward Digital Intermediation in the Information Society'. *Journal of Economic Issues*, 33/2: 383–91.

Hazell, R. (1996). 'Could London Survive without Lloyd's?'. Conference address by EPG Insurance Systems presented at Competition in the Global Insurance Market Conference, Brighton, June.

Healy, D. (1997). 'Cyberspace and Place: The Internet as Middle Landscape on the Electronic Frontier', in D. Porter (ed.), *Internet Culture*. New York: Routledge, 55–68.

Heidegger, M. (1962). *Being and Time*, trans. J. Macquarrie and E. Robinson. New York: Harper & Row.

Helpman, E. (1998) (ed.). *General Purpose Technologies and Economic Growth*. Cambridge, MA: MIT Press.

Her Majesty's Treasury (1996). *Invisible Earnings: The UK's Hidden Strength*. London: HM Treasury.

Hirschman, A. O. (1984). 'Against Parsimony: Three Easy Ways of Complicating some Categories of Economic Discourse'. *American Economic Review*, 74/2: 88–96.

Hofstader, R., and Lipset, S. M. (1968). *Turner and the Sociology of the Frontier*. New York: Basic Books Inc.

Hollanda, E. (1993). 'Usuário deve ter Direito de Escolha', *Gazeta Mercantil*, 15 Oct.

Homans, G. (1961). *Social Behaviour: Its Elementary Forms*. New York: Harcourt.

Houston, F. S., and Gassenheimer, J. B. (1987). 'Marketing and Exchange'. *Journal of Marketing*, 51 (Oct.): 3–18.

Howell, J. (1999). 'GIS Implementation at the IEC'. Paper presented at the Earth Data Information Systems Conference, CSIR, Pretoria, 12–14 July.

HSBC (1995). *Annual Report*. London: HSBC.

Hudson, H. (1997). *Global Connections: International Telecommunications Infrastructure and Policy*. New York: Van Nostrand Reinhold.

Hughes, T. P. (1987). 'The Evolution of Large Technological Systems', in W. E. Bijker, T. P. Hughes, and T. J. Pinch (eds.), *The Social Construction of Technological Systems: New Directions in the Sociology and History of Technology*. Cambridge, MA: MIT Press, 51–82.

Hunt, C. S., and Aldrich, H. E. (1998). 'The Second Ecology: Organizational Communities', in B. M. Staw and L. L. Cummings (eds.), *Research in Organizational Behavior*. Greenwich, CT: JAI Press Inc., 267–301.

IEEE Transactions on Communications (1976). 'Special Issue on Telecommunications in Developing Countries'. *IEEE Transactions on Communications*, 24/7: 673–782.

Independent Commission for WorldWide Telecommunication Development (1984). *The Missing Link*. Report of the Independent Commission for WorldWide Telecommunication Development. Geneva: International Telecommunication Union.

Industry Canada (1997). *Preparing Canada for the Digital World: Final Report of the Information Highway Advisory Council*. Ottawa: Communications Branch, Industry Canada.

Information Infrastructure Task Force (1997). *A Framework for Global Electronic Commerce*. Washington: Information Infrastructure Task Force, 7 Jan.

Innis, H. A. (1950). *Empire and Communication*. Toronto: Oxford University Press.

—— (1951). *The Bias of Communication*. Toronto: University of Toronto Press.

INPE (1976–97): Instituto Nacional de Pesquisas Espaciais, *Annual Reports*. São José dos Campos, São Paulo: Instituto Nacional de Pesquisas Espaciais.

Iridium (1996). 'Reaching a Global Objective'. *Iridium Today*, 3/1: 9–11.

ITU (1996): International Telecommunication Union, *World Telecommunication Policy Forum. Final Report*. Geneva: International Telecommunication Union.

—— (1997). *World Telecommunication Development Report 1996/97: Trade in Telecommunications—World Telecommunication Indicators*. Geneva: International Telecommunication Union.

—— (1999a). *Challenges to the Network—Internet for Development*. Geneva: International Telecommunication Union.

—— (1999b). *Internet ISO 3166-Based Top Level Domains Survey*. Geneva: International Telecommunications Union, www.itu.int/net/cctlds (accessed 27 Apr. 1999).

Jackson, M. (1975). *Principles of Program Design*. New York: Academic Press.

James, J. (1994). 'Breaking the Speed Limit'. *Computing*, 6 Oct., 32.

Jarque, C. M. (1997). 'An Application of New Technologies: The National Geographic Information System of Mexico', in D. Rhind (ed.), *Framework for the World*. Cambridge: GeoInformation International, 63–70.

Jessop, B. (1997). 'The Governance of Complexity and the Complexity of Governance: Preliminary Remarks on Some Problems and Limits of Economic Guidance', in A. Amin and J. Hausner (eds.), *Beyond Market and Hierarchy*. Cheltenham: Edward Elgar, 95–128.

Johnson, B. (1995). 'Banking on Multimedia'. *McKinsey Quarterly*, 2: 94–106.

Johnson, D. R., and Post, D. G. (1996*a*). 'And How Shall the Net be Governed? A Meditation on the Relative Virtues of Decentralized, Emergent Law'. Paper presented at the Coordination and Administration of the Internet Workshop, Kennedy School of Government, Harvard University, Boston, 8–10 Sept., **www.cli.org/emdraft.html**. (accessed 2 June 1997).

—— —— (1996*b*). 'Law and Borders—the Rise of Law in Cyberspace'. *Stanford Law Review*, 48/1367, **www.cli.org/X0025_LBFIN.html** (accessed 10 May 1999).

Johnson, D. S., and Grayson, K. (1998). 'Sources and Dimensions of Trust in Service Relationships'. Working Paper No. 98-503. London: Centre for Marketing, London Business School.

Johnson, R. J. (1997). 'Spatial Science and Spatial Analysis'. *Futures*, 29/4–5: 323–36.

Jonscher, C. (2000). *Wiredlife: Who Are We in the Digital Age?* London: Anchor Books.

Jussawalla, M., and Tehranian, J. (1993). 'The Economics of Delayed Access. Developing Nations and Satellite Technology'. *Telecommunications Policy*, 17/7: 517–28.

Kalakota, R., and Whinston, A. (1997). *Electronic Commerce: A Manager's Guide*. New York: Addison-Wesley Longman Inc.

—— Robinson, M., and Tapscott, D. (2000). *E-Business 2.0*. 2nd edn. New York: Addison-Wesley.

Katz, J. (1997). 'The Digital Citizen'. *Wired*, 5/12: 60, 275.

Kelly, K. (1997). 'The New Rules of the New Economy'. *Wired*, 5/9: 140, 275.

—— (1998). *New Rules for the New Economy: Ten Radical Strategies for a Connected World*. New York: Viking.

Kiefl, B. (1996). 'Measuring the Use of the Internet: The Future of New Media'. Paper presented at the Conference of the Institute for International Research, Chicago, 24 Apr.

Kim, A. J. (2000). *Community Building on the Web: Secret Strategies for Successful Online Communities*. Berkeley, CA: Peachpit Press.

Kim, W. C., and Mauborgne, R. (1998). 'Procedural Justice, Strategic Decision Making, and the Knowledge Economy'. *Strategic Management Journal*, 19/4: 323–38.

Kling, R. (1993). 'Organizational Analysis in Computer Science'. *Information Society Journal*, 9/2: 71–87.

—— and Lamb, R. (1999). 'IT and Organizational Change in Digital Economics: A Socio-Technical Approach'. Paper presented at a conference on Understanding the Digital Economy: Data, Tools and Research, Washington, 25–26 May, **mitpress.mit.edu/UDE/kling.rtf** (accessed 17 Sept. 2000).

—— Crawford, H., Rosenbaum, H., Sawyer, S., and Weisband, S. (1999). 'Information Technologies in Human Contexts: Learning from Organizational and Social Informatics'. Manuscript prepared by the Center for Social Informatics, Indiana University, Bloomington, IN, May.

Kollock, P. (1994). 'The Emergence of Exchange Structures: An Experimental Study of Uncertainty, Commitment, and Trust'. *American Journal of Sociology*, 100/2: 313–45.

—— (1999). 'The Production of Trust in Online Markets', in E. J. Lawler, M. Macy, S. Thyne, and H. A. Walker (eds.), *Advances in Group Processes*. Greenwich, CT: JAI Press, **www.sscnet.ucla.edu/soc/faculty/kollock/papers/online_trust.htm** (accessed 4 July 2000).

KPMG (2000). *The Impact of the New Economy on Poor People and Developing Countries: Draft Final Report*. London: KPMG report prepared for the Department of International Development, 12 July.

Kracaw, T. S. C. (1980). 'Information Production, Market Signalling, and the Theory of Financial Intermediation'. *Journal of Finance*, 35/4: 863–83.

Kraut, R., Steinfield, C., Chan, A., Butler, B., and Hoag, A. (1998). 'Coordination and Virtualization: The Role of Electronic Networks and Personal Relationships'. *Journal of*

Computer Mediated Communication, 3/4, www.ascusc.org/jcmc/vol3/issue4/kraut.htm (accessed 1 Sept. 2000).

Krueger, R. A. (1994). *Focus Groups: A Practical Guide for Applied Research*. 2nd edn. Thousand Oaks, CA: Sage Publications.

Kurzweil, R. (1999). *The Age of Spiritual Machines: How we will Live, Work and Think in the New Age of Intelligent Machines*. London: Phoenix Orion Books.

Lamberton, D. M. (1971) (ed.). *The Economics of Information and Knowledge*. Harmondsworth: Penguin.

—— (1983). 'Information Economics and Technological Change', in S. Macdonald, D. M. Lamberton, and T. Mandeville (eds.) *The Trouble with Technology: Explorations in the Process of Technological Change*. London: Pinter Publishers, 75–92.

Lasswell, H. D., and Kaplan, A. (1950). *Power and Society: A Framework for Political Inquiry*. New Haven: Yale University Press.

Latour, B. (1999). *Pandora's Hope*. Cambridge, MA: Harvard University Press.

—— and Woolgar, S. (1986). *Laboratory Life: The Construction of Scientific Facts*. Princeton: Princeton University Press.

Lave, J., and Wenger, E. (1991). *Situated Learning: Legitimate Peripheral Participation*. Cambridge: Cambridge University Press.

Leadbeater, C. (1999). *Living on Thin Air: The New Economy*. London: Viking.

Lee, Y. S. F. (1998). 'Intermediary Institutions, Community Organizations, and Urban Environmental Management: The Case of Three Bangkok Slums'. *World Development*, 26/6: 993–1011.

LeFebvre, H. (1991). *The Production of Space*, trans. D. Nicholson-Smith. Oxford: Blackwell Publishers. Originally published in 1974.

Lessig, L. (1998a). 'Governance'. Paper presented at the Computer Professionals for Social Responsibility (CPSR) Annual Meeting, One Planet, One Net: The Public Interest in Internet Governance, Massachusetts Institute of Technology, Cambridge, MA, cyber.harvard.edu/works/lessig/Ny_q_d1.pdf (accessed 11 May 1999).

——(1998b). 'The Laws of Cyberspace', Taiwan Net '98, Taipei, Mar., cyber.harvard.edu/works/lessig/laws_cyberspace.pdf (accessed 11 May 1999).

—— (1999). *Code and Other Laws of Cyberspace*. New York: Basic Books.

Lester, K. (1999). 'Using GIS to Establish a Corporate Database'. Paper presented at the Earth Data Information Systems Conference, CSIR, Pretoria, 12–14 July.

Levinson, P. (1997). *The Soft Edge: A Natural History and Future of the Information Revolution*. London: Routledge.

Lewis, D. J., and Weigert, A. (1985). 'Trust as a Social Reality'. *Social Forces*, 63/4: 967–85.

Lipsey, R. G. (1991). *Economic Growth: Science and Technology and Institutional Change in a Global Economy*. CIAR Publication No. 4. Toronto: Canadian Institute for Advanced Research.

——(1994). 'Sustainable Growth, Innovation, Competitiveness and Foreign Trade'. Mimeo, Whistler, British Columbia, 14 Aug.

Lissoni, F., and Metcalfe, J. S. (1994). 'Diffusion of Innovation Ancient and Modern: A Review of the Main Themes', in M. Dodgson and R. Rothwell (eds.), *The Handbook of Industrial Innovation*. Cheltenham: Edward Elgar, 106–41.

Lombard, M., and Ditton, T. (1997). 'At the Heart of it All: The Concept of Presence'. *Journal of Computer Mediated Communication*, 3/2: 44 pages, www.ascusc.org/jcmc/vol3/issue2/lombard.html (accessed 19 Sept. 2000).

Luhmann, N. (1988). 'Familiarity, Confidence, Trust: Problems and Alternatives', in D. Gambetta (ed.), *Trust: Making and Breaking of Cooperative Relations*. Oxford: Blackwell, 94–107.

Lukes, S. (1974). *Power: A Radical View*. London: MacMillan.

Lundvall, B.-Å. (1992*a*). 'Introduction', in B.-Å Lundvall (ed.), *National Systems of Innovation: Towards a Theory of Innovation and Interactive Learning*. London: Pinter Publishers, 1–19.

—— (1992*b*) (ed.). *National Systems of Innovation: Towards a Theory of Innovation and Interactive Learning*. London: Pinter Publishers.

Lyon, D. (1994). *The Electronic Eye: The Rise of the Surveillance Society*. Cambridge: Polity Press.

Macdonald, S. (1992). 'Formal Collaboration and Informal Information Flow'. *International Journal of Technology Management*, 7/1–3: 49–60.

MacKenzie, D. (1996*a*). 'Economic and Sociological Explanations of Technological Change', in D. MacKenzie (ed.), *Knowing Machines: Essays on Technical Change*. Cambridge, MA: MIT Press, 49–66.

—— (1996*b*). *Knowing Machines: Essays on Technical Change*. Cambridge, MA: MIT Press.

—— and Wajcman, J. (1999) (eds.). *The Social Shaping of Technology*. 2nd edn. Milton Keynes: Open University Press.

MacLean, D. (1996). 'Global Mobile Personal Communications by Satellite: Regulatory Risk vs. Sovereignty Risk at the ITU World Telecommunication Policy Forum'. Paper prepared for Mobile Satellite Communications Global Conference, London, 17–19 June.

McLuhan, H. M. (1962). *The Gutenberg Galaxy: The Making of Typographic Man*. Toronto: University of Toronto Press.

—— (1964). *Understanding Media: The Extensions of Man*. New York: Penguin.

Macneil, I. R. (1980). *The New Social Contract*. New Haven: Yale University Press.

Maculan, A. M. (1981). 'Processo Decisório no Setor de Telecomunicações'. Unpublished M.Sc. thesis, Instituto Universitário de Pesquisas do Rio de Janeiro, Rio de Janeiro.

Maguire, D. J. (1991). 'An Overview and Definition of GIS', in D. J. Maguire, M. F. Goodchild, and D. W. Rhind (eds.), *Geographical Information Systems—Principles and Applications*. Harlow: Longman Scientific & Technical, 9–20.

Mahoney, D. L. (1988). 'The Cost of Doing Business'. Paper delivered at a conference on Networking at Lloyd's, London, 10–11 Oct.

Makridakis, S. (1995). 'The Forthcoming Information Revolution: Its Impact on Society and Firms', *Futures*, 27/8: 799–821.

Malkin, G. (1994). *RFC 1718: The Tao of IETF—A Guide for New Attendees of the Internet Engineering Task Force*. Nov., www.ietf.org/rfc/rfc1718.txt (accessed 22 Nov. 1999)

Malone, T. W., and Rockart, J. F. (1991). 'Computers, Networks and the Corporation'. *Scientific American*, 265/3: 92–9.

Mansell, R. (1996), 'Communication by Design?', in R. Mansell and R. Silverstone (eds.), *Communication by Design: The Politics of Information and Communication Technologies*. Oxford: Oxford University Press, 15–43.

—— (2001). 'Issues Paper'. Paper Prepared for the OECD Emerging Market Economy Forum, Dubai, 16–17 Jan.

—— and Silverstone, R. (1996) (eds.). *Communication by Design: The Politics of Information and Communication Technologies*. Oxford: Oxford University Press.

—— and Steinmueller, W. E. (2000). *Mobilizing the Information Society: Strategies for Growth and Opportunity*. Oxford: Oxford University Press.

—— and Wehn, U. (1998) (eds.). *Knowledge Societies: Information Technology for Sustainable Development*. Oxford: Published for the United Nations Commission on Science and Technology for Development by Oxford University Press.

—— Schenk, I., and Steinmueller, W. E. (2000). 'Net Compatible: The Economic and Social Dynamics of E-commerce'. *Communications & Strategies*, 38/2: 241–76.

Marcell, P. (1988). 'What the Network Will Mean for the Insurer'. Paper delivered at a conference on Insurance Electronic Networks—Progress and Prospects, London, 8 Nov.

March, J. G., and Simon, H. (1958). *Organizations*. London: Wiley.

Margherio, L., Henry, D., Cooke, S., and Montes, S. (1998). 'The Emerging Digital Economy'. Washington: Department of Commerce, **www.ecommerce.gov/emerging.htm** (accessed 23 June 2000).

Markus, M. L., and Bjorn-Andersen, N. (1987). 'Power over Users: Its Exercise by Systems Professionals', *Communications of the ACM*, 30/6: 498–504.

Martin, L. (1999). 'An IEC Case Study—the Technology behind SA's Democracy'. *Network Times*, 7: 3–4.

Martin-Barbero, J. (1993). *Communication, Culture and Hegemony: From the Media to Mediations*. London: Sage Publications.

Marvin, C. (1988). *When Old Technologies Were New: Thinking about Electric Communication in the Late Nineteenth Century*. Oxford: Oxford University Press.

Masser, I. (1999). 'All Shapes and Sizes: The First Generation of National Spatial Data Infrastructures'. *International Journal of Geographical Information Science*, 13/1: 67–84.

—— and Onsrud, H. J. (1993) (eds.). *Diffusion and Use of Geographic Information Technologies*. London: Kluwer Academic Publishers.

Mathiason, J. R., and Kuhlman, C. C. (1998a). *An International Communication Policy: The Internet, International Regulation and New Policy Structures*, **www.intlmgt.com/ITSpaper.html** (accessed 26 May 1998).

—— —— (1998b). 'International Public Regulation of the Internet: Who will give you your Domain Name?'. Paper prepared for the Panel on Cyberhype or the Deterritorialization of Politics: The Internet in a Post-Westphalian Order, International Studies Association, Minneapolis, **www.intlmgt.com/pastprojects/domain.html** (accessed 26 May 1998).

Mattos, H. (1984). *Brasilsat: O Satélite Brasileiro de Telecomunicações*. Brasília, DF: Ministério das Comunicações, Sept.

Mauss, M. (1969). *The Gift: Forms and Functions of Exchange in Archaic Societies*. London: Routledge & Kegan Paul.

Mayer, R. C., David, J. H., and Schoorman, F. D. (1995). 'An Integrative Model of Organisational Trust'. *Academy of Management Review*, 20/3: 709–34.

Melbin, M. (1978). 'Night as Frontier'. *American Sociological Review*, 43/1: 3–22.

Melody, W. H. (1973). 'The Role of Advocacy in Public Policy Planning', in G. Gerbner, L. Gross, and W. Melody (eds.), *Communication Technology and Social Policy*. New York: John Wiley & Sons, 165–81.

—— (1985). 'The Information Society: Implications for Economic Institutions and Market Theory'. *Journal of Economic Issues*, 19/2: 523–39.

—— (1987). 'Information: An Emerging Dimension of Institutional Analysis'. *Journal of Economic Issues*, 21/3: 1313–39.

—— (1997) (ed.). *Telecom Reform: Principles, Policies and Regulatory Practices*. Lyngby: Technical University of Denmark.

Merton, R. K. (1987). 'The Focused Interview and Focus Groups: Continuities and Discontinuities'. *Public Opinion Quarterly*, 51/4: 550–6.

—— Fiske, M., and Kendall, P. L. (1956). *The Focused Interview: A Manual of Problems and Procedures*. Glencoe, IL: Free Press.

Metcalfe, B. (2000). *Internet Collapses and Other InfoWorld Punditry*. Foster City, CA: IDG Books Worldwide Inc., an International Data Group Company.

Metcalfe, J. S. (1988). 'The Diffusion of Innovations: An Interpretative Survey', in G. Dosi, C. Freeman, R. Nelson, G. Silverberg, and L. Soete (eds.), *Technical Change and Economic Theory*. London: Pinter Publishers, 560–89.

—— and Miles, I. (1994). 'Standards, Selection and Variety: An Evolutionary Approach'. *Information Economics and Policy*, 6/3–4: 243–68.

Meyrowitz, J. (1985). *No Sense of Place: The Impact of Electronic Media on Social Behaviour*. Oxford: Oxford University Press.

—— (1994). 'Medium Theory', in D. Crowley and D. Mitchell (eds.), *Communication Theory Today*. Stanford, CA: Stanford University Press, 50–77.

Miles, I., and Thomas, G. (1995). 'User Resistance to New Interactive Media: Participants, Processes and Paradigms', in M. Bauer (ed.), *Resistance to New Technology*. Cambridge: Cambridge University Press, 255–75.

—— Bolisani, E., Scarco, E., and Boden, M. (1999). 'Electronic Commerce Implementation: A Knowledge-Based Analysis'. *International Journal of Electronic Commerce*, 3/3: 53–69.

Millar, J. (1996). 'Interactive Learning in Situated Software Practice: Factors Mediating the New Production of Knowledge during iCASE Technology Interchange'. Unpublished D.Phil. thesis, SPRU, University of Sussex, Brighton.

Miller, D., and Slater, D. (2000). *The Internet: An Ethnographic Approach*. Oxford: Berg.

Miller, L. (1995). 'Women and Children First: Gender and the Settling of the Electronic Frontier', in J. Beck and I. A. Boal (eds.), *Resisting the Virtual Life: The Culture and Politics of Information*. San Francisco: City Lights, 49–57.

Ministério das Comunicações (1995). *Recovery and Expansion Program for Telecommunication and Postal Systems: PASTE*. Brasília, DF: Ministério das Comunicações, Nov.

—— (1996). *Projeto de Lei Geral das Telecomunicações Brasileiras*. Brasília, DF: Ministério das Comunicações, Dec.

Misztal, B. A. (1996). *Trust in Modern Societies: The Search for the Bases of Social Order*. Cambridge: Polity Press.

Mitchell, W. J. (1996). *City of Bits: Space, Place and the Infobahn*. Cambridge, MA: MIT Press

—— (1999). *e-topia: Urban life, Jim—But not as we Know it*. Cambridge, MA: MIT Press.

Mitter, S., and Bastos, M.-I. (1999) (eds.). *Europe and Developing Countries in the Globalised Information Economy: Employment and Distance Education*. London: Routledge for UNU/INTECH Studies in New Technology and Development.

Molina, A. (1997). 'Issues and Challenges in the Evolution of Multimedia: The Case of the Newspaper'. *Futures*, 29/3: 193–212.

Moody, F. (1995). *I Sing the Body Electronic: A Year with Microsoft on the Multimedia Frontier*. London: Hodder & Stoughton.

Morris, G. H., and Hopper, R. (1980). 'Remediation and Legislation in Everyday Talk: How Communicators Achieve Consensus'. *Quarterly Journal of Speech*, 66/3: 266–74.

Morrison, D. E. (1998). *The Search for a Method: Focus Groups and the Development of Mass Communication Research*. Luton: University of Luton Press.

Mueller, M. (1997). 'Internet Governance in Crisis: The Political Economy of Top-Level Domains'. Paper presented at INET '97, Kuala Lumpur, www.isoc.org/isoc/whatis/conferences/inet/97/proceedings/B5/B5_1.htm (accessed 3 Mar. 1998).

—— (1998). 'The Battle over Internet Domain Names: Global or National TLDs'. *Telecommunication Policy*, 22/2: 89–107.

Muller, G. (1996). 'Secure Communication: Trust in Technology or Trust with Technology?' *Interdisciplinary Science Reviews*, 21/4: 336–47.

Mumford, L. (1934). *Technics and Civilization*. London: Routledge & Kegan Paul.

—— (1964). *The Myth of the Machine: The Pentagon of Power*. New York: Harcourt Brace Jovanovich, Inc.

Myers, G. J. (1975a). *Reliable Software through Composite Design*. New York: Van Nostrand Reinhold Company.

—— (1975b). *Software Reliability: Principles and Practices*. New York: John Wiley & Sons.

—— (1976). *Composite/Structured Design*. New York: Van Nostrand Reinhold Company.

Naming Committee Information Sheet (1999). *Application for a Domain Name with co.uk Domain*. London: Naming Committee.

Nardi, B. A. (1997) (ed.). *Context and Consciousness: Activity Theory and Human–Computer Interaction*. Cambridge, MA: MIT Press.

Nass, C., and Steuer, J. (1993). 'Voices, Boxes, and Sources of Messages: Computers and Social Actors'. *Human Communication Research*, 19/4: 504–27, SRCT Paper No. 107, **www.cyborganic.com/_People/jonathan/Academia/Papers/Web/casa-hcr1.html** (accessed 1 Oct. 2000).

—— —— Henriksen, L., and Dryer, D. C. (1994). 'Machines, Social Attributions, and Ethopoeia: Performance Assessments of Computers Subsequent to "Self-" or "Other-" Evaluations', *International Journal of Human–Computer Studies*, 40/3: 543–59, SRCT Paper No. 106, **www.cyborganic.com/People/jonathan/Academia/Papers/Web/casa-ijmms1.html** (accessed 1 Oct. 2000).

—— Lombard, M., Henriksen, L., and Steuer, J. (1995). 'Anthropocentrism and Computers'. *Behaviour and Information Technology*, 14/4: 229–38.

National Research Council (1993). *National Collaboratories: Applying Information Technology for Scientific Research*. Washington: National Academy Press.

Neef, D. (1998) (ed.). *The Knowledge Economy*. Boston: Butterworth-Heinemann.

Negroponte, N. (1995). *Being Digital*. New York: Vintage Books.

Neice, D. (1996). 'Information Technology and Citizen Participation'. SRA Working Paper 167. Ottawa: Department of Canadian Heritage, Aug.

—— (1998a). 'ICTs and Dematerialization: Some Implications for Status Differentiation in Advanced Market Societies'. FAIR Working Paper No. 43. Brighton: SPRU, University of Sussex, Mar.

—— (1998b). 'Measures of Participation in the Digital Techno-Structure: Internet Access'. FAIR Working Paper No. 44. Brighton: SPRU, University of Sussex, Mar.

—— (2000). 'Conspicuous Contributions: Signs of Social Esteem on the Internet'. Unpublished D.Phil. thesis, SPRU, University of Sussex, Brighton.

Nelson, R. (1993) (ed.). *National Innovation Systems: A Comparative Analysis*. New York: Oxford University Press.

—— and Winter, S. (1982). *An Evolutionary Theory of Economic Change*. Cambridge, MA: Belknap Press.

Nemzow, M. (1999). 'ECommerce "Stickiness" for Customer Retention'. *Journal of Internet Banking and Commerce*, 4/1, **www.arraydev.com/commerce/jibc/9908-03.htm** (accessed 12 July 2000).

Nettleton, G. S., and McAnany, E. G. (1989). 'Brazil's Satellite System: The Politics of Applications Planning'. *Telecommunication Policy*, 13/2: 159–66.

Neuberger, C., Tonnemacher, J., Biebl, M., and Duck, A. (1998). 'Online—The Future of Newspapers? Germany's Dailies on the World Wide Web'. *Journal of Computer Mediated Communication*, 4/1, **www.ascusc.org/jcmc/vol4/issue1/index.html** (accessed 5 June 2000).

NewsCentral (2000). 'NewsCentral Home Page'. 5 July, www.all-links.com/newscentral (accessed 30 July 2000).

Newspaper Society (2001a). 'Total Regional Press Publishers' (July 2001 data), available at www.newspapersoc.org.uk/total-regional.html (accessed 1 Aug. 2001).

—— (2001b). 'Ownership Changes, Mergers & Acquisitions', available at www.newspapersoc.org.uk/factsfigures-frameset.html (accessed 1 Aug. 2001).

Noble, D. F. (1998). *The Religion of Technology: The Divinity of Man and the Spirit of Invention.* New York: Alfred A. Knopf.

Nonaka, I. (1994). 'A Dynamic Theory of Organizational Knowledge Creation'. *Organization Science,* 5/1: 14–37.

—— and Takeuchi, H. (1995). *The Knowledge-Creating Company: How Japanese Companies Create the Dynamics of Innovation.* New York: Oxford University Press.

Nooteboom, B. (1996). 'Trust, Opportunism and Governance: A Process and Control Model'. *Organization Studies,* 17/6: 985–1010.

Norris, M., and West, S. (2001). *eBusiness Essentials: Technology and Network Requirements for Mobile and Online Markets.* 2nd edn. Chichester: John Wiley & Sons.

Nourozi, A., and Blonz, T. (1998). *LEOs, MEOs and GEOs: The Market Opportunity for Mobile Satellite Services.* 2nd edn. London: Ovum Consulting.

NSIF (1998a): *National Spatial Information Framework, Interim Policy Guidelines.* Pretoria, www.nsif.org.za (accessed 21 June 2000).

—— (1998b). *The Way Forward: Developing a Framework Facilitating the Exchange and Utilization of Geospatial Information.* Pretoria, www.nsif.org.za (accessed 30 July 1999).

Nuttall, C., and Tunstall, D. (1996). *AFRICAGIS'95—Inventory of GIS Applications in Africa.* Johannesburg: United Nations Institute for Training and Research (UNITAR), World Resources Institute (WRI), Observatoire du Sahara et du Sahel (OSS), UNITAR/95/9, Jan.

OECD (1996) (ed.). *Employment and Growth in the Knowledge-Based Economy.* Paris: OECD.

—— (1997). *Electronic Commerce: Opportunities and Challenges for Government* (The Sacher Report). Paris: OECD.

—— (1999). *The Economic and Social Impact of Electronic Commerce: Preliminary Findings and Research Agenda.* Paris: OECD.

—— (2000). 'E-Commerce: Impacts and Policy Challenges'. ECO/WKP(2000)25. Working Paper No. 252. Paris: OECD, 23 June.

Oliveira, E. Q. (1992). *Renascem as Telecomunicações. 1 Construindo as Base.* Paraná: Editora Editel.

Oliveira, F. (1991). *Caminhos para o Espaço: 30 Anos do INPE.* São Paulo: Contexto.

Onsrud, H. (1999). *Survey of National Spatial Data Infrastructures—Compiled Responses by Question for Selected Countries,* no city, www.spatial.main.edu/~onsrud/GSID.htm (accessed 21 June 2000).

—— and Pinto, J. K. (1991). 'Diffusion of Geographic Information Innovations'. *International Journal of Geographical Information Systems,* 5/4: 447–67.

—— and Rushton, G. (1992). *NCGIA Research Initiative 9: Institutions Sharing Geographic Information.* Technical Report 92-5. Johannesburg: National Center for Geographic Information and Analysis, June.

—— —— (1995) (eds.). *Sharing Geographic Information.* New Brunswick, NJ: Center for Urban Policy Research.

Open GIS Consortium (1999). *Open GIS Consortium Inc.* Open GIS Consortium Inc., www.ogis.org (accessed 21 June 2000).

Openshaw, S., Charlton, M., and Carver, S. (1991). 'Error Propagation: A Monte Carlo Simulation', in I. Masser and M. Blakemore (eds.), *Handling Geographical Information: Methodology and Potential Applications.* New York: Longman Scientific & Technical, 78–101.

Orlikowski, W. J. (1999). 'The Truth is Not Out There: An Enacted View of the "Digital Economy" '. Paper presented at a conference on Understanding the Digital Economy: Data, Tools and Research, Washington, 25–26 May, **mitpress.mit.edu/UDE/orlikowski.rtf** (accessed 17 Sept. 2000).

Overby, B. A. (1996). 'Identification and Validation of a Societal Model of USENET'. Unpublished Ph.D. thesis, San Jose State University, **www.well.com/user/deucer/thesis/html** (accessed 30 Sept. 2000).

Peak, S., and Fisher, P. (1999). *The Media Guide 2000*. London: Fourth Estate.

Pearsall, J., and Trimble, B. (1996) (eds.). *The Oxford English Reference Dictionary*. 2nd edn. Oxford: Oxford University Press.

Pennings, J. M., and Harianto, F. (1992). 'The Diffusion of Technological Innovation in the Commercial Banking Industry'. *Strategic Management Journal*, 13/1: 29–46.

Perez, C. (1983). 'Structural Change and Assimilation of New Technologies in Economic and Social Systems'. *Futures*, 15/5: 357–75.

—— (1985). 'Microelectronics, Long Waves and World Structural Change: New Perspectives for Developing Countries'. *World Development*, 13/3: 441–63.

—— and Soete, L. (1988). 'Catching up in Technology: Entry Barriers and Windows of Opportunity', in G. Dosi, C. Freeman, R. R. Nelson, G. Silverberg, and L. Soete (eds.), *Technical Change and Economic Theory*. London: Pinter Publishers, 459–79.

Perri 6, Lasky, K., and Pletcher, A. (1998). *The Future of Privacy*. 2 vols. London: Demos.

Pessini, J. (1993). 'Competitividade da Indústria de Equipamentos de Telecomunicações', in L. Coutinho, J. C. Ferraz, A. Santos, and P. M. Veiga (eds.), *Estudo da Competitividade Brasileira, Relatório Técnico*. Rio de Janeiro and Campinas: Universidade Estadual de Campinas, Instituto de Economia, Universidade Federal do Rio de Janeiro, Instituto de Economia Industrial, FUNCEX, 20–50.

Pitt, W. (1988). 'Children of the Revolution'. *Reinsurance* (Feb.), 37–43.

Polanyi, M. (1966). *The Tacit Dimension*. London: Routledge & Kegan Paul.

Porter, M. E. (1991). 'Towards a Dynamic Theory of Strategy'. *Strategic Management Journal*, 12 (Special Issue, Winter), 95–117.

Post, D. G., and Johnson, D. R. (1997a). 'Borders, Spillovers, and Complexity: Rule-Making Processes in Cyberspace (and Elsewhere)'. Paper prepared for Olin Law and Economics Symposium on 'International Economic Regulation', Georgetown University Law Centre, Washington, 5 Apr.

—— —— (1997b). 'The New Civic Virtue of the Net: A Complex Systems Model for the Governance of Cyberspace'. Paper presented at the Annual Review of the Internet as a Platform, The Aspen Institute, Aspen, **www.cli.org/paper4.htm** (accessed 7 Mar. 1997).

Postel, J. (1994). *RFC 1591: Domain Name System Structure and Delegation*. ISI, Mar., **info.internet.isi.edu:80/in-notes/rfc/files/rfc1591.txt** (accessed 5 Nov. 1999).

Poster, M. (1997). 'Cyberdemocracy: Internet and the Public Sphere', in D. Porter (ed.), *Internet Culture*. New York: Routledge, 201–17.

Primo Braga, C. A. (2000). 'The Networking Revolution: Opportunities and Challenges for Developing Countries'. *info*Dev Working Paper. Washington: Global Information and Communication Technologies Department, The World Bank Group (and others).

Quah, D. T. (1993). 'Empirical Cross-Section Dynamics in Economic Growth'. *European Economic Review*, 37/2–3: 426–34.

—— (1996). 'The Invisible Hand and the Weightless Economy'. Occasional Paper No. 12. London: Programme on National Economic Performance, LSE Centre for Economic Performance, Apr.

—— (1997). 'Increasingly Weightless Economies'. *Bank of England Quarterly Bulletin* (Feb.): 49–56.

Quintas, P. (1996). 'Software by Design', in R. Mansell and R. Silverstone (eds.), *Communication by Design: The Politics of Information and Communication Technologies*. Oxford: Oxford University Press, 75–102.

Raab, C. D., and Bennett, C. J. (1998). 'The Distribution of Privacy Risks: Who Needs Protection?'. *Information Society*, 14/4: 263–74.

Ramirez, R. (1999). 'Value Co-Production: Intellectual Origins and Implications for Practice and Research'. *Strategic Management Journal*, 20/1: 49–65.

Raymond, E. S. (1999). *The Cathedral and the Bazaar: Musings on Linux and Open Source by an Accidental Revolutionary*. Sebastopol, CA: O'Reilly & Associates, Inc.

Raynes, H. E. (1964). *A History of British Insurance*. London: Pitman & Sons.

Reagle, J. M. J. (1996). 'Trust in Electronic Markets'. *First Monday*, 1/2, **www.firstmonday.dk/issues/issue2/markets/index.html** (accessed 6 July 2000).

Reeve, C. (1998). 'Presence in Virtual Theatre'. Paper prepared for the workshop 'Presence in Shared Virtual Environments', BT Laboratories, Martlesham Heath, Ipswich, 10–11 June, **www.eimc.brad.ac.uk/research/presence.htm** (accessed 16 June 1998).

Reeves, B., and Nass, C. (1996). *The Media Equation: How People Treat Computers, Television, and New Media like Real People and Places*. Palo Alto, CA: CSLI Publications.

Reidenberg, J. R. (1996). 'Governing Networks and Cyberspace Rule-Making'. *Emory Law Journal*, 45, **www.law.emory.edu/ELJ/volumes/sum96/reiden.html** (accessed 10 May 1999).

—— (1998). 'Lex Informatica: The Formulation of Information Policy Rules Through Technology'. *Texas Law Review*, 76/553, **www.epic.org/misc/gulc/materials/reidenberg2.html** (accessed 10 May 1999).

Rengger, N. (1997). 'The Ethics of Trust in World Politics'. *International Affairs*, 73/3: 469–87.

Rheingold, H. (1993). *The Virtual Community: Homesteading on the Electronic Frontier*. Reading, MA: Addison-Wesley Publishing Co.

—— (1995). *The Virtual Community: Finding Connection in a Computerized World*. London: Minerva.

—— (2000). *The Virtual Community: Homesteading on the Electronic Frontier*. Rev. edn. Cambridge, MA: MIT Press. First published by Addison-Wesley Publishing Co., 1993.

Rice, R. E. (1993). 'Media Appropriateness: Using Social Presence Theory to Compare Traditional and New Organizational Media'. *Human Communication Research*, 19/4: 451–84.

—— (1999). 'Artifacts and Paradoxes in New Media'. *New Media & Society*, 1/1: 24–32.

Rifkin, J. (1995). *The End of Work—The Decline of the Global Labor Force and the Dawn of the Post-Market Era*. London: G. P. Putnam's Sons.

Ring, P. S., and Van de Ven, A. H. (1994). 'Developmental Processes of Cooperative Interorganizational Relationships'. *Academy of Management Review*, 19/1: 90–118.

Robins, K., and Webster, F. (1999). *Times of the Technoculture*. London: Routledge.

Roche, E. M. (2000). 'Information Technology and the Multinational Enterprise', in E. M. Roche and M. J. Blaine (eds.), *Information Technology in Multinational Enterprises*. Cheltenham: Edward Elgar, 57–89.

Rochlin, G. I. (1997). *Trapped in the Net: The Unanticipated Consequences of Computerization*. Princeton: Princeton University Press.

Rogers, D. (1993). *The Future of American Banking: Managing for Change*. New York: McGraw-Hill Inc.

Rogers, E. M. (1968). *Diffusion of Innovations*. New York: Free Press.

Rogers, E. M. (1982). 'Information Exchange and Technological Innovation', in D. Sahal (ed.), *The Transfer and Utilization of Technical Knowledge*. Lexington, MA: Lexington Books, 105–23.

—— (1993). 'The Diffusion of Innovations Model: Keynote Address', in I. Masser and H. J. Onsrud (eds.), *Diffusion and Use of Geographic Information Technologies*. London: Kluwer Academic Publishers, 9–24.

—— (1995). *Diffusion of Innovations*. 4th edn. New York: Free Press.

Romer, P. (1986). 'Increasing Returns and Long-Run Growth'. *Journal of Political Economy*, 94: 1002–37.

—— (1990). 'Endogenous Technological Change'. *Journal of Political Economy*, 98/5, pt. 2: 71–102.

—— (1993). 'Ideas and Things'. *The Economist*, 11 Sept., 6–7.

—— (1994). 'The Origins of Endogenous Growth'. *Journal of Economic Perspectives*, 8/1: 3–22.

—— (1995). 'Beyond the Knowledge Worker'. *World Link* (Jan.–Feb.), 56–60.

Romm, C., Pliskin, N., and Clarke, R. (1997). 'Virtual Communities and Society: Toward an Integrative Three Phase Model'. *International Journal of Information Management*, 17/4: 261–70.

Rony, E., and Rony, P. (1998). *The Domain Name Handbook: High Stakes and Strategies in Cyberspace*. Lawrence, KS: R&D Books.

Rotter, J. B. (1967). 'A New Scale for the Measurement of Interpersonal Trust', *Journal of Personality*, 35/4: 651–5.

Ruggie, J. G. (1975). 'International Responses to Technology: Concepts and Trends'. *International Organization*, 29 (Summer): 557–83.

Ruggles, R. L. (1997) (ed.). *Knowledge Management Tools*. New York: Butterworth-Heinemann.

Rush, H., Hobday, M., Bessant, J., and Arnold, E. (1995). 'Strategies for Best Practice in Research and Technology Institutes: An Overview of a Benchmarking Exercise'. *R&D Management*, 25/1: 17–31.

Rutkowski, A. M. (1998*a*). *Competing Models of Internet DNS Service Governance*, 5 Apr. **www.wia.org/pub/models.html** (accessed 24 May 1999).

—— (1998*b*). 'The Internet: Governance for Grabs?'. *World Internet Alliance*, n.d., **www.wia.org/pub/forgrabs.html** (accessed 24 May 1999).

Sanchez, R., and Mahoney, J. T. (1996). 'Modularity, Flexibility, and Knowledge Management in Product and Organization Design'. *Strategic Management Journal*, 17 (Winter Special Issue), 63–76.

Sarkar, J. (1998). 'Technological Diffusion: Alternative Theories and Historical Evidence'. *Journal of Economic Surveys*, 12/2: 131–76.

Sarkar, M., Butler, B., and Steinfield, C. (1998). 'Cybermediaries in Electronic Marketspace: Toward Theory Building'. *Journal of Business Research*, 41/3: 215–21.

Saxenian, A. L. (1994). *Regional Advantage: Culture and Competition in Silicon Valley and Route 128*. Cambridge, MA: Harvard University Press.

Schmid, B. F., and Lindemann, M. A. (1998). *Elements of a Reference Model for Electronic Markets*. Proceedings of the 31st Annual Hawaii International Conference on System Sciences (HICSS), IEEE Computer Society, Honolulu, Hawaii, **www.computer.org/proceedings/hicss/8242/8242toc.htm** 824201193 (accessed 4 July 2000).

Schmittbeck, R. (1994). 'Intermediation Environments of West-German and East-German Voters—Interpersonal-Communication and Mass-Communication during the 1st All-German Election Campaign'. *European Journal of Communication*, 9/4: 381–419.

Schneider, F. B. (1999). *Trust in Cyberspace*. Washington: National Research Council (US), Committee on Information Systems Trustworthiness, National Academy Press.

Schneier, B. (1996). *Applied Cryptography: Protocols, Algorithms, and Source Code in C*. New York: John Wiley & Sons.

Schön, D. (1991). *The Reflective Practitioner: How Professions Think in Action*. London: Basic Books Inc.

Schramm, W. (1968). *Communication Satellites for Education, Science and Culture*. Reports and Papers on Mass Communication No. 53. Paris: United Nations Educational, Scientific, and Cultural Organization (UNESCO).

Schumpeter, J. (1947). *Capitalism, Socialism and Democracy*. 2nd edn. New York: Harper & Row.

—— (1961). *The Theory of Economic Development: An Inquiry into Profits, Capital, Credit, Interest and the Business Cycle*. Oxford: Oxford University Press

Schwabe, C., O'Leary, B., and Sukai, S. B. (1998). 'Putting all the Facts on the Map'. *In Focus Forum*, 5/3: 4–7.

Scott, J. (1996). *Stratification and Power: Structures of Class, Status and Command*. Cambridge: Polity Press.

Seabrook, J. (1997). *Deeper: A Two Year Odyssey in Cyberspace*. London: Faber & Faber.

Senker, J. (1995). 'Tacit Knowledge and Models of Innovation'. *Industrial and Corporate Change*, 4/2: 425–47.

Shannon, C. E., and Weaver, W. (1949). *The Mathematical Theory of Communication*. Urbana: University of Illinois Press.

Shapiro, A. L., and Leone, R. C. (1999). *The Control Revolution: How the Internet is Putting Individuals in Charge and Changing the World We Know*. Washington: A Century Foundation Book.

Shapiro, C., and Varian, H. (1998). *Information Rules: A Strategic Guide to the Network economy*. Boston: Harvard Business School Press.

Shapiro, S. P. (1987). 'The Social Control of Impersonal Trust'. *American Journal of Sociology*, 93/3: 623–58.

Sharpe, R. (1994). 'Quick Off the Mark'. *Computer Weekly*, 8 Sept., 30.

Sheridan, T. B. (1992). 'Musings on Telepresence and Virtual Presence'. *Presence: Teleoperators and Virtual Environments*, 1/1: 120–6.

Short, J., Williams, E., and Christie, B. (1976). *The Social Psychology of Telecommunications*. Chichester: Wiley.

Sieger, S. (1998). 'EC in the (Re-)Insurance Industry'. *EM—Electronic Markets*, 8/1: 7–9, www.electronicmarkets.org/netacademy/publications.nsf/all_pk/1116 (accessed 14 Apr. 1999).

Silverstone, R. (1994). 'Domesticating the Revolution: Information and Communication Technologies and Everyday Life', in R. Mansell (ed.), *Managing Information and Communication Technologies: Emerging Patterns of Control*. London: ASLIB, 221–33.

—— (1999). *Why Study the Media?* London: Sage Publications.

—— (forthcoming). 'Mediation and Communication', in C. Galhoun, C. Rojek, and B. S. Turner (eds.), *The International Handbook of Sociology*. London: Sage Publications.

—— and Haddon, L. (1996). 'Design and the Domestication of Information and Communication Technologies: Technical Change and Everyday Life', in R. Mansell and R. Silverstone (eds.), *Communication by Design: The Politics of Information and Communication Technologies*. Oxford: Oxford University Press, 44–74.

—— and Hirsch, E. (1992) (eds.). *Consuming Technologies: Media and Information in Domestic Spaces*. London: Routledge.

Simon, H. (1973). 'The Structure of Ill Structured Problems'. *Artificial Intelligence*, 4: 181–201.

Simonson, H. P. (1963). *Frederick Jackson Turner: The Significance of the Frontier in American History*. New York: Frederick Unger Publishing Co.

Smith, M. A., and Kollock, P. (1999) (eds.). *Communities in Cyberspace*. London: Routledge.

Smith, R. (1998). 'Southern African Metadata Consortium (SAM)'. Paper presented at the National Spatial Information Framework Workshop, Pretoria, 11 Feb.

Smith System Engineering (1997). *Feasibility of Censoring and Jamming Pornography and Racism in Informatics*. Guildford: Final report prepared for the European Parliament, Scientific and Technical Options Assessment (STOA).

Smythe, D. W. (1957). *The Structure and Policy of Electronic Communications*. Chicago: University of Illinois Bulletin.

——(1962). 'Considerations on a Worldwide Communications Satellite System'. *Telecom Journal*, 29/9: 264–73.

——(1977). 'Communications: Blindspot of Western Marxism'. *Canadian Journal of Political and Social Theory*, 1/3: 1–27.

——(1981). *Dependency Road: Communications, Capitalism, Consciousness and Canada*. Norwood, NJ: Ablex Publishers.

Stefik, M. (1997). 'Trusted Systems'. *Scientific American*, 276/3: 78–81.

——(1999). *The Internet Edge: Social, Technical and Legal Challenges for a Networked World*. Cambridge, MA: MIT Press.

Steinmueller, W. E. (1992). 'The Economics of Production and Distribution of User-Specific Information via Digital Networks', in C. Antonelli (ed.), *The Economics of Information Networks*. Amsterdam: North Holland, 173–94.

——(2000). 'Will New Information and Communication Technologies Improve the "Codification" of Knowledge?' *Industrial and Corporate Change*, 9/2: 361–76.

Sterling, B. (1992). *The Hacker Crackdown: Law and Disorder on the Electronic Frontier*. London: Penguin Books.

Sternberg, R. J., and Lubart, T. I. (1999). 'The Concept of Creativity: Prospects and Paradigms', in R. J. Sternberg (ed.), *Handbook of Creativity*. Cambridge: Cambridge University Press, 3–15.

Steuer, J. (1992). 'Defining Virtual Reality: Dimensions Determining Telepresence'. *Journal of Communication*, 42/4: 73–93.

——(1995). 'Defining Virtual Reality: Dimensions Determining Telepresence', in F. Biocca and M. R. Levy (eds.), *Communication in the Age of Virtual Reality*. Hillsdale, NJ: Lawrence Erlbaum, 33–56.

Stevens, W. P., Myers, G. J., and Constantine, L. L. (1974), 'Structured Design'. *IBM Systems Journal*, 13/2: 115–39. Repr. in P. Freeman and A. I. Wasserman (eds.), *Tutorial on Software Design Techniques*. 4th edn. Silver Spring, MD: IEEE Computer Society Press, 1983, 328–52.

Stewart, D. W., and Shamdasani, P. N. (1990). *Focus Groups: Theory and Practice*. Newbury Park, CA: Sage Publications.

Stiglitz, J. (1999). 'Knowledge as a Global Public Good', in I. Kaul, I. Grunberg, and M. Stern (eds.), *Global Public Goods: International Cooperation in the 21st Century*. New York: Published for the United Nations Development Program by Oxford University Press, 308–26.

Strausak, N. (1998). 'Resumee of Votalk'. *Virtual Organization-Net Newsletter*, **www.virtual-organization.net/mailinglist/resum997-398.pdf** (accessed 6 June 1999).

Strauss, A., and Corbin, J. (1990). *Basics of Qualitative Research: Grounded Theory, Procedures and Techniques*. London: Sage Publications.

Summers, R. (1997). *Secure Computing: Threats and Safeguards*. New York: McGraw-Hill Publications.

Swarts, M. (1998). 'Nutshell'. *In Focus Forum*, 5/3: 2.

Tapia, J. (1995). 'O Desenvolvimento de Sistemas de Produtos Complexos: O Caso do Satélite Brasileiro', in L. Coutinho, J. Cassiolato, and A. Silva (eds.), *Telecomunicacões, Globalização e Competitividade*. Campinas, São Paulo: Papirus, 222–69.

Tapscott, D. (1995). *The Digital Economy: Promise and Peril in the Age of Networked Intelligence*. New York: McGraw-Hill.

——Ticoll, D., and Lowy, A. (2000). *Digital Capital: Harnessing the Power of Business Webs*, Boston: Harvard Business School Press.

Taylor, J. (1998). 'Inaugural President's Address, Engineering the Information Age'. Institution of Electrical Engineers, Savoy Place, London, 1 Oct.

Telebras (1995). *ECCO System Description, Telebras, CCI, E-Systems, Bell Atlantic, INPE, Lockheed Martin, Texas Instruments, BR-Industrial Companies*. A Telebras System Operator Publication. Brasília, DF: Telebras, Feb.

Telecomunicações Brasileiras SA (1974–97). *Telebras System*. Annual reports, various years. Brasília, DF: Telecomunicações Brasileiras SA.

——(1994). *MoU Telebrás and Bell Atlantic Enterprises International Corporation and Constellation Communications Inc. (CCI)*. Brasília, DF: Telecomunicações Brasileiras SA, Oct.

Terracine, E. (1997). *Contribuição das Telecomunicações para as Áreas da Educação e da Saúde*. Relatório No. 7. Brasília, DF: Agência Espacial Brasileira (AEB).

Thompson, J. B. (1995). *The Media and Modernity: A Social Theory of the Media*. Cambridge: Polity Press.

Tidd, J., Bessant, J., and Pavitt, K. (1997). *Managing Innovation: Integrating Technological, Market and Organisational Change*. Chichester: John Wiley & Sons

Timmers, P. (1999). *Electronic Commerce: Strategies and Models for Business to Business Trading*. Chichester: Wiley.

Townsend, R. (1988). 'Back Office Processing'. Paper delivered at a conference on Networking at Lloyd's, London, 10–11 Oct.

Tremblay, G. (1995). 'The Information Society: From Fordism to Gatesism'. *Canadian Journal of Communication*, 20/4: 461–82.

——and Lacroix, J.-G. (1997). 'The "Information Society" and Cultural Industries Theory'. *Current Sociology (Monograph)*, 45/4; 1–155.

Turkle, S. (1996). *Life on the Screen: Identity in the Age of the Internet*. London: Weidenfeld & Nicolson

Turner, F. (1999) 'Cyberspace as the New Frontier? Mapping the Shifting Boundaries of the Network Society'. Paper presented at the International Communication Association Conference, San Francisco, 29 May.

UBS (1995): Union Bank of Switzerland, *Annual Report*. Zurich: UBS.

Uchôa, C. (1997). *A Globalização das Telecomunicações: Sistemas Globais de Comunicação Pessoal Via Satélite (GMPCS)*. Iridium SudAmerica Service Provider Publication.

UNCTAD (2000). *Building Competence: Electronic Commerce and Development*. Geneva: UN Conference on Trade and Development, www.unctad.org (accessed 2 July).

Upton, D. M., and McAfee, A. (1995). 'The Real Virtual Factory'. *Harvard Business Review*, 74/4: 123–33.

Valente, A. C. (1998). *Solicitação de Outorga de Permissão para Exploração de Serviço Móvel Global por Satélites Não-Geostacionários pela Iridium-Brasil S.A.—Processo No. 53500.000001/97 de 05.11.1997, Análise No. 11/98-GCAV*. Brasília, DF: Agência Nacional de Telecomunicações (Anatel).

van Dusseldorp, M. (1998). *The Future of the Printed Press: Challenges in a Digital World*. The European Journalism Centre, www.ejc.nl/hp/fpp/contents.html (accessed 28 Nov. 1998).

van Helden, P. (1999). 'GIS for Development: An Analysis of GIS Diffusion in Government Organisations in South Africa'. Paper prepared for the Earth Data Information Systems Conference, CSIR, Pretoria, 12–14 July.

Veblen, T. (1899/1994). *The Theory of the Leisure Class*. New York: Dover Publications.

Verhoest, P., and Hawkins, R. W. (2000). 'A Transaction Structure Approach to Assessing the Dynamics and Impacts of "Business-to-Business" Electronic Commerce'. A report prepared for the Impacts and Perspectives of Electronic Commerce (IPEC) Project of the Telematica Instituut by TNO Institute for Strategy, Technology and Policy (STB) and SPRU, Delft/Brighton, 17 Apr.

Vianna, C. (1993). *Privatização das Telecomunicações*, Rio de Janeiro: Notrya.

von Hippel, E. (1994).' "Sticky Information" and the Locus of Problem Solving: Implications for Innovation'. *Management Science*, 40/4: 429–39.

—— (1998). 'Economics of Product Development by Users: The Impact of "Sticky" Local Information'. *Management Science*, 44/5: 629–44.

Vygotsky, L. S. (1978). *Mind in Society: The Development of Higher Mental Processes*. Cambridge, MA: Harvard University Press.

Wallis, J., and Dollery, B. (1997). 'Autonomous Policy Leadership: Steering a Policy Process in the Direction of a Policy Quest'. *Governance*, 10/1: 1–22.

Waugh, E. (1962). *Black Mischief*. Middlesex: Penguin Books.

Webster, F. (1995). *Theories of the Information Society*. London: Routledge.

—— (2001) (ed.). *Culture and Politics in the Information Age*. London: Routledge.

Wehn de Montalvo, U. (2001). 'Crossing Organizational Boundaries: Prerequisites for Spatial Data Sharing in South Africa'. Unpublished D.Phil. thesis, SPRU, University of Sussex, Brighton.

Weick, K. E. (1976). 'Educational Organizations as Loosely Coupled Systems'. *Administrative Science Quarterly*, 21/1: 1–19.

Wendt, A. E. (1987). 'The Agent-Structure Problem in International Relations Theory'. *International Organization*, 41/3: 335–70.

—— and Duvall, R. (1989). 'Institutions and International Order', in E.-O. Czempiel and J. N. Rosenau (eds.), *Global Changes and Theoretical Challenges: Approaches to World Politics for the 1990s*. Lexington, MA: Lexington Books, 51–73.

Wertsch, J. V. (1985) (ed.). *Culture, Communication and Cognition: Vygotskian Perspectives*. Cambridge: Cambridge University Press.

White, H. S. (1995). 'Information Intermediary—a Fancy Name for Reference Work'. *Library Journal*, 120/5: 44–5.

Williams, R., and Edge, D. (1992). 'Social Shaping Reviewed: Research Concepts and Findings in the UK'. Edinburgh PICT Working Paper No. 41. Edinburgh: Research Centre for Social Sciences, Edinburgh University.

—— and Slack, R. S. (1999) (eds.). *Europe Appropriates Multimedia: A Study of the National Uptake of Multimedia in Eight European Countries and Japan*. Report No. 42. Trondheim: Norwegian University of Science and Technology, Centre of Technology and Society.

Wilson, J. D. (1999). 'Stellar Performers—Top Companies Vie for Position in the GIS Universe'. *GEOEurope* (Mar.), 36–41.

Winn, J. K. (1998a). 'Couriers without Luggage: Negotiable Instruments and Digital Signatures', *South Carolina Law Review*, 49/4, www.smu.edu/~jwinn/ecouriers.html (accessed 4 July 2000).

—— (1998b). 'Open Systems, Free Markets, and Regulation of Internet Commerce'. *Tulane Law Review*, 72/4, www.smu.edu/~jwinn/esig.htm (accessed 4 July 2000).

—— (1999). 'The Hedgehog and the Fox: Distinguishing Public and Private Sector Approaches to Managing Risk for Internet Transactions'. *ABA Administrative Law Review*, 51/3, www.smu.edu/~jwinn/hedgehogfox.htm (accessed 4 July 2000).

Winner, L. (1977). *Autonomous Technology—Technics-out-of-Control as a Theme in Political Thought*. Cambridge, MA: MIT Press.

—— (1986). *The Whale and the Reactor: A Search for Limits in an Age of High Technology*. Chicago: University of Chicago Press.

Winslow, C. D., and Bramer, W. L. (1994). *Futurework: Putting Knowledge to Work in the Knowledge Economy*. New York: Free Press.

Wohlers, M. (1996). 'A Guerra das Telecomunicações: Internacionalização, Privatização e Novas Oportunidades'. Unpublished D. Phil. thesis, Instituto de Economia, Universidade Estadual de Campinas, Campinas.

Wood, J., and Silver, D. (1995). *Joint Application Development*. New York: John Wiley & Sons Inc.

Woolgar, S. (forthcoming) (ed.). *Virtual Society? Get Real!* Oxford: Oxford University Press.

Woolley, B. (1993). *Virtual Worlds: A Journey in Hype and Hyperreality*, London: Penguin Books.

World Bank (1998). *World Development Report 1998/99: Knowledge for Development*. Washington: World Bank.

Wright, C., and Fayle, E. (1988). *A History of Lloyd's*. London: Macmillan & Co.

Wright, D. (1997). 'Obtaining Global Market Access for GMPCS', *Telecommunications Policy*, 21/9–10: 775–82.

Wrobel, D. W. (1993). *The End of American Exceptionalism: Frontier Anxiety from the Old West to the New Deal*. Lawrence, KA: University Press of Kansas.

Wyatt, S. (1998). 'From Metaphor to Reality: Images of the Internet and Change'. Paper prepared for the Cultural Politics of Technology Workshop, London, June.

—— (2000). 'Talking about the Future: Metaphors of the Internet', in N. Brown, B. Rappert, and A. Webster (eds.), *Contested Futures*. Aldershot: Ashgate, 109–26.

—— Henwood, F., Miller, N., and Senker, P. (2000) (eds.). *Technology and In/Equality: Questioning the Information Society*. London: Routledge.

Yanelle, M. O. (1989). 'The Strategic Analysis of Intermediation'. *European Economic Review*, 33/2–3: 294–301.

Yates, J. (1997). 'Early Interactions between the Life Insurance and Computer Industries: The Prudential's Edmund Berkeley and the Society of Actuaries Committee, 1946–1952'. *IEEE Annals of the History of Computing*, 19/3: 60–73.

Yourdon, E., and Constantine, L. L. (1978). *Structured Design*. New York: Yourdon Press.

Zerdick, A., Picot, A., Schrape, K., Artope, A., Goldhammer, K., Lange, U. T., Vierkant, E., Lopez-Escobar, E., and Silverstone, R. (2000) (eds.). *E-conomics: Strategies for the Digital Marketplace*. Berlin: Springer.

Zorkoczy, P., and Heap, N. (1995). *Information Technology: An Introduction*. 4th edn. London: Pitman Publishing.

Zuboff, S. (1988). *In the Age of the Smart Machine: The Future of Work and Power*. London: Heinemann Professional Publishing.

Zucker, L. G. (1986). 'Production of Trust: Institutional Sources of Economic Structure', in B. M. Staw and L. L. Cummings (eds.), *Research in Organizational Behavior*. Greenwich, CT: JAI Press Inc., 53–112.

Author Index

Abbott, J. 190, 191, 196, 200
Agre, P. 83, 134, 258, 264 n.
Ainley, R. 61, 63
Ajzen, I. 199
Åkesson, K.-P. 261
Alasuutari, P. 61, 85
Albitz, P. 207 n., 208 n.
Aldrich, H. E. 221 n.
Allen, F. 193
Alter, C. 187
Amaral, L. 238
Ancori, A. 24 n.
Anderson, C. 7
Andrade, M. 240
Angell, I. 8, 257
Antonelli, C. 187
Arafat, Y. 126
Armstrong, A. 8, 35, 260
Arroio, A. C. 225, 229 n., 238
Arrow, K. 187

Bachrach, P. 220
Baier, A. 121 n., 125
Bambace, L. A. 235
Baratz, M. S. 220
Barbour, R. S. 59 n.
Barley, S. R. 220
Barnatt, C. 88, 260
Barnett, G. A. 187
Barras, R. 143, 147, 157
Barrett, M. I. 135
Bashe, C. J. 129
Bastos, C. A. 235
Bastos, M.-I. 258
Baudrillard, J. 17 n.
Baum, M. 112 n., 122
Beamish, R. 94
Bell, M. 230, 244, 245
Bell, T. W. 219 n.
Beniger, J. R. 24
Benjamin, R. I. 168
Bennett, C. J. 258
Bensimon, C. 239
Berkowitz, D. 105
Bessant, J. 86
Bewley, T. F. 123 n.
Biddle, C. B. 111, 122
Bijker, W. E. 2, 4, 139 n.
Biocca, F. 92

Bjorn-Andersen, L. 134
Blackler, F. 169
Blonz, T. 227, 228, 242
Boden, Margaret 172
Boden, Mark 2
Boisot, M. H. 3 n., 56, 153 n., 156, 157, 158, 159,
    261–2
Bolter, J. D. 4, 260
Borsook, P. 259
Bothamly, J. 252 n., 255 n.
Bourdieu, P. 71 n.
Boyle, J. 219 n.
Bradach, J. L. 124 n.
Bradner, S. 210 n.
Bramer, W. L. 162
Braverman, H. 258
Brisolla, S. 232
Brown, J. S. 5, 145, 261
Bryan, I. 145
Buber, M. 258
Bultje, R. 90
Bunker, N. 131, 141
Bureth, A. 24 n.
Burns, T. 166 n.
Burrough, P. A. 189 n., 192, 195
Butler, D. 8, 252 n.
Byrne, J. A. 90

Cairncross, F. 5, 128, 251
Campbell, H. 186
Cappelli, P. 148
Carter, R. L. 130, 141
Carver, S. 190
Cassiolato, J. 240
Castells, M. 2, 3, 10, 11, 56, 128, 251,
    253–4, 261
Cawson, A. 9 n.
Cazale, L. 221 n.
Ceballos, D. 235
Charlton, M. 190
Chelmsford, J. 135
Chesborough, H. W. 90
Chiesa, V. 229 n.
Christ, H. 189
Christiansen, T. 189
Christie, B. 140
Ciborra, C. U. 3 n.
Citicorp 149
Clarke, D. G. 192, 199

Clarke, R. 8
Clarke, S. 98
Cleaver, A. 94
Clemente, P. C. 61
Cochrane, P. 256–7
Cohendet, P. 7, 24 n.
Cole, M. 169–70
Coleman, W. E. 4, 252 n.
Constantine, L. L. 134 n.
Cooley, T. F. 4, 252 n.
Coombs, R. 86, 187 n.
Coppock, J. T. 190
Corbin, J. 59, 173 n.
Cosimano, T. F. 4, 252 n.
Coughlan, P. 229 n.
Cowan, R. 7, 24 n., 153 n.
Coyle, D. 72
Credé, A. 144 n., 146 n., 147
Crompton, R. 71 n.

Dahl, R. A. 220
Daley, W. M. 133
Dalum, B. 230
Dasgupta, P. 50
Dataquest 186 n., 189
Davenport, T. H. 10, 85, 162, 166, 168, 170 n.
David, J. H. 121 n.
David, P. A. 7, 23, 24 n., 50, 153 n.
Davidow, W. H. 23 n., 85, 88
De Bony, E. 117 n.
Decol, R. 242 n.
DeLisi, P. S. 89
Denis, J.-L. 221 n.
Department of Commerce (US) 3 n., 66
Department of Trade and Industry (UK)
        66 n., 129 n., 138 n., 256
Dess, G. G. 89–90
Diamond, D. 147, 152
Dibona, C. 51 n.
Dickenson, P. 59, 66 n.
Ditton, T. 72, 139, 259, 261
Dizard, W. 3 n., 8
Dollery, B. 221 n.
Dosi, G. 11, 139 n., 244 n.
Drenth, H. 193
Drucker, P. F. 85–6, 88–9
Dudman, J. 172, 173 n.
Duff, A. S. 3 n.
Duguid, P. 5, 145, 261
Durlach, N. 92, 93 n.
Dutton, W. H. 3 n., 8, 10, 257
Duvall, R. 220

Easterbrook, S. 168
Easterwood, J. C. 8, 252 n.

Easton, D. 220
Eccles, R. G. 124 n.
Edge, D. 9, 255
Edquist, C. 86, 262 n.
Egri, C. P. 220–1
Elg, U. 220 n.
Eliasson, G. G. 7, 56, 145
Emmott, S. J. 259
Erber, F. 238
Evans, E. 228

Falush, P. 130, 141
Fayle, E. 136 n.
Feghhi, J. and Feghhi, J. 112, 122
Feigenbaum, J. 111
Ferné, G. 138 n.
Ferreira Silva, L. F. 236, 238, 240–25 n.
Finholt, T. A. 51 n.
Fiol, C. M. 221 n.
Fisher, P. 94
Fiske, M. 59
Flaaten, P. O. 135
Flexner, S. B. 4, 255 n.
Flowerdew, R. 190
Foray, D. 7, 23, 24 n., 138 n., 153 n.
Ford, W. 112 n., 122
Foster, W. A. 219
Fourie, H. 193, 196
Freeman, C. 2, 6, 7, 9, 11, 72, 86, 230,
        244 n., 245 n., 251, 253
Freitas, O. 241
Froomkin, M. A. 112
Frost, P. J. 220–1
Furiati, G. 242 n.

Galaskiewicz, J. 187
Gane, C. 134 n.
Garcia, D. L. 7
Garfinkel, S. 110 n., 112 n.
Garnham, N. 8, 17–18, 269
Gassenheimer, J. B. 124 n.
Gauntlett, D. 3
Gavin, E. 196–7
Gibbons, M. 3, 5, 88, 166 n., 167
Gibson, W. 26, 61 n.
Gilder, G. 228
Gillett, S. E. 219
Gilster, P. 56, 61
Glaser, B. G. 59, 173 n.
Goffman, E. 4
Goggin, G. 13
Goldhaber, G. M. 187
Goldstein, A. 10
Göranson, B. 238
Gould, M. 207 n., 219

Grayson, K. 122
Green, M. 190
Greenwood, D. J. 111
Gristock, J. 86, 87, 93, 94 n.
Grusin, R. 4, 260
Guthke, K.S. 63

Haas, E. B. 262
Haddon, L. 5, 9 n.
Hafner, K. 210 n.
Hage, J. 187
Hagel III, J. 8, 35, 260
Hagstrom, W. O. 58
Hammer, M. 165
Handy, C. 89
Hanseth, O. 210 n.
Hansmann, B. 189
Hanson, H. 47 n.
Hardin, R. 121 n.
Harianto, F. 146
Harrigan, K. R. 187
Harris, R. C. 90
Hart, P. 123
Hatling, M. 210 n.
Hauck, L. C. 4, 255 n.
Hawkins, R. W. 9, 138 n., 193
Hazell, R. 131
Healy, D. 61
Heap, N. 259
Heidegger, M. 269
Helpman, E. 7 n.
Hirsch, E. 93
Hirschman, A. O. 125 n.
Hofstader, R. 65
Hollanda, E. 240 n.
Homans, G. 124 n.
Hopper, K. 4, 252 n.
Houston, F. S. 124 n.
Howell, J. 196
Hudson, H. 247 n.
Hughes, T. P. 4
Hunt, C. S. 221 n.

Innis, H. 253 n.
International Telecommunication Union
 206, 227, 228, 246

Jackson, M. 134 n.
James, J. 173 n.
Jarque, C. M. 192
Jessop, B. 245
Johansson, U. 220 n.
Johnson, B. 146, 230
Johnson, D. R. 219, 220 n., 256
Johnson, D. S. 122

Johnson, R. J. 98
Jonscher, C. 3
Jussawalla, M. 247 n.

Kalakota, R. 7
Kaplan, A. 220
Kapor, K. 219
Katz, J. 59
Kelly, K. 82, 128, 251, 256
Kendall, P. L. 59
Kiefl, B. 70 n.
Kim, A. J. 5
Kim, T. 92
Kim, W. C. 123–4
Kitzinger, J. 59 n.
Kling, R. 10, 26 n., 133
Kollock, P. 5, 56, 124 n.
KPMG 10, 258
Kracaw, T. S. C. 144
Kraut, R. 252 n., 259
Krueger, R. A. 59 n.
Kuhlman, C. C. 219
Kurzweil, R. 260

Lacroix, J.-G. 56
Lamb, R. 133
Lamberton, D. M. 56, 145
Langley, A. 221 n.
Lasky, K. 258
Lasswell, H. D. 220
Latour, B. 58, 74, 261 n.
Lave, J. 24 n., 170, 171, 261 n.
Law, J. 139 n.
Leadbeater, C. 5, 256
Lee, Y. S. F. 193
Lefebvre, H. 4
Leone, R. C. 5, 138, 251
Lessig, L. 8 n., 56, 138 n., 219–20, 257, 260 n.
Lester, K. 196
Levinson, E. 168
Levinson, P. 4, 252
Levy, M. R. 92
Lewis, D. J. 121 n.
Lindemann, M. A. 124
Lipset, S. M. 65
Lipsey, R. 7 n.
Lissoni, F. 187 n.
Liu, C. 207 n., 208 n.
Ljungberg, F. 261
Lombard, M. 72, 139, 259, 261
Lowy, A. 5, 23 n.
Lubart, T. I. 172
Luhmann, N. 123
Lukes, S. 220
Lundvall, B.-Å. 85, 86, 153 n., 169, 230, 245 n.

Lyon, D. 257
Lyon, M. 210 n.

McAfee, A. 89
McAnany, E. G. 231, 240
Macdonald, S. 187
McDonnell, R. A. 189 n., 192, 195
MacKenzie, D. 2, 9, 139 n., 255
MacLean, D. 228
McLuhan, H. M. 18, 93
Macneil, I. R. 124 n.
Maculan, A. M. 237 n.
Maguire, D. J. 194 n.
Mahoney, D. L. 132
Mahoney, J. T. 167 n.
Makridakis, S. 88
Malerba, F. 11
Malkin, G. 210 n.
Malone, M. S. 23 n., 85, 88
Malone, T. W. 88
Mansell, R. 3 n., 4, 5, 7, 8, 9, 10, 11, 23 n., 30 n., 50,
     56, 138 n., 193, 227, 245, 252, 253 n., 255,
     257, 258, 265
Marcell, P. 132–3
March, J. G. 160
Margherio, L. 128, 129, 251
Markus, M. L. 134
Martin, L. 196
Martin-Barbero, J. 259
Marvin, C. 252
Marx, K. 83
Masser, I. 186, 195, 199
Mathiason, J. R. 219
Mattos, H. 239
Mauborgne, R. 124
Mauss, M. 74
Mayer, R. C. 121 n.
Melbin, M. 63
Melody, W. H. 269
Merton, R. K. 59
Metcalfe, B. 257
Metcalfe, J. S. 139 n., 187 n.
Meyrowitz, J. 4
Miles, I. 2, 9 n., 139 n.
Millar, J. 166, 173 n., 184
Miller, D. 259 n.
Miller, L. 61
Misztal, B. A. 121 n.
Mitchell, W. J. 4, 5, 34 n., 49
Mitter, S. 258
Molina, A. 96
Monteiro, E. 210 n.
Moody, F. 61
Morgan, G. E. 8, 252 n.
Morris, A. 193

Morris, G. H. 4, 252 n.
Morrison, D. E. 59 n.
Mueller, M. 206, 219
Muller, G. 124
Mumford, L. 256
Myers, G. J. 134 n.

Nam, K. 4, 252 n.
Nardi, B. A. 259
Nass, C. 139, 140 n.
Neef, D. 7, 260
Negroponte, N. 5, 56, 257
Neice, D. 60 n., 63, 66 n., 67 n., 74 n.
Nelson, R. 158, 244 n.
Nemzow, M. 123 n.
Nettleton, G. S. 231, 240
Neuberger, C. 96
Newman, W. H. 187
Noble, D. F. 258
Nonaka, I. 7, 145
Nooteboom, B. 124 n.
Norris, M. 7
Nourozi, A. 227, 228, 242
Nuttall, C. 189

Ockman, S. 51 n.
O'Connor, D. 10
OECD 7, 128, 129, 142, 258
O'Leary, B. 186, 188
Oliveira, E. Q. 236
Oliveira, F. 231
Olson, G. M. 51 n.
Onsrud, H. 186, 191, 194
Openshaw, S. 190
Orlikowski, W. J. 133
Overby, B. A. 30 n.

Pavitt, K. 86
Pavitt, P. 230, 244, 245
Peak, S. 94
Pearsall, J. 4
Pearson, K. 252 n.
Pennings, J. M. 146
Perez, C. 2, 6, 9, 11, 72, 244 n., 251, 253
Perri 6 258
Pessini, J. 236 n., 238
Pinch, T. J. 4
Pinto, J. K. 186
Pitt, W. 138
Pletcher, A. 258
Pliskin, N. 8
Polanyi, M. 160
Porter, M. E. 187
Post, D. G. 219, 220 n., 256
Postel, J. 217 n.

Poster, M. 72
Primo Braga, C. A. 10
Prusak, L. 10, 85, 162, 168, 170 n.

Quah, D. T. 72, 128, 251
Quintas, P. 135, 219 n.

Raab, C. D. 258
Ramirez, R. 124 n., 125 n.
Raymond, E. S. 31 n., 50 n., 51–2
Raynes, H. F. 129
Rayport, J. F. 260
Reagle, J. M. J. 125 n.
Reeve, C. 261
Reeves, B. 139, 140 n.
Reidenberg, J. R. 219
Rengger, N. 124
Rheingold, H. 5, 8, 34, 35–7, 56, 61, 73
Rhind, D. W. 190
Rice, R. E. 26 n., 260
Rifkin, J. 258
Ring, P. S. 123, 124 n.
Robins, K. 8
Robinson, M. 7
Roche, E. M. 259
Rochlin, G. I. 257
Rockart, J. F. 88
Rogers, D. 152 n.
Rogers, E. M. 186, 187
Romer, P. 7, 23
Romm, C. 8
Rony, E. and Rony P. 207 n., 210 n.
Roome, N. 98
Rotter, J. B. 121 n.
Ruggie, J. G. 262
Ruggles, R. 7, 10
Rush, H. 229 n.
Rushton, G. 194 n.
Rutkowski, A. M. 219

Sanchez, R. 167 n.
Santomero, A. M. 193
Sarkar, J. 4, 187 n.
Sarkar, M. 8, 252 n.
Sarson, T. 134 n.
Saunders, C. 123
Saviotti, P. 86, 187 n.
Saxenian, A. L. 56
Schenk, I. 9, 265
Schmid, B. F. 124
Schmittbeck, R. 193
Schneider, F. B. 122
Schneier, B. 112 n.
Schön, D. 167
Schoorman, F. D. 121 n.

Schramm, W. 225 n.
Schuler, D. 134, 258
Schumpeter, J. 6
Schwabe, C. 186, 188
Sciadas, G. 59, 66 n.
Scott, J. 56, 71 n.
Seabrook, J. 73
Senker, J. 153 n., 262 n.
Shamdasani, P. N. 59 n.
Shannon, C. E. 260 n.
Shapiro, A. L. 5, 128, 251
Shapiro, C. 4, 7
Shapiro, S. P. 121 n.
Sharpe, R. 173 n.
Sheridan, T. B. 260
Short, J. 140
Sieger, S. 142
Silver, D. 166, 173 n.
Silverstone, R. 3 n., 4, 5, 8, 10, 56, 93, 99,
    258, 259 n.
Simon, H. 160, 167 n.
Simonson, H. P. 65 n.
Slack, R. S. 9, 255
Slater, D. 259 n.
Slater, M. 92, 93 n.
Smith, A. 83
Smith, M. A. 56
Smith, R. 196 n.
Smythe, D. W. 225 n., 269
Soete, L. 6, 7, 9, 11, 86, 230, 244 n., 245 n.,
    251
Spafford, G. 110 n., 112 n.
Stalker, G. 166 n.
Stanton, S. 165
Stefik, M. 122
Steinfield, C. 8, 252 n.
Steinmueller, W. E. 3 n., 7, 8, 9, 11, 23 n.,
    27 n., 30 n., 35 n., 50, 138 n., 193, 252,
    253 n., 257, 265
Sterling, B. 61
Sternberg, R. J. 172
Steuer, J. 92, 140, 260
Stevens, W. P. 134 n.
Stewart, D. W. 59 n.
Stiglitz, J. 7 n.
Stone, M. 51 n.
Strausak, N. 90
Strauss, A. L. 59, 173 n.
Sukai, S. B. 186, 188
Summers, R. 112
Swarts, M. 188
Szapiro, M. 240

Takeuchi, H. 7
Tapia, J. 234

Tapscott, D. 5, 7, 23 n.
Taylor, J. 257
Teece, D. J. 90
Tehranian, J. 247 n.
Terracine, E. 231, 239, 240
Thomas, G. 9 n.
Thompson, J. B. 259 n.
Ticoll, D. 5, 23 n.
Tidd, J. 86
Timmers, P. 7, 138 n.
Tolbert, P. S. 220
Townsend, R. 138
Tremblay, G. 56
Trimble, B. 4
Tseng, G. 193
Tunstall, D. 189
Turkle, S. 56, 59, 258
Turner, F. 61, 65 n.

Uchôa, C. 242
UNCTAD 258
Upton, D. M. 89

Valente, A. C. 242
Van de Ven, A. H. 123, 124 n.
Van Dusseldorp, M. 96
Van Helden, P. 189, 195
Van Wijk, J. 90
Varian, H. 4, 7
Veblen, T. 57, 74
Verhoest, P. 9
Vianna, C. 235
Von Hippel, E. 123 n.
Voss, C. 229 n.
Vygotsky, L. S. 169

Wajcman, J. 139 n.
Wallis, J. 221 n.
Walsh, V. 86, 187 n.
Walsham, G. 135

Waugh, E. 225
Weaver, W. 260 n.
Weber, M. 83
Webster, F. 8, 56
Wehn de Montalvo, U. 5, 8, 50, 200 n., 227, 258
Weick, K. E. 167
Weigert, A. 121 n.
Wendt, A. E. 220
Wenger, E. 24 n., 170, 171, 261 n.
Wertsch, J. V. 169, 259
West, S. 7
Whinston, A. 7
White, H. S. 4, 193, 252 n.
Williams, E. 140
Williams, P. 112, 122
Williams, R. 9, 255
Wilson, J. D. 122, 189 n.
Winn, J. K. 110, 111, 117 n., 122
Winner, L. 221 n., 256 n.
Winslow, C. D. 162
Winter, S. 158
Wohlers, M. 237
Wood, J. 166, 173 n.
Woolgar, S. 3, 8, 10, 58, 74
Woolley, B. 85, 109
World Bank 7, 256
Wright, C. 136 n.
Wright, D. 228
Wrobel, D. W. 65
Wyatt, S. 57, 255, 257, 261 n.

Yanelle, M. O. 4, 252 n.
Yates, J. 129
Yourdon, E. 134 n.

Zaheer, A. 187
Zerdick, A. 5
Zorkoczy, P. 259
Zuboff, S. 159
Zucker, L. G. 123

Subject Index

access:
  Internet 66–71
  negotiated access to data 153–4, 156
accreditation 111, 116
accuracy, data 190–2
acquisitions and mergers 94, 145
Adobe 82
advertising 40, 47–8
alliances 90
ambiguity 255–8
America Online (AOL) 32, 35, 70
Anatel (Brazilian National
    Telecommunications Agency) 227,
    242, 248
Andreeson, Marc 73
anonymity 110
Apache 31
arbitrary rules 212–14
architecture, Internet 219–20, 221–3
Associated New Media 105
Assurance Companies Act 1910 136
attitude 200, 201, 202
authentication:
  data 153, 156–7, 160
  pre-transaction 120
  tacit 160
  trust services 112, 114–15, 120
  user authentication 45
authority, procedural 24–5

banking industry 14–15, 144–62, 263, 265
  credit procedures 150–2
  handling of codified data 152–5
  nature of commercial banking 145–7
  repersonalization and banking intermediation
    155–60
  selection of ICTs 147–50
  trust services 121
behaviour:
  and motive in virtual communities 34–9
  self-regulation 175–6
  theory of planned behaviour 199–204
Berners-Lee, Tim 73
Black, William 215–16, 217, 222
Brazil:
  building technological and institutional
    capabilities 229–35, 238–9, 247
  ECO-8 233–5, 240–1, 242–3, 246–7, 249
  Embratel 236, 237, 238–9, 240, 242, 247

Equatorial Constellation Communications
    system (ECCO) 228, 241, 242–3
  government 229, 231–7, 240–1, 246–7, 248–9
  market liberalization 239–43
    satellite 231, 239–40, 241, 246
  Telebras 229, 235, 236–7, 241–3, 246–7
  telecommunications 16–17, 225–50, 266, 268
British Chambers of Commerce (BCC) 120, 121
British Telecom 114–15, 119, 120, 121
  BT Internet services 209, 209 n., 215–6
broadcast, interactive 27
  channel guides 41
business process re-engineering 165–6
business professionals 173–83

capabilities:
  for knowledge exchange 15–18, 266–8
  technological and institutional in
    telecommunications 229–35, 244–7
Cerf, Vinton 61, 73
certificates, digital 14, 110–27, 263, 265
certification practice statements (CPS) 116–17
chat circles 49
claims handling 136–7
clearing-house arrangements 194–5
club goods 46, 51–2
co.uk sub-domain 208–9, 211, 212–13
co-design 15, 165–85, 266, 267
  in communities of practice 167–8, 173–83
  learning and production of new design
    knowledge 168–73
  outcomes 183–5
codified data 152–5
collaboration 50–2
collective action 37–8
commercial banks, see banking industry
communication:
  banking industry 148–9, 159–60
  computer-mediated 25–7, 259–60
  inter-user 35, 37, 39–40
  tacit dimension 159–60
community:
  communities of practice 173–83
  organizational 90–1, 93, 98–103
  virtual, see virtual communities
community developers 60
  digital divide metaphor 68–9
  frontier metaphor 62–3
  network peer reciprocity 76, 77, 79, 79–80, 81

computer-mediated communication 25–7,
  139–41, 259–60
confidentiality 153–4
conflict 176–8, 183
conspicuous consumption 57–8, 81–2
consumption 57–8, 81–2
control 23–4
  of information 78
  perceived behavioural control 200, 201, 203–4
cooperation 23–4, 186–205
coordination 23–4
  failures 29, 40–4
costs 132
  congestion costs 40–4
creativity 80–1, 172
credit review process 150–2
  see also banking industry
cyberspace:
  metaphors for cyber-experience 56–8
  virtual communities, *see* virtual communities

data:
  accuracy 190–2
  banks 153–4, 156
  complexity 190–2
  handling codified data 152–5
  repersonalization of 14–15, 155–60, 161–2,
    263, 265
  sharing spatial data 15–16, 186–205, 266, 267
  standards 194–5
  see also information
dematerialization 72
developing countries 10
  development projects 189–90
  Geographic Information Systems 189–90, 192
  satellites and the access gap 225, 227–9
diffusion of innovations 187
digital divide metaphor 57, 66–71, 257–8
disintermediation 144–5, 193
dystopian vision 257–8

earth station 236, 240
economizing 156, 158
education, distance 231
electronic commerce 37, 264–6
  digital certificates, trust and 13–14, 110–27,
    263, 265
  London insurance market 14, 128–43, 263,
    265
  regional newspaper sites 96–7
electronic data interchange (EDI) 129
Electronic Frontier Foundation (EFF) 61
electronic signature legislation 117
email 148, 149
esteem, social 71–4, 81–2

European Union 66, 117
exchange:
  gifts 36–7, 74, 82
  information 253–4
  knowledge, *see* knowledge exchange
  market forms and the Internet 82–3
  network peer reciprocity 74–5, 82–3
  reducing costs of 23
  trust services and digital certificates 124–5
  virtual communities 36–7, 51
    expertise, respect for 79–80

facilitation 88, 177–8, 179–80
facsimile machine 159
focused interviewing 58–9, 60–1
freedom 80–1

generic business processes 137–8
geographic information systems (GIS) 15–16,
    186–205
  core data-sets 194–5
  use and spread 188–90
  see also spatial data
gifts 36–7, 74, 82
global financial centres 161
global information society 8, 254
globalization 141–2
governance:
  Brazilian telecommunications 243–7
  Internet, *see* Internet governance
  virtual communities 29–33
graffiti attacks 40–4

hostility 176–8
human capital 7
humility, self-effacing 77–8
humour 181

Independent Commission for Worldwide
    Telecommunication Development
    226 n., 227 n.
indwelling 160
informating 159
information:
  asymmetries 146–7
  exchange 253–4
  and knowledge 261–2
  network peer reciprocity 78–9
  technologies, organization and 88–9
  see also data
information-processing view 145, 146–7,
    157, 260
information systems professionals 173–83
innovation 87
  diffusion of innovations 187

newspaper products 103–8
  social innovation 22
  technology-push and market-pull 86
institutions 266–8
  authority 25
  capabilities 229–35, 244–7
  social 38–9, 261
insurance industry 14, 128–43, 263, 265
  development of general-purpose underwriting
    system 135–41
integrated Computer-Aided Software Engineering
  (iCASE) 172
interactive Joint Application Development (JAD)
  172–3
intermediation 4–6, 87
  banking industry 144–5, 146
    credit decisions 151–2
    repersonalization and 155–60
  co-design 168–9, 170–1
  institutional 226
  spatial data sharing 193–5
  see also mediation
International Telecommunication Satellite
  Organization (INTELSAT) 236
Internet 13, 142, 255
  digital divide metaphor 57, 66–71
  domain name registry system 206–18
  experienced Internet-users 58–82, 263, 264
  free serve ISPs 70, 82
  frontier metaphor 57, 61–3
  Internauts 60
  Internet service providers (ISPs) 41, 70, 82
  governance 16, 206–24, 266, 267–8
  network peer reciprocity 74–81
    contribution 58, 73–4, 81–2
  Nominet UK 206, 207, 214–18, 222–3
  UK Naming Committee 209–18, 221–3
  see also virtual communities
Inter-user communication 35, 37, 39–40
intimacy 34–9
  defining virtual communities 36–9
investors 146
Iridium 228, 241, 242

Joint Academic Network (JANet) 208–9

knowledge:
  exchange 15–18, 102–3, 266–8
  information and 261–2
  Mode 1 and Mode 2 167–8, 174, 183–5
  novelty 172
  sharing 15, 165–85
knowledge management 7
  information-processing view 145, 260
  mechanistic 174, 175–8, 183–5

organic 174, 178–85
  and virtual organization 13, 85–109, 263, 264
knowledge production 7, 167–73

learning 15, 165–85
  building capabilities for knowledge exchange
    266–8
  and production of new design knowledge
    168–73
  strong learning 172, 178–83
  tacit authentication 160
  weak learning 172, 175–8
liability 117–18
liberalization of markets 239–43
LIMNET (London Insurance Market Network)
  129, 131–42, 263, 265
Linux 31
localization 41
London Insurance Market 14, 128–43, 263, 265
  Lloyd's of London 130–1, 136, 138
  Marine, Aviation and Transport (MAT)
    insurance 136
London Internet Exchange (LINX) 209–10
low earth orbiting satellites (LEOS) 16–17,
  225–50, 266, 268
'lurkers' 38

management techniques 183, 184
markets:
  entry strategies 113–15
  liberalization 239–43
  market-making strategies 118–21
  market-pull model 86
  market societies 71, 74
matrices 99–102
mediation 17, 254–5, 269–70
  cooperation 186–205
  knowledge management and the virtual
    organization 86–7, 92–3, 102–8
  knowledge sharing and co-design 165–85
  mediated learning 174–83
  see also intermediation
merchant bankers 144
mergers and acquisitions 94, 116
metadata 192, 194–5
microelectronics technologies 7, 9–10
MP3 files 31–2
multi-user dungeon (MUD) 49
music 31–2, 33

Napster 31–2, 33
negotiating capability 229–30, 243–7,
  248–50
Netscape 82
network externalities 74

new economy 6, 8
  virtual communities and 12, 21–54
    basis for new economy 33–4
new media developers 60, 63, 68, 75, 77, 79–81
newspaper industry 13, 85–109
  analysis of web sites 95–7
  new media developers 98, 107–8
  regional daily newspapers 98, 107–8
night 63, 64–5
Nominet UK 206, 207, 214–18, 222–3
non-profit organizations 42, 43
norms 38–9, 123, 266–8

objectification 255–8
Office of Telecommunications (UK) 116
Open GIS Consortium 194
open-source software 31, 33, 51–2, 54
organization-space 90–1, 98–102
  organizational virtuality 103–8
  see virtual organization
organizations 22–5, 88–9
outsourcing 90

parastatals 197, 198
peer interaction 74–6
peer-to-peer file exchange 31–2, 33
physical meetings 36
planned behaviour, theory of 199–204
policy:
  Internet as digital divide 67, 69–71
  priorities 269–70
  TSPs and business policies 115–18
politics of Internet governance 219–23
Postel, Jon 73
practices 266–8
  TSPs and 115–18
presence 92–3, 103, 139–41, 260–1
private sector 197, 198
privatization 241–3
procedures 115–18
public goods 46, 51–2
public key infrastructures (PKIs) 110–11, 113, 121
  see also trust services
public sector 195, 197, 198

Rapid Application Development (RAD) 172–4
rapid prototyping 172
reality:
  virtual communities and 36
  and the virtual organization 89–90
reciprocity:
  network peer reciprocity 57–8, 71–82
  trust 123–6

recruitment 29–33
  user recruitment 47–8
registration:
  digital certificates 114–15
  domain name registry system 206, 208–18, 221–3
regulation:
  deregulation 132
  satellite telecommunications 243–7
  self-regulatory initiatives 116
  see also Internet governance
relationships, *see* interaction
remote sensing satellites 234
repersonalization of data 14–15, 155–60, 161–2, 263, 265
reputation capital 50–2
rules 38–9
  implementation of arbitrary rules 212–14

satellite communications, *see* Brazilian telecommunications
scalability 119–20
science networks 58, 74
search engines 39–40
  news archive 96–7
secrecy culture 153–4
self-regulatory initiatives 116
serendipity 79
shopping 96–7
skills:
  intellective skills 159
  Internet users and respect for 79–80
  interpersonal 203
  and spatial data sharing 203
social distinctions 13, 55–84, 263, 264
  research on intensive Internet users 58–61, 83–4
  values and social esteem 71–4, 81–2
social inclusion 269–70
social presence theory 139–41
social pressure 200, 201, 202–3
software development 133–9, 168
South Africa 195–204
  Chief Directorate of Surveys and Mapping 196, 198
  Chief Surveyor General 198
  Committee for Spatial Information 196
  Department of Land Affairs 196, 198
  determinants of spatial data sharing 199–204
  Independent Electoral Commission (IEC) 196
  National Spatial Information Framework (NSIF) 196–9, 205
space 90–1, 98–103
space research 230–3, 246
spatial data 15–16, 186–205, 266, 267
  characteristics 190–2

distribution in South Africa 195–204
importance 188–92
intermediation 193–5
national spatial data infrastructures 194–5
standardization 38–9, 137–8
data standards 194–5
digital certificates 116–17
status, social, *see* social distinctions
sticky trust 123
surveys 59
sustainability 29–33

tacit authentication 160
tacit knowledge 157, 159–60
technical systems:
ambiguity 255–8
emergent regimes 253–5
mediating social and technical relationships
4–6, 12–15, 262–6
technological capabilities 229–35,
244–7
techno-economic paradigm 6–12
guiding principles and new 253–5
technological change 6
enacted view 133–4
mediation and 258–62
technology-push model 86
telecommunications:
access gap 225, 227–9
Brazilian, *see* Brazilian telecommunications
telepresence 93, 260–1
theory of planned behaviour 199–204
time:
as frontier 63, 64–5
organization-space 90–1, 93, 98–103
Torvalds, Linus 73
trust 13–14, 110–27, 263, 265
building 121–6
interactive process 123–6
tacit authentication 160
trust service providers (TSPs) 110–27
VeriSign 112, 113, 115, 116–17, 119

.uk domain 216–17, 221–3
UK Education and Research Networking
Association (UKERNA) 209, 215
UK Naming Committee 208–18, 221–3
USENET 29–31, 33, 35
user authentication 45
user recruitment 47–8
utopian vision 257–8

validation 153, 156–7
value added networks (VANs) 129
virtual communities 12, 21 54, 93, 263, 264, 270
affinity-based virtual communities 44–5,
46–51, 54
and brand name 41, 44, 46–51, 54
and cyberspace 25–34
closed 43, 44–51
collective production services 46, 51–2
convenors 49–50
dynamics of virtual-community types 46–52
information services and resources 46–51
market for 39–45
motive and behaviour 34–9
open 44–5, 46–51
purpose-built virtual communities 45, 46,
51–2, 54
typology 44–5
virtual organization 13, 85–109, 263, 264
concept of virtuality 93–102
defining 89–90
limits of virtuality 90–3, 103–8
virtual reality 92
voluntary associations 24, 27–9, 33, 37, 39
voluntary licensing 111, 116

Wales 103 5, 107
Whole Earth 'Lectronic Link (WELL) 34, 35–6, 42

Yahoo 41

zone of proximal development (ZPD) 169–70, 181
zones of virtuality 98–100